Analytical Methods for Environmental Monitoring

Edited by

RAFI AHMAD
MICHAEL CARTWRIGHT
FRANK TAYLOR

Prentice
Hall

An imprint of **Pearson Education**

Harlow, England · London · New York · Reading, Massachusetts · San Francisco · Toronto · Don Mills, Ontario · Sydney
Tokyo · Singapore · Hong Kong · Seoul · Taipei · Cape Town · Madrid · Mexico City · Amsterdam · Munich · Paris · Milan

Pearson Education Limited
Edinburgh Gate
Harlow
Essex CM20 2JE
England

and Associated Companies throughout the world

Visit us on the World Wide Web at:
http://www.pearsoneduc.com

First published 2001

ISBN 0 582 25357 8

The programs in this book have been included for their instructional value. The publisher does not offer any warranties or representations in respect of their fitness for a particular purpose, nor does the publisher accept any liability for any loss or damage (other than for personal injury or death) arising from their use.

Many of the designations used by manufacturers and sellers to distinguish their products are claimed as trademarks. Pearson Education Limited has made every attempt to supply trademark information about manufacturers and their products mentioned in this book. A list of trademark designations and their owners appears on p xix.

British Library Cataloguing-in-Publication Data
A catalogue record for this book is available from the British Library

Library of Congress Cataloging-in-Publication Data
Analytical methods for environmental monitoring / edited by Frank Taylor, Mike
Cartwright, Rafi Ahmad.
       p.   cm.
    Includes bibliographical references.
    ISBN 0–582–25357–8 (pbk.)
    1. Environmental monitoring—Methodology.   I. Taylor, Frank, 1956–   II. Cartwright,
    Mike, 1940–   III. Ahmad, Rafi, 1944–
    QH541.15.M64   A53   2001
    577.27′028′7—dc21   00–032402

10 9 8 7 6 5 4 3 2 1
05 04 03 02 01

Typeset by 35 in 9.5/11pt Times New Roman
Produced by Pearson Education Asia Pte Ltd.
Printed in Malaysia ,LSP

Brief contents

Contents

Contents

Contents

Contents

List of contributors

Dr Rafi Ahmad
Department of Environmental and Ordnance Systems
Cranfield University
Royal Military College of Science
Shrivenham
Swindon
SN6 8LA

Dr Michael Cartwright
Department of Environmental and Ordnance Systems
Cranfield University
Royal Military College of Science
Shrivenham
Swindon
SN6 8LA

Frank Taylor
Cranfield Biotechnology Centre
IBST
Cranfield University
Cranfield
Bedford
MK43 0AL

Dr Jeff Newman
Cranfield Biotechnology Centre
IBST
Cranfield University
Cranfield
Bedford
MK43 0AL

Dr Ibtisam Tothill
Cranfield Biotechnology Centre
IBST
Cranfield University
Cranfield
Bedford
MK43 0AL

Dr Brian Dacre
Department of Environmental and Ordnance Systems
Cranfield University
Royal Military College of Science
Shrivenham
Swindon
SN6 8LA

Dr Sarah Stephens
Genetix Ltd
63–69 Somerfield Rd
Christchurch
Dorset
BH23 3QA

Preface and acknowledgements

The delicate balance and interactions between the three extensive components of our environment, namely air, water and land, is vital to maintain the ecosystem capable of sustaining life on Earth. Any imbalances created either gradually by prolonged or repeated pollution incidents or through sudden major incidents can have catastrophic environmental consequences. The growth in humanity's agricultural, industrial and urban activities has resulted in the appearance of a number of problems associated with apparently harmless activities of ordinary people. The atmospheric degradation of Freon and the interactions which have reduced the upper atmosphere ozone levels are a case in point. Hence there is a need for effective pollution control. An essential component of any control programme is adequate monitoring to provide early warning of pollution incidents.

The range of technologies available for monitoring is very extensive. Each is characterised by the nature of the sample presentation, the species to be monitored, the sensitivity and precision of the method. Equally important is the time scale for the measurement. In many pollution incidents the speed of response can be critical. Many pollution incidents occur in remote locations requiring long range monitoring technologies, which do not require contact with the sample. Analytical methods for environmental monitoring thus cover a complete range of methods from simple physical methods using the absorption and emission of electromagnetic radiation, through classical wet chemical methods to microbiological cultivation techniques.

There is a need for a textbook which covers all aspects of environmental monitoring detailing the basic principles of the techniques. This volume attempts to meet this need by providing details of the analytical methods available, with discussion of their capabilities and limitations, and the options for interpretation of the data produced. The opening three chapters deal with the environment and the nature of pollution, statistical methods of data analysis followed by sample collection and preservation. These chapters are not intended to be exhaustive but to give adequate information for applications in most laboratory analysis. The following eight chapters detail classical chemical analysis, instrumental methods, chromatography, tandem techniques, remote sensing, biochemical techniques, electrochemical methods and finally radiochemical techniques. The book also contains appendices, which provide details on the Red List of chemical pollutants and an indication of some of the limits proposed for some of these materials in environmental water samples. Other topics covered are SI units, standard materials and sources of certified standards.

The idea for writing this book arose from the realisation of the need for a teaching manual to support instruction on pollution and its detection as part of two master's degree

Preface and acknowledgements

courses in environmental diagnostics and water pollution control technology. The principal aim was to provide an introductory text for people with a limited background in the relevant science and technologies, while at the same time providing a source book for people involved in day-to-day analysis of environmental samples. The bibliographies at the end of most chapters steer the reader to more advanced specialist sources.

Whilst we have endeavoured to provide as complete a picture as possible, within the confines of a single volume, we would welcome any comments and suggestions from readers.

Finally, the authors are very grateful to the publishers, Pearson Education Limited, particularly Alexandra Seabrook, for allowing us the opportunity to publish this work and also to Tina Cadle, Managing Editor and their colleagues. The authors also gratefully acknowledge the following organisations through assistance with permission to reproduce figures from the sources listed: Royal Society of Chemistry, American Chemical Society, Dionex Incorporated, Varian Ltd, VCH IM Publications, Lab Connections Inc., John Wiley & Sons Inc., Perkin Elmer Ltd.

Other figures have been produced either by the publishers or in house and we acknowledge the assistance of Mrs J. Mosely of the Cranfield design staff, whose ability to turn very rough draft diagrams into camera-ready copy has been outstanding. We also acknowledge helpful discussions with colleagues and the students who provided the inspiration for the book.

Finally we are indebted to our families who have given us support and graciously tolerated our tantrums during the long gestation period for the manuscript.

Rafi Ahmad
Michael Cartwright
Frank Taylor
Cranfield University, April 2000

Publisher's acknowledgements

The publishers wish to thank the following for permission to reproduce the material:

Figure 5.17 from V.A. Fassel, *Science*, 1978, 202, 185 (1978) published by the American Association for the Advancement of Science; Figures 5.14 (a and b), 5.15 and 5.21 (a and b) from *Principles of Instrumental Analysis*, Fourth Edition by Douglas A. Skoog and James L. Leery, copyright © 1992, by Saunders College Publishing, reproduced by permission of the publisher; Figure 9.1 from UK SCA, 1981, crown copyright reproduced with the permission of the Controller of Her Majesty's Stationery Office; Figure 9.4 from *Genes V* reprinted by permission of Oxford University Press (Lewin, 1984).

While every effort has been made to obtain permission in respect of Figures 8.4, 8.5, 8.6 and 8.9 from the *RCA Electro-Optics Handbook (Technical Series EOH-11)*, the publishers have been unable to contact the copyright holders. However, we would be grateful for any information that will enable us to do so.

Trademark notice

The following designations are trademarks or registered trademarks of the organisations whose names appear in brackets:

Freon, Mylar, Teflon (Du Pont U.K. Ltd.); Ascarite II (Thomas Scientific Inc.); Carbasorb (BDH (Merck) Ltd.); Corex (Corning Ltd.); OPTO8, OPTO Hamamatsu (Hamamatsu Photonics U.K. Ltd.); Lambda 9 (Perkin Elmer Inc.); Apiezon (Apiezon Products Ltd.); Cellusolve (Union Carbide Inc.); Image isocon, Orthicon (English Electric Valve Co. Ltd.); Low'Tran-6 (Air Force Systems Command, USAF); Analar, Volucon (BDH (Merck) Ltd.); MetPAD (Group 206 Technologies Inc.); ECHA (Microbiology Ltd.); COMPUT-OX, DODDX (N-CON Systems Inc.); Merit 20 (Yorkshire Water plc.); ARAS BOD (Dr. Bruno Lange GmBH); LUMIS-tox, Microtex (Beckman Inc.); Mutatox (Microbics U.K. Ltd.); Microwell (Nalge Nunc Inc.); Biacore, Captagene GCN4 (Pharmacia); Polytox (Polybac Inc.); Genetip (Affymetrix Inc.); Hybrid capture (Digene Inc. USA); Coulter counter (Coulter Electronic Ltd.); Taq Man (Perkin Elmer Applied Biosystems Inc.); RiboPrinter (Qualicon Inc.); SOS Chromatest (Environmental Biodetection Products Inc.).

1

The environment and the nature of pollution

1.1 Introduction

There has been a trend in legislation spanning two centuries designed to protect our natural environment, from the earliest Acts of Parliament (e.g. Alkali Act 1863) to control the emissions of noxious pollutants to the atmosphere to the Clean Air Act 1993. This trend arguably stemmed from a public desire to protect people in the first instance from the widespread activities of an industrialised community. The desire to protect people and their environments in specific terms has grown and evolved into a desire to protect the population at large, the workplace and the natural environments by the gradual evolution of integrated management systems.

Measurement and monitoring of the state of health of the natural and workplace environments is a core activity in any environmental management system. The purpose of this book is to introduce students and practitioners to current and novel methods for environmental monitoring. To place this contribution in context we must first examine the broader picture with regard to what the health of the environment is. From this understanding, the position held by analysis and monitoring will be more clearly appreciated.

1.2 Ecosystems and the effects of pollution

1.2.1 What is an ecosystem?

Odum states: 'Living organisms and their nonliving (i.e. abiotic) environment are inseparably inter-related and interact upon each other. Any unit that includes all of the organisms [...] in a given area interacting with the physical environmental so that a flow of energy leads to clearly defined trophic (levels of feeding) structure, biotic diversity, and material cycles [...] within the system is an ecological system or *ecosystem*' (Odum, 1971). An ecosystem is the smallest unit to which a biological community may be reduced. Upon Earth, life is contained within the **biosphere** but the sources of materials required for biotic function are to be found also in the other three global arenas: the **hydrosphere**, the **geosphere** and the **atmosphere**. Interactions taking place between these spheres are fundamental to life on Earth (Figure 1.1). An example of an ecosystem would be a watershed, where the hydrosphere commences with evaporation of water from the oceans followed by precipitation onto dry land. Water then moves under gravity taking with it inorganic (and, to some extent, organic) nutrients, to return again to the oceans. During this cycle there are

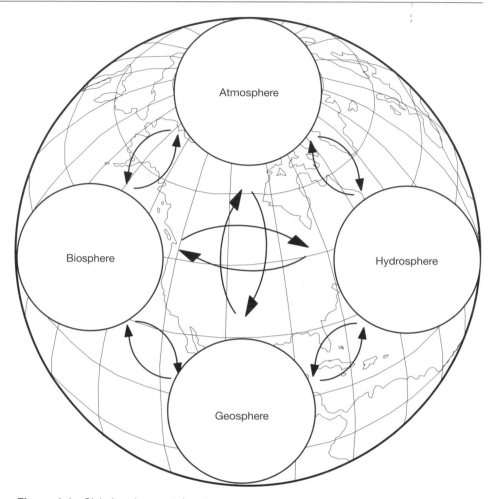

Figure 1.1 Global environmental systems

interactions between the hydrosphere and the geosphere, the atmosphere and the biosphere. Parallel interactions take place between all components of the global environmental systems, modelled by the elemental cycles. Thus, carbon is fixed by plants growing on the land, fed from the precipitated water. The plants in turn feed animals and micro-organisms, and these in turn produce waste and die and decay, returning carbon and nutrients to the oceans and so on.

The interactions which take place between the various parts of the global environmental system are complex, but they can be modelled at the elemental or chemical level (e.g. the carbon, nitrogen and sulphur cycles), the physical level (e.g. the hydrological cycle and sedimentary cycles), the biochemical level (e.g. nutrient cycles) and the energy level. Collectively, these interactions are known as **biogeochemical cycles**. Energy flows may be superimposed on all other biogeochemical cycles and are inseparable from

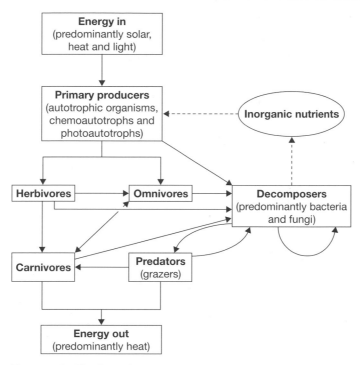

Figure 1.2   The flow of energy through an ecosystem

them. Figure 1.2 describes a schematic representation of energy flow through a food web or an ecosystem.

1.2.2   What is pollution?

To define pollution may appear straightforward at first glance, e.g. 'too high a concentration of any chemical material in any location'. However, scientists, politicians, legislators, enforcers and the public freely and frequently use the term, and the perception of its meaning is as diverse as the people who use it. There is one fundamental property of pollution, however, which is that it has an adverse effect, directly or indirectly, on life or on quality of life. If it does not affect life in any way then it is not pollution. However, since the biosphere interacts with all other quarters of the globe (see Figure 1.1), pollution may be anything which affects the hydrosphere, the geosphere and the atmosphere – or water, soil and air, i.e. pollution acts detrimentally upon an ecosystem. In consequence, the exercise of monitoring the natural environment is applicable to all environmental media. This has far-reaching implications for sampling and analytical protocols.

Pollutants, then, are materials that affect the global environment, with a detrimental effect on the biosphere. They may be chemical in nature, they may be a purely physical phenomenon, e.g. noise and vibration, or they may be chemical in nature but result in a harmful physical phenomenon, e.g. greenhouse gases. Those pollutants of a chemical nature are the major remit of this book although measurement, monitoring and protection of noise, etc., are covered to a lesser extent.

1.3   Atmospheric pollution

The atmosphere is a **reservoir pool** for a number of elements essential to life on Earth. A reservoir pool is that part of a biogeochemical cycle which contains the bulk of an element on the planet and which interacts slowly with the other sectors of the global environmental system (Odum, 1971). The remaining complement of an element is accounted for in a faster-moving and generally smaller **exchange pool**. Many pollutants may affect the atmosphere and those with the greatest impact are those which affect the atmosphere as a reservoir pool. An impact on an exchange pool affects those ecosystems associated with that pool. An impact on the reservoir pool pervades the entire global environmental system. A notable example of this is the oxides of nitrogen. Atmospheric nitrogen chemistry is complex and by its nature it is essential to life on Earth. Nitrogen is an element, required by biota, for the synthesis of structures such as proteins, nucleic acids, cofactors and many derivatives of carbohydrates. So nitrogen is fundamental to most classes of biochemical compounds, secondary only to carbon which is fundamental to all. The oxides of nitrogen are important in the movement of nitrogen from its reservoir pool to biota but in addition, if in excess, contribute to acid rain. With the Industrial Revolution came legislation to control the emissions to the atmosphere resulting directly from commercial activity. The concept of **best practical means** (BPM) was introduced and embodied in a local bye-law in Leeds to prevent or abate a nuisance arising from smoke emissions. Failure to do so resulted in the offender being fined. The concept of BPM was included in the 1874 amendment to the 1863 Alkali Act, which aimed to prevent the emission of noxious gases from alkali works.

1.4   Aqueous pollution

Water is the basis of all life. If a pollutant enters the hydrosphere then it enters the reservoir and/or exchange pools of all of the major elements. The hydrosphere is the fastest medium of transmission to the biosphere of most elements. Therefore any impact on it impacts on the biosphere both quickly and in a widespread fashion. The legislation in the UK protecting water goes back as far as the Middle Ages where regulatory controls were put into place to prevent the tipping of waste into public watercourses. Indeed, even from Roman times laws prevented the fouling of potable water sources.

Thus, industrial humanity realised long ago that pollutants in air and water were a great threat to humanity itself. However, it has since become apparent that water-borne pollutants affect a much wider range of biota than humans alone. Pollution liberated, for example into the atmosphere, has a direct impact on the hydrosphere, and reduction in air quality can bring about a reduction in plant growth (i.e. the biosphere) as well as impacting on human health. Impact is even apparent on inert structures such as building materials. So pollution control has mirrored these observations with the evolution of integrated pollution control systems, which take into account the impact of human activity on all environmental media. Thus, the impacts of a process emitting persistent and volatile hydrocarbons in an aqueous effluent, for example, are considered and controlled in air, soil and water.

The impact of different toxicants in aqueous media need not be investigated in detail to highlight their ubiquity. The effects of mercury on the Japanese fishing population and the effects of petrochemical spillages at sea are well known and are detailed in the public as well as the scientific press. The range of chemical types encountered in water is greater than that encountered in any other environmental medium and their toxic effects on the biosphere have a rapid impact. For this reason analytical chemistry has evolved most of its methodology, protocols and standards to describe the quality of water. All well-tried toxicity-testing methods occur in aqueous media. All biodegradability-testing methods relate to an aqueous medium intended to simulate the hydrosphere.

Many aqueous toxicants are carbon based but there are many inorganic pollutants with high priority. Heavy metals are a notable example. Chelates, complexes of metals with organic structures, and organometallics may be particularly hazardous to the environment, e.g. tributyl tin.

1.5   Pollution of soils and sediments

1.5.1   General considerations

The presence of toxicants in soil and sediments may pose a great and prolonged threat to life. Sediments in lakes and oceans have been demonstrated to be the site of considerable accumulation of harmful chemicals, both organic and inorganic. Metals such as chromium, iron, mercury, tin and lead have all been highlighted in lake sediments as a legacy of previous industrial activity. Wheal Jane tin mine in Cornwall is still the cause of great concern despite the cessation of extraction for some decades. This is due in part to leachate emanating from spoil heaps and in part to the discontinuation of pumping out shafts, which results in the accumulation of contaminated water and subsequent over-spill into the surrounding hydrogeological systems. Organic materials too have been seen to contaminate mineral extraction sites for many years. The range of monocyclic and polycyclic aromatic hydrocarbons, which pollute aquifers underlying coal-gasification sites in the US, is huge.

1.5.2   Contaminated land

The UK has recently witnessed a growth in the legislation to protect and remediate land contaminated by the products of commerce. The following effects may arise from contamination of land: inhibition of the growth and development of plants; contamination of associated surface and groundwater; direct uptake by animal life by absorption, ingestion or inhalation; indirect uptake of toxicants via the human food web; chemical degradation of structures; fire and explosion; odour nuisance. The Environment Act 1995 places upon local authorities the new duty to inspect land in their area of jurisdiction for contamination and to identify any 'special sites'. Under the Act remediation notices may be issued under the authority of the Environment Agency, which are intended to secure the assessment and clean-up of contaminated land. The remedial treatment of contaminated land may be carried out in one of four ways: removal of the contamination to other specially designated areas ('dig and dump'); on-site retention; dilution; elimination by biological, chemical or physical treatments. All of these treatments, especially the last two, require analytical protocols to ensure that standards are met and that remedial action is effective.

1.6   Origins of pollution – natural pollution versus anthropogenic pollution

There are three principal areas of human activity which result in the liberation of pollutants into the natural environment: industry; agriculture; domestic life. Some activities pervade all of these areas and have immense impact, such as transport.

1.6.1 Industry

The industrial sector manufactures, utilises and discards more types of pollutant than any other sector of human activity. The breadth of toxic materials used by industry is almost impossible to catalogue. However, the controls placed upon industry by legislation are extensive and continue to grow. Growth in environmental law relating to industry should logically outstrip growth in industrial activities, since new activities are legislated for and old ones continue to be further legislated for owing to increasing public pressure for industry to 'clean up its act'. New understanding of the effect of pollution on life and the natural environment has never led to a relaxation of pollution quality standards, so the trend is set. Recent legislation under the UK Environmental Protection Act 1990 prescribes certain substances as harmful and detrimental to the health of the environment. Processes which generate or utilise such prescribed substances are prescribed as activities with the potential to cause harm to the environment. Prescribed processes and substances, under the Act, must be entered into a public register and it is required to monitor their emissions to air, soil and water. Manufacturing industry has a need to monitor its emissions to the environment, in part, for reasons of process control and optimisation. More and more legal requirements are placed on manufacturing industry to monitor its emissions to meet environmental standards. These two principles result in the two major areas of environmental monitoring: process monitoring and compliance monitoring.

1.6.2 Agriculture

Agriculture has an immediate effect on the natural environment and on the biosphere because it takes place, by definition, predominantly within the biosphere. Modern agricultural practice utilises many chemical agents for fertilisation of soil, protection of crops against vertebrate, invertebrate and microbial spoilage, protection of cattle against insect infestation and microbial infection, and in waste treatment and hygiene. In the UK environmental protection in the agricultural arena is embodied in Acts of Parliament and published Codes of Practice. Examples are as follows: Clean Air Act, 1993 parts I and III; Environmental Protection Act 1990, parts I, II, and III; The Sludge (Use in Agriculture) Regulations 1989; Code of Practice for the Safe Use of Pesticides on Farms and Holdings 1990, under review 1995 etc. In addition to UK legislation further controls are placed upon agriculture by the EC Framework Directive on Waste.

In England and Wales two Codes of Practice cover most agricultural activity: Code of Good Agricultural Practice for the Protection of Air, 1992; Code of Good Agricultural Practice for the Protection of Soil, 1993. In Scotland these two codes of practice are combined in a single unit, The Code of Good Practice on the Prevention of Environmental Pollution from Agricultural Activity, 1992, which was due to be revised in 1997 and which revision was still awaited at the time of going to press. The Code of Good Agricultural Practice for the Protection of Water was published in July 1997. The aforementioned codes place requirements on farmers to control emissions to all environmental media. The place of environmental monitoring within this framework is largely the remit of the enforcement body (the Environment Agency and the Ministry of Agriculture, Fisheries and Food (MAFF)), but extensive costs of monitoring activity will be borne by the farmer in accordance with the **polluter-pays principle**. The polluter-pays principle states that the originator of pollution will pay for its abatement or clean-up and in sufficient amount. This includes activities such as analysis and monitoring, which are associated with continued emission of pollution under licence, abatement and remediation to agreed standards.

1.6.3   Domestic impact on the environment

Domestic waste is covered by various pieces of legislation, largely aimed at the waste handling industry, such as landfill operators. The emissions to the environment from this activity largely emanate from landfill gas and odours and from leachate. The emission of smoke and other nuisance from domestic dwellings is now controlled by the Environmental Protection Act 1990 and its subsequent amendments. This followed on from the Clean Air Act 1956, which dealt with smoke emissions in particular, subsequent to the London smog disaster of December 1952. Considerable requirements are placed upon waste handling operators to monitor their activities, but on the private householder the only requirements fall to local authorities. The cost of their monitoring activities is, of course, ultimately borne by the tax-payer.

The above introduction is not intended to serve as a comprehensive treatise on the law governing the generators of pollution but rather to give sufficient insight to identify where responsibility lies for the various areas of environmental protection.

1.7   Toxicological aspects of pollution

Pollution may have many effects on life. The degree to which xenobiotic materials affect the biosphere is determined by toxicological and ecotoxicologial examination. Without toxicological information, risk assessments for the protection of the natural environment and for the workplace environment (for example under the Control of Substances Hazardous to Health Regulation 1985, or COSHH) become impossible to perform. In this text we have incorporated in the appendices lists of priority pollutants and associated toxicological data. In addition, data in the form of occupational exposure limits are presented for many common reagents and chemicals. The main purpose for this inclusion is to highlight and assist in the procedural aspects of analysis and monitoring designed to protect the analyst in the laboratory. The ecotoxicological information required to carry out environmental risk assessments is beyond the scope of the present volume. Furthermore, a risk assessment for environmental management purposes is founded upon the entire knowledge base, which is a dynamic entity, changing extensively and continuously. Any attempt to provide such information to an environmental analyst would be fruitless. However, it should be borne in mind that there is a potential risk of harm from any contaminated environmental sample, and risk assessments should be performed on all tasks and operations involved in environmental monitoring.

1.8   Environmental quality management – the place of environmental monitoring

1.8.1   General considerations

The natural and workplace environments are protected by standards, which in turn arise from legislation. The standards are demonstrated to have been met, or otherwise, by the implementation and maintenance of an environmental monitoring programme, to be carried out by or on behalf of the generators of the pollution, following the precept of 'the polluter pays'. Where standards are not met the polluter is brought to task and, ideally, compelled to act appropriately. So environmental monitoring is the penultimate link in a chain of responses by humankind to unacceptable levels of pollution in the natural environment.

Environmental monitoring is therefore of equal importance to environmental protection as are environmental management systems and other management tools (e.g. environmental impact assessment, or EIA, life-cycle analysis, risk assessment), environmental law, remediation technologies etc.

To teach or instruct students in the methods for environmental monitoring is no easy task. There are two major approaches: to offer guidance on the methods appropriate to each environmental medium in turn or to instruct in the methods of analytical chemistry, physics and biology and then subsequently to identify where each method may be applied to environmental problems. The editors prefer the latter approach, because all analytical methods have found application in the monitoring of the natural environment. In almost all cases the only components of an analytical protocol which varies depending upon the environmental source of the sample are the sampling strategy, sample storage and sample pre-treatment. For this reason analytical techniques have not been presented as groups of applications, but rather as discrete entities in their own right. Wherever possible examples have been cited within each chapter of applications to specific environmental media. This is sometimes easy, such as the application of remote sensing techniques to atmospheric monitoring. Sometimes, however, an analytical technique has such broad application that specific examples become meaningless, such as gas chromatography–mass spectrometry (GC–MS). This exceptionally powerful technique can be applied to air monitoring by entrapment of airborne pollutants on adsorbent material followed by automated thermal desorption (ATD) and GC–MS. It may be applied to water monitoring either by direct injection or by solvent extraction followed by GC–MS. Finally, it may be applied to soil monitoring by solvent extract or headspace analysis followed by GC–MS. The latter three examples are not intended to be exhaustive.

1.8.2 Environmental standards

Environmental standards are not merely the establishment of an acceptable level of a pollutant in a given medium. They are intended to incorporate an understanding of the capabilities of analytical science. Consequently standard methods are born out of environmental legislation and standards. Wherever possible, practitioners of environmental monitoring should refer to standard methods, if they exist, and the standard operating procedures (SOPs) contained within them. SOPs should contain all of the component parts of an analysis, from sampling methodology to statistical treatment of results. Where a standard method does not exist then the analyst must assemble all of these component parts. It is hoped that this book will provide the tools for students and practitioners of environmental analysis and monitoring to do just that.

1.9 Using *Analytical Methods for Environmental Monitoring*

This book is intended to serve as an introduction to analytical techniques, which are extensively used in environmental protection. The present chapter introduces the reader to the fundamental reasons, both scientific and legal, for environmental protection and, within that discipline, for environmental monitoring. The fundamentals of analysis are dealt with in Chapter 2, providing statistical methodology upon which all analysis is based.

Sampling of environmental media is the first step in a monitoring programme, and the reader should seek advice from Chapter 3 to develop a sampling programme to suit his or her needs.

Chapters 4 through 11 provide a discipline by discipline account of the available techniques for analysis of environmental samples which should be available in most 'well-found' laboratories, i.e. those which are equipped to perform all fundamental procedures in analytical chemistry such as titrimetric and gravimetric analysis with pre-treatment protocols where appropriate, such as distillation. Basic instrumentation is also implied: UV–visible spectrophotometry; IR specrophotometry; pH, dO_2 and redox measurement.

Chapter 4 provides an introduction to small-scale test kits, which are designed for use in the field and which provide rapid and inexpensive analysis of water, soil and gases. It should be borne in mind, however, that speed and economy are almost universally associated with decreases in accuracy, precision and limits of detection.

The appendices are intended as a ready reference for that information required to establish and develop analytical protocols.

1.10   Reference

Odum E P 1971 *Fundamentals of ecology*, Saunders, Philadelphia, PA

1.11   Bibliography

Department of the Environment 1992 *The UK environment*, HMSO
Methods for the examination of waters and associated materials, published in parts, HMSO
Greenberg A E, Clesceri L S and Eaton A D (eds) 1995 *Standard methods for the examination of water and wastewater*, 19th edn, American Public Health Association
McEldowney J F and McEldowney S 1996 *Environment and the law*, Addison Wesley Longman, London

2

Essentials of analysis

2.1 Introduction

Environmental analysis is primarily concerned with the measurement of parameters and their changes relevant to the environment. This, however, constitutes a very wide field, encompassing parameters of the landmass, water bodies and the atmosphere. Global assessment of the environment is normally carried out from the analysis of data recorded by remote sensing. This includes studies of soil and rock, water content in soil, flood and drainage of both surface and underground waters, the ozone loading in the upper atmosphere and aerosol loading including weather in the lower atmosphere. The ecosystem maintains a delicate balance to sustain life on Earth and the pollution loading will destroy this balance. Any species or parameter (e.g. temperature, humidity, pH) which disturbs the delicate balance in the ecosystem and poses a physiological hazard can be considered as pollution. Monitoring of pollution, therefore, is an integral part of environmental diagnostics. The confidence and the effectiveness of such a diagnosis, however, depend critically on the evaluation of limitations of instruments and methods, statistics of signals, efficiency of signal retrieval from noise and, finally, the interpretation of results through data reductions and manipulations.

Laboratory measurements of parameters are preceded by the collection (sampling), preparation and presentation of samples and are followed by data analysis. The latter depends on the method of measurements and the nature of the parameter measured. In the conventional wet chemical analysis, the constituents of a sample are usually measured by gravimetric or by titration methods and often samples are separated by chromatographic techniques. Nowadays, different kinds of spectroscopic and physical techniques are extensively used for such analysis. These techniques are explained in detail in the subsequent chapters; the methods of data analysis, however, pertain to all analytical measurement techniques and are described below.

2.2 Data analysis

Confidence in experimentally obtained data is defined in terms of instrumental precision, accuracy of observation and reproducibility under similar conditions. Whatever be the instrument and the mode of operation, the experimental results will be fundamentally inaccurate to some degree. The magnitude of this uncertainty, i.e. the size of the error in the final value of the parameter of interest, must be known to quantify the degree of confidence in the results.

The precision of a measuring device and the accuracy with which a parameter has been measured jointly determine the degree of reproducibility of a set of measurements of the same parameter under the same conditions. The degree of reproducibility, on the other hand, is governed either intrinsically, i.e. by the manufacturer's prescribed accuracy of the measuring device, or extrinsically, i.e. by the scatter from the mean value of the data in repeated measurements. The latter depends on the operator's skill and the variation in the operating conditions.

Most measurements, particularly those in chemical analysis, rely on visual inspection with respect to a scale. Whether this be a metre rule, a measuring flask or a moving-coil meter is of no consequence; only the markings made by the manufacturer of the scale are important. The accuracy of the measurement will, therefore, be defined by the half fraction of the smallest calibration unit of the scale. For example, if we are measuring the length of an object by a ruler calibrated and marked in mm units, the length of the object should be quoted as, for example, 80.45 ± 0.05 cm. In this, it is assumed that the initial value is certain, i.e. one end is positioned exactly on to a line marking, and that the eye is capable of judging a ±0.05 cm difference. The same is applicable in the measurement of weight, i.e. the initial weight is assumed to be zero with no error when balanced. Measurements of parameters with digital electronics offer much better accuracy than visual measurements from markings.

If, however, the measurements are repeated, e.g. a set of 10 measurements of the length of the same object are taken, the results will not always be the same. There will be a scatter in the value, either due to the variation in observation of the scale in the ruler or as a result of uncertainty in the positioning of the object on to the ruler or due to some changes in the object itself during the course of the measurement. Let us assume that the mean of the 10 repeated measurements of length of the object that we considered before is 80.42 cm. Let us also take it that the maximum scatter values, i.e. the maximum deviations from the mean value, are +0.1 (higher) and −0.05 (lower). In this case, it can be said that the 'acceptable' value for the length of the object is 80.42 cm and it is measured with an accuracy of ±0.1 cm.

It should be noted that the precision of the measuring tool (±0.05) is much higher than the maximum deviation of data from their mean (±0.1). It is obvious that, in this circumstance, accuracy of measurement cannot be improved by simply using a more precise ruler, i.e. a ruler with finer markings. The measurement accuracy is, therefore, limited by the random distribution of data (scatter). The measurement uncertainty or accuracy governed by the statistical nature of a set of data is known as random error. The statistical analysis of random errors is an essential element of data analysis in most physical analytical measurements.

The accuracy of a set of measurements, not limited by the quoted precision of the device, is normally expressed in terms of either absolute error or relative error. If the value is known precisely (accepted value) then the absolute error $E_a = x \pm \bar{x}$, where x is the accepted value and $\bar{x}$ is the mean of the set of measurements. The absolute error, therefore, could be positive if more or negative if less than the accepted value. It is, however, customary to express the accuracy in terms of relative error as a percentage as follows:

$$\text{relative error} - \frac{x \pm \bar{x}}{x} \times 100\% \qquad (2.1)$$

Table 2.1 *Example of scatter in data: repeated measurements of mass of a substance in an analytical experiment*

Measurement number	Recorded mass (g)	Measurement number	Recorded mass (g)
1	12.010	21	12.022
2	12.016	22	12.050
3	12.026	23	12.008
4	12.034	24	12.038
5	12.40	25	12.005
6	12.021	26	12.027
7	12.046	27	12.020
8	12.031	28	12.015
9	12.035	29	12.037
10	12.029	30	12.034
11	12.025	31	12.009
12	12.051	32	12.044
13	12.041	33	12.023
14	12.013	34	12.028
15	12.026	35	12.040
16	12.026	36	12.031
17	12.017	37	12.045
18	12.036	38	12.018
19	12.034	39	12.011
20	12.024	40	12.021

2.3   Statistical analysis of random errors

For a large number of repeated measurements of a parameter the calculation of error using equation (2.1) gives only the limit but not the distribution of this error. It is desirable to know how often a value within a set range (sampling interval) has been recorded. This is done by sampling the data and generating a frequency spectrum and is explained by an example as follows.

Let us consider a set of 40 repeated measurement data for the mass of a sample as shown in Table 2.1. The random variation in this collection of data could be attributed to small errors of judgement at the limits of definition of the balance, accidental errors due to small disturbances and the consequent fluctuating conditions in the environment and the sample etc.

The mean or average of the 40 values of mass is 12.028 g. The lowest and the highest values recorded are 12.005 and 12.051 g respectively. For most practical purposes the accepted value of the mass will be 12.028 g with a spread of ±0.027 g, i.e. all the values lie in the range 12.028 ± 0.027 g. However, it is accepted that most appropriate value of the mass will be that which has been recorded (within a small range) the majority of times out of the total number of trials. This is best analysed by setting data groups of equal sizes called cells or sampling intervals and recording the number of times the observed data points fall within each cell, as shown in Table 2.2.

Table 2.2 *Distribution of data within intervals and the frequency of occurrence*

Range of values (sampling interval)	Number of times recorded (number of samples)	Frequency F of occurrence out of 40 samples (%)
12.001–12.005	1	2.5
12.006–12.010	2	5.0
12.011–12.015	3	7.5
12.016–12.020	4	10.0
12.021–12.025	6	15.0
12.026–12.030	7	17.5
12.031–12.035	6	15.0
12.036–12.040	5	12.5
12.041–12.045	3	7.5
12.046–12.050	2	5.0
12.051–12.055	1	2.5

For most practical purposes, the random errors set a limit to the accuracy of a measurement. In principle, the mean value of a very large number of replicated data points may be considered to be the true value of the parameter. In practice, however, it may not be possible to take a very large number of measurements. In this case, the data set is usually treated using the laws of statistics with some restrictions.

The laws of statistics apply strictly to a very large number (infinity, ∞) of replicated data points, known as the 'population'. A finite number of data points is called a 'sample' of the theoretically possible population. In statistical terms, the mean (μ – pronounced as 'mew') of a population is defined as:

$$\mu = \lim_{N \to \infty} \frac{\sum_{i=1}^{N} x_i}{N} \tag{2.2}$$

i.e. the mean μ is the sum of the values of the parameter for a large number of trials, divided by the number of trials.

The data in Table 2.2 are plotted as a histogram with the frequency of occurrence against the sampling interval, as shown in Figure 2.1. If the sampling intervals were made very small, then a smooth curve can be drawn (as shown). This smooth curve, for a large number of data points, would represent the frequency distribution of the data. For random errors, such a curve is most likely to be a Gaussian type and is called a 'normal error curve'. From this curve two major conclusions can be drawn as follows:

(a) The most frequently observed value in the range interval is the mean, μ, of the data points.
(b) Large variations from the mean occur less frequently and symmetrically about the mean.

Intuitively then, it is understood that, for a very large number of data points, the curve in Figure 2.1 will be sharply defined and the mean value will approach the true value. This will be achieved only in the absence of a bias, i.e. systematic errors.

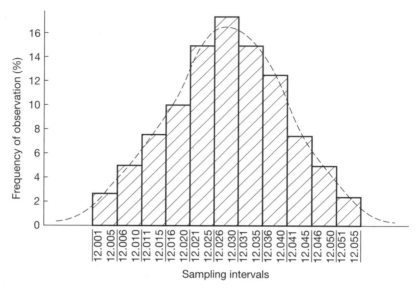

Figure 2.1   Histogram plot of frequency of occurrence of events against sampling interval of hypothetical data points

2.3.1   Systematic error

This type of error could be either subjective or objective and gives rise to a consistent bias. The subjective bias is due to personal judgement in making measurements such as reading the needle of a meter consistently to higher or lower values at the limit of accuracy, variation in reaction time of the observer in the case of timing an event etc. Whatever be the reason, some people will record a parameter consistently higher and others may record consistently lower than the true value.

The objective bias comes from inherent misadjustment, misalignment or wrong calibration of instruments. The instrumental bias could also arise from drifts in electronic measuring circuits due to temperature rise (or fall), fluctuation in leakage current, decrease in voltage in the battery etc.

2.3.2   Standard deviation and variance

Statistically, the accuracy of the value of a parameter in relation to the calculated mean (nearly true value for a very large number of samples) is quantified by a parameter called the standard deviation, σ. It is essentially a measure of the maximum variation of a set of data points about the mean and is defined as:

$$\sigma = \sqrt{\lim_{N \to \infty} \frac{\sum_{i=1}^{N}(\bar{x} - x_i)^2}{N}} \qquad (2.3)$$

where x_i is the value of the ith sample (i.e. the ith trial). In other words, the standard deviation is the square root of the mean of the squares of all individual deviations from the mean value, $\bar{x}$.

Sometimes the accuracy of measurements is quantified by a parameter called the variance, which is simply the square of the standard deviation, i.e. σ^2. The advantage of using variance is that it combines additively. For example, if a parameter is evaluated from two or more sets of replicated measurements, e.g. $y = x + z$, then the variance of the parameter is given as:

$$\sigma_y^2 = \sigma_x^2 + \sigma_z^2$$

The standard deviation is defined for $N \approx \infty$, i.e. for a very large sample; therefore it is an unknown or unobtainable quantity. When the number of replicated data points to be averaged is small, e.g. <30, a practical standard deviation, S, is then defined as:

$$S = \sqrt{\frac{\sum\limits_{i=1}^{N} (\bar{x} - x_i)^2}{N-1}} \tag{2.4}$$

From the statistical analysis it is found that the above expression provides a better estimate of the standard deviation than that obtained using equation (2.3). In other words, the use of equation (2.4) allows statistical inferences (conclusions) about the parameter of the population from which the samples are derived.

2.4   Probability distribution

In the previous section the mean and standard deviation were defined for a sample data set. Since the laws of statistics applies to populations, the analysis of sample data must be subject to some uncertainty. To take account of this the laws of probability are invoked.

In these analyses, a probability value is attributed to each occurrence of a data point so that a distribution of probabilities will arise in the set of data points. In the case of a discrete variable which can have certain distinct random values, e.g. 1, 2, 3, ... , in a given range, there could be several types of probability distributions. Most important among these are the binomial and Poisson distributions.

The binomial distribution is characterized by a fixed number of trials (n), giving one or the other of two outcomes such as 0 or 1 in a binary result, head or tail in the tossing of a coin. For an unbiased condition each will have equal probability ($p = 0.5$) of occurrence. The variable, x, for having, for example, a head in the tossing of a coin n times will have a binomial distribution, $B(n, 0.5)$, given by the general formula:

$$P(x) = \frac{n!}{x!(n-x)!} p^x (1-p)^{n-x} \tag{2.5}$$

where $P(x)$ is the probability of obtaining x heads for a total number of n trials and the symbol '!' stands for factorial, i.e. $n! = n \times (n-1) \times (n-2) \times \ldots \times [n - (n-2)]$. For this distribution the mean and standard deviation are given by statistical laws as follows:

$$\text{mean} = n \times p \quad \text{and} \quad \text{standard deviation} = [n \times p(1-p)]^{-1/2}$$

The Poisson distribution applies to theoretically unlimited discrete random events occurring per unit of time or space. Examples are β decay in radioactivity and photoelectric emission in photomultiplier tubes. The probability of such events per unit time (or space) is given by:

$$P(x) = \exp(-m) \times m^x / x! \tag{2.6}$$

where m is the mean number of events per unit of time (or space) and the standard deviation is $m^{1/2}$.

For continuous random variables which can have any value in a range, the normal distribution applies. This occurs when a variable is measured for a large number of identical objects or samples or by a large number of trials on the same sample. The statistical variation could be due to slight differences in the sample or measurement errors. The properties of the normal distribution allow conclusions to be drawn from a sample data set about the population from which the sample is taken. In addition, the analysis of the normal distribution allows a better understanding of the significance of the standard deviation. Therefore the analysis of the normal distribution is described in detail below.

2.4.1   Normal distribution

The normal distribution of a population is defined jointly by the mean, μ, and the standard deviations, σ, of the population. If the data are collected over a fixed sampling interval then the relative number of data points contained in this interval, dN/N (for $N \to \infty$), can be termed the relative frequency of occurrence, F. From the theory of statistics, the normal frequency distribution is expressed as:

$$F = \frac{\Delta x}{\sigma\sqrt{2\pi}} \exp\left[-0.5\left(\frac{x-\mu}{\sigma}\right)^2\right] \tag{2.7}$$

The population mean and standard deviation are theoretical quantities and are always unknown, as the number of data points, in reality, is never infinite. In practice, therefore, one has to be content with a sample; the larger the sample the closer will be these parameters to the true values.

One way to estimate the values of μ and σ from a relatively large sample (>30 data points) is to plot the equation for various combinations of values of the pair of parameters μ and σ. From the best fit to the experimental frequency distribution, these parameters can be obtained. This, however, leaves an uncertainty as to the definition of 'best fit' and does not provide a confidence level or limit of the error. Besides, this could be a long numerical mathematical process.

A convenient way to use statistical methods to analyse the normal distribution is to standardise it. The exponential parameter in the equation, i.e. $(x - \mu)/\sigma$, is simply the deviation of the variable (x) from the mean, per unit of standard deviation. This, therefore, can be replaced by only one parameter, say $z = (x - \mu)/\sigma$, so that the normal distribution in equation (2.7) can now be written as:

$$f(z) = \frac{\Delta z}{\sigma\sqrt{2\pi}} \exp(-0.5z^2) \tag{2.8}$$

This equation can be plotted as the distribution of f against z values for small intervals Δz with normalisation so that $\mu = 0$ when $\sigma = 0$. Such a plot of this equation, as shown in Figure 2.2, is known as a standard normal curve. Certain characteristic features of a standard normal curve are:

- The maximum frequency occurs when the deviation is zero and the deviations are symmetrically distributed on both sides of the mean (i.e. positive and negative). The frequency decreases rapidly with increasing deviation, reaching the z-axis asymptotically. The latter means that the large errors occur much less frequently than the small ones.

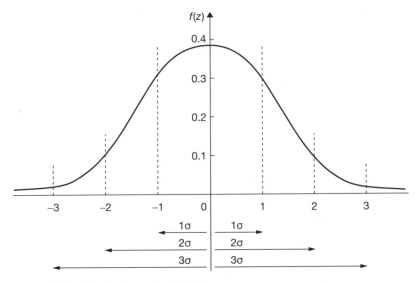

Figure 2.2 Standard normal distribution

• It can be mathematically shown that the area under the normal distribution curve between $z = +\infty$ and $z = -\infty$ is a square unit representing the total population. Furthermore, it can be shown that approximately 68%, 95% and 100% of the total area lie between $\pm 1\sigma$, $\pm 2\sigma$ and $\pm 3\sigma$ respectively. This means that if, circumstantially, the standard deviation of a sample is considered to be a good match to that of the population from which the sample is taken, then about 68% of the population will be contained within $\pm 1\sigma$ and almost 100% of the data points will lie within $\pm 3\sigma$.

From the theory of probability the above can be interpreted in another way. Suppose the sample is large and the value of this sample is accepted as the correct one. It can, then, be inferred that the value of a single data point in the sample will have 68% chance of being accurate within $\pm 1\sigma$ and 100% chance of being accurate within $\pm 3\sigma$. In other words, the population mean will be $\pm \sigma$ or $\pm 3\sigma$ with probabilities of 0.68 and 1 respectively. With the help of statistical tables (e.g. Neave, 1992) any level of probability statement on the errors of a normally distributed data set can be made. This type of probability statement or statistical inference is often called a confidence interval or confidence limit over an interval. For example, one can say with 95% confidence (0.95 probability) that the mean of a single sample, $\bar{x}$, will lie within two standard deviations ($\pm 2\sigma$) of the population from which the sample has been chosen randomly.

If r such samples are chosen from a population, the means of the individual samples will approach a normal distribution as r increases, according to central limit theory. The theorem also tells that the mean of the sampling distribution of means, X, is equal to the mean of the population, μ, i.e. $\mu = X$, and the standard deviation of the sampling distribution of means, S, is related to the standard deviation of the population, σ:

$$S = \frac{\sigma}{\sqrt{N}} \tag{2.9}$$

where N is the number of data points in the sample.

Returning to the concept of confidence interval, the population mean is then related to the mean of a single sample, X_1, as follows:

$$\mu = X_1 \pm tS \tag{2.10}$$

where t is the multiplying factor corresponding to the desired confidence limit for the mean of the sample having a standard deviation S. From equation (2.8) we can write

$$\mu = \bar{X}_1 \pm \frac{t\sigma}{\sqrt{N}} \tag{2.11}$$

For large samples, e.g. $N > 30$, statistical analysis and everyday experience tell us that it is usual to assume that $\sigma = S$, i.e. the population mean is practically the same as the sample mean. Therefore, it can be inferred that

$$\mu = \bar{x}_1 \pm \frac{tS}{\sqrt{N}} \tag{2.12}$$

This equation gives a neat way of estimating the confidence limits and level in the correct value of the mean of a large sample. The value of t is chosen by the confidence level desired, e.g. $t \sim 1$ for 68% confidence, $t \sim 3$ for 99.7% confidence etc. The inference, therefore, is that, for a large enough sample, approximately 99.7% of the data points will lie within $\pm 3\sigma$ for $t = 3$.

For small samples ($N < 30$), the multiplying factor can be obtained from standard statistical tables given in most textbooks on statistics. For this it must be assumed that the distribution is approximately normal. In addition, one needs to evaluate a parameter, $\alpha = (1 - C)/2$, where C is the confidence level desired, e.g. $C = 0.95$ for 95% confidence limit, and a value for the degrees of freedom, $N - 1$.

The probability distribution of a large sample having a mean μ and a standard deviation S can be represented by what is known as a 't' distribution:

$$t = \frac{X - \mu}{S/N^{1/2}} \tag{2.13}$$

Inference on the statistical parameters such as confidence interval estimation of a population from a sample can also be carried out by a method called hypothesis testing using the above equation. In this the 'confidence limit' is replaced by the concept of 'significance level'. The test involves calculating the value of t from equation (2.13) using known (calculated) values of the other parameters (null hypothesis) and assuming a normal distribution and comparing this with the value obtained from the standard statistical tables for a chosen significance level (usually 5%) and alternative parameters (alternative hypothesis). The test will allow conclusions to be drawn: for example, if the calculated value of t is less than the value obtained from the table for, say, a significance level of 5%, the assumed hypothesis is valid within 5% risk.

However, there are some hypothesis tests which do not require rigorous assumptions about the distribution of the population. These are called non-parametric hypothesis tests. Among these, the sign test, the Wilcoxon signed rank test and the Mann–Whitney U test are commonly used for both small and large sample sizes. These tests make less rigorous assumptions than the t tests. If the assumptions of the t tests are found to be valid, the non-parametric tests are not needed as these are less powerful. Details of both parametric and non-parametric tests can be found in most standard books on statistics and are not included here.

2.5   Regression analysis (least-squares fit)

In experiments where two interdependent variables are measured it may be a requirement to establish a mathematical correlation between them. An equation representing such a correlation is called a regression equation and provides a prediction of data beyond the limits of measured values. By the process of regression analysis, which is essentially finding a plot which is the best fit to the data points, one can obtain various constants of the equation. Such analyses are widely used in situations where linear relationships are expected or the plotted data points show a linear trend. In cases where the data points do not show a linear trend these can be transformed into a linear relation by mathematical manipulation. The most commonly used transformations are 'logarithm' (log or ln) in cases of exponentially varying data points, the 'square root' transformation (for data which follow Poisson distribution) and the 'arcsine' transformation (for data which follow binomial distribution). The validity of such transformations can be justified by various tests known as 'goodness-of-fit tests' or in most cases by a parameter called the 'correlation coefficient value'.

In cases where a simple transformation is not possible, other complex procedures are available known as non-linear regression analysis. This is beyond the scope of this chapter and the readers are referred to relevant textbooks. However, these analytical procedures are now available as computer software and could be readily utilised for the analysis of experimental data. For a basic understanding of the procedure, the method of linear regression analysis is explained below.

All experiments are prone to both subjective and objective uncertainties. Therefore, even for an expected linear relationship between two variables, e.g. x and y, the discrete data points (x_i, y_i) may not lie exactly on a straight line, although they may show a tendency to follow one. A common practice is to draw a straight line through the points so that, on the average, half of the points at the lower values of the axes fall on one side of the line and the higher values lie on the opposite side. This is rather a crude method and is not satisfactory. A statistical method is, therefore, used to draw a line of 'best fit' through randomly scattered data points, or mean values of such points when replicated measurements of the dependent parameter are carried out. This method is known as the 'least-squares fit' and is based on numerical linear regression of the independent variable, x, on the dependent one, y. The scheme for such analysis is widely available in modern pocket calculators and in statistics software packages for computers. However, for effective utilisation of such a scheme it is necessary to understand the basic principle of the methodology and its limitations.

The least-squares method is based on some fundamental assumptions, which are as follows:

1. The random or statistical errors in the independent variable x_i ($i = 1, 2, 3$, etc.) either are negligible or do not exist. For example, in an experiment to determine the dependence of absorbance (y) on the concentration (x) of a sample, replicated measurements of absorbance are made on each sample concentration, which is assumed to be measured with no errors.
2. The errors in the value of the dependent parameter in replicated measurements corresponding to each independent parameter are normally distributed. That is, if the absorbance of a sample of concentration c is measured repeatedly in a spectrometer, the frequency distribution of the values will be a Gaussian curve as described earlier.
3. The errors in the values of the dependent variable are not related to the values of the independent variable. This means that, as far as the magnitude of the error is concerned, all data points are considered with equal importance, in other words the

Table 2.3 *Example of typical fluorescence intensity versus concentration of species data for least-squares analysis*

Index (*i*)	Concentration (x_i) (mM)	Measured fluorescence intensity (y_i) (arbitrary units)	Fluorescence intensity from the 'best-fit' data (y_i')	$y_i - y_i'$
1	0	0.1	0.418	−0.318
2	0.5	3.0	2.410	0.590
3	1.0	3.8	4.721	−0.921
4	1.5	8.0	7.501	0.499
5	2.0	10.0	9.861	0.139
6	2.5	12.2	12.612	−0.412
7	3.0	14.4	15.000	−0.600
8	3.5	19.1	17.824	1.276
9	4.0	20.7	20.139	0.561
10	4.5	22.6	23.000	−0.400
11	5.0	26.9	25.279	1.621
12	5.5	27.8	27.922	−0.122
13	6.0	29.1	30.418	−1.318

errors are unweighted. Although this assumption is not always valid, it simplifies the calculations. In the case where the error increases in proportion to the value of the independent variable, a more complicated algorithm based on weighted regression analysis is needed.

On the basis of the above simplifying assumptions, the principle of the least-squares method is described as follows.

The equation for a straight line is given as:

$$y = mx + c \tag{2.14}$$

where m is the slope and c is the y intercept. Let us consider the results of a series of fluorescence intensity data for some discrete values of concentrations as shown in Table 2.3 and plotted in Figure 2.3. To fit a 'best' straight line through the data points in the intensity (y) against concentration (x) plot, appropriate values of the slope and the intercept need to be found. This is done by first assigning a parameter called the 'y residual'. If the solid line on the graph is the best fit and y_i' represents the values of absorbances on the straight line for the corresponding values of x_i ($i = 1, 2, 3$, etc.) then the absolute value of $y_i - y_i'$ is called the y residual. It is obvious that the value of this residual could be positive or negative. In the least-squares method, the sum of the squares of these residuals is minimised. Since $y_i' = mx_i + c$, the sum of the squares of the residuals is given as:

$$S = y_i - y_i' = \sum_{i=1}^{n} (y_i - mx_i - c)^2 \tag{2.15}$$

If the above equation is differentiated separately with respect to m and c, the minimum in these values will occur when the results are independently set equal to zero, i.e. when $dS/dm = dS/dc = 0$. From such mathematical manipulations, expressions for m and c in terms of mean values of x and y are obtained, which are given as follows:

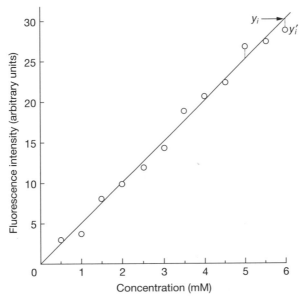

Figure 2.3   Typical plot of fluorescence intensity against concentration of species

$$m = \frac{S_{xy}}{S_{xx}} = \frac{\displaystyle\sum_{i=1}^{n}(x_i - \bar{x})(y_i - \bar{y})}{\displaystyle\sum_{i=1}^{n}(x_i - \bar{x})^2} \tag{2.16}$$

and

$$c = \bar{y} - m\bar{x} \tag{2.17}$$

where

$$\bar{x} = \sum_{i=1}^{n} x_i/n \quad \text{and} \quad \bar{y} = \sum_{i=1}^{n} y_i/n$$

The straight line in Figure 2.3 is actually drawn as a least-squares fit of the data in Table 2.3, using equations (2.16) and (2.17).

A general property of the least-squares calculation based on the above equation is that the residual sum must be zero (allowing for rounding errors) if

$$\sum_{i=1}^{n}(y'_i - \bar{y}) = 0$$

and the line of 'best fit' must pass through the centroid, i.e. the point $(\bar{x}, \bar{y})$.

In applying this method it is advisable to look at the data points on the graph and if one or two points appear to have large y residuals a gross error should be suspected and such points should be verified or rejected as outliers. Otherwise such points (even a single outlier) may substantially affect the regression analysis.

In some experiments it is reasonable to assume that the line should pass through the (0, 0) point. For example, in absorption measurements, the absorbance should be ideally zero when there is no target sample in the solution and the absorption by the solvent has been compensated. Therefore, in such situations, equation (2.14) can be written simply as $y = mx$. However, stray light, background electronic noise and a host of other experimental factors may give rise to a detectable signal even in the absence of the absorbing species. Therefore, it is not a good practice to force the line through the origin (0, 0). If this is done, the least-squares line may no longer pass through the centroid $(\bar{x}, \bar{y})$.

2.6   Noise analysis

Modern detection and measurement techniques are based on either electronic (e.g. ammeter, voltmeter) or electro-optic (e.g. photodiode, photomultipliers) detection devices. Noise is an integral part of signals in all such detection devices and puts a limit to sensitive detections. Sources of noise are diverse and their contribution to signals depends on the characteristics of the instruments and their environment. Noise arising from environmental effects is diverse and will depend on the experimental conditions, whereas that arising from instruments may be classified and quantified as follows.

2.6.1   Thermal noise (Johnson noise)

This is generated in electronic components and electrochemical cells, transducers, etc. owing to random agitation of electrons or other charge carriers at the ambient temperature. The random motion (agitation) of the charge carriers gives rise to a voltage fluctuation in the electronic monitoring device, which appears as noise or statistical fluctuation in the readout.

To carry information as electric signals or to respond faithfully to a sudden change in signal intensity an instrument must have a finite bandwidth. The bandwidth, i.e. the frequency interval within which such response is possible, is determined by the overall resistive and capacitive elements in the instrument. The bandwidth, Δf (hertz), is inversely related to the response time of the device, t_r (seconds). The latter is defined as the time required for the output to rise from 10% to 90% of its maximum value.

The root-mean-square voltage V_{rms} at the output of a device without any signal input can readily be calculated from a simple relation involving the bandwidth and temperature as follows:

$$V_{rms} = \sqrt{4kTR\,\Delta f} \tag{2.18}$$

where k is the Boltzmann constant ($=1.38 \times 10^{-23}$ J K^{-1}), T is the absolute temperature in kelvins and R is the resistance of the component across which voltage signal is measured.

It is obvious from the above equation that the thermal noise can be minimised either by lowering the operating temperature or by reducing the bandwidth. The implication of the latter is that the instrument becomes slower to respond to sudden signal changes.

2.6.2   Flicker noise (1/f noise)

The long-term drift observed in dc amplifiers, ammeters, galvanometers etc. is a manifestation of this type of noise. The cause of this type of noise in analytical measurements is not clearly understood but could be due to a slow increase in temperature in the electronic components. The main characteristic of this type of noise is that its magnitude is inversely proportional to the frequency of the signal and becomes significant at frequencies lower than 100 Hz. The flicker noise can be reduced by using metallic film resistors in place of commonly used carbon resistors.

2.6.3 Background noise

This is inherent in all electrical or optical detection. In voltage- or current-measuring devices external electromagnetic pick-up and the random fluctuation in the main supply voltage may give rise to the background noise. In photodetectors, in addition to the current due to cosmic radiation, thermal fluctuation etc., there may be noise due to the leakage of light other than that being detected, as a result of inefficient baffling or filtering. These, in general, give rise to a dc level of voltage and can usually be biased off in most cases. However, the statistical fluctuation in this noise (discussed below) will remain and may limit the detection sensitivity.

2.6.4 Shot noise

This is inherent in all measurements involving transducers whether these are photoemissive devices such as vacuum photodiodes or photomultipliers or semiconductor detectors. The shot noise arises as a result of the granular (quantised) nature of light (flow of photons) and the photon-induced current (flow of electrons). The photoelectric emission is a random process, i.e. the time between successive photoelectric events is statistically uncertain. This type of process is analysed by probability theory. From the analysis it has been found that photoelectric emission obeys Poisson's statistics, i.e. the uncertainty in photoelectric events is proportional to the square root of the rate of events. The root-mean-square current fluctuation i_{rms} associated with the average direct current I (proportional to electron flow rate) is given by:

$$i_{rms} = \sqrt{2Ie\,\Delta f} \tag{2.19}$$

where e is the charge on the electron and Δf, as before, is the bandwidth of the detection system.

It is obvious from the above equation that the shot noise will be minimised by detecting weaker signals and by reducing the bandwidth of the system. If the magnitude of the fluctuation in current is considered to be the measure of uncertainty, termed 'noise' in practice, then the normalised error will decrease as averages are taken of more and more replicated measurements. This is explained in the next section.

2.7 Noise reduction

For sensitive detection of weak electrical signals generated by external stimuli (e.g. light, heat) the associated noise, which could sometimes be orders of magnitude higher than the signal, must be reduced. The extent by which the average level of the noise is reduced in comparison with the signal intensity is quantified by a parameter called the signal-to-noise ratio or S/N. In most measurements the average value (intensity) of the noise, N, is constant and independent of the intensity of the signal. Therefore, the effect of noise becomes progressively more pronounced as weaker signals are measured. Therefore, S/N is a more appropriate 'figure of merit' than the quotation of the level of the noise alone.

There are many electronic and electro-optical techniques for the improvement of S/N in analytical measurements. The frequency distribution of a set of replicated measurements is a Gaussian curve (Figure 2.2). The statistical analysis shows that for a large number of measurements the mean of this distribution can be taken to be true value, i.e. the error will be very small. In a replicated measurement of a parameter (X) such as mass or temperature, $S/N = X/\sigma_s$. Since the standard deviation, σ_s, decreases with the number of measurements, S/N increases as the sample becomes larger. Electronically the data averaging is done on a series of pulses generated by a modulated (pulsed) stimuli or on a series of pulses sampled electronically from a constant dc level of the signal. Since the random shot noise will be positive or negative with respect to the mean, the average of such random noise will approach zero, because the signal, however small, will always have the same polarity and its mean will approach the true value on prolonged averaging, i.e. for a large number of replicated measurements.

There are many different techniques for data averaging and signal retrieval (see Chapter 5) depending on the types and levels of signals and noise. For example, the technique known as phase-sensitive detection by a device called 'lock-in' amplifier is widely used for signal retrieval by averaging the signal at only the modulating frequency. Gated charge integration, box-car averaging, photon counting etc. are some of the techniques currently being used for S/N improvements through data averaging. An example of the effect of data averaging on the Raman spectrum of nitrogen in air is demonstrated in Chapter 5 (Figure 5.13).

2.8   Reference

Neave H R 1992 *Elementary statistical tables*, Routledge, London

2.9   Bibliography

Nolan B 1994 *Data analysis – an introduction*, Polity Press, Cambridge

3

Sampling methods

3.1 General introduction

Sampling holds a position of great importance in an analytical protocol. The only absolutely certain way to ascertain the quality of, for example, a body of water is to analyse the entire body. This would clearly be impractical. The aim of sampling is to acquire for analysis a very small portion of the whole which is representative of the whole. In very few instances is this as simple as when the whole body to be examined is homogeneous or approximates to homogeneity.

There are three main categories of sampling: grab (or spot), composite and continuous sampling. The first of these, grab sampling, may be either random or systematic. Where the determined parameter varies considerably then grab sampling is an inappropriate strategy if the reason for sampling is to determine process efficiency. Where the purpose of sampling is to determine compliance with a consent limit issued by a regulatory body, then the variability of the determinand is irrelevant since compliance must be achieved at all times, regardless of scatter. Thus most regulatory bodies such as MAFF and the Environmental Agency (EA) take random grab samples. Systematic grab sampling is applied to the analysis of a medium which is subject to regular changes in composition, influx of materials etc. Thus, where for example a weekly cycle is encountered in the quality of an aqueous effluent, then a systematic grab sampling programme may be instituted to avoid the loss of data from the extremes experienced.

Composite sampling employs a system for pooling several samples taken over a predetermined time as grab samples. This strategy is employed where information is required for process control or for long-term environmental studies and where continuous methods are unavailable for the determination to be carried out. The effect of composite sampling is to smooth out the extremes of determinand value so that corrective action can be taken to improve process performance.

Continuous sampling is the strategy to be employed wherever possible since, in terms of environmental study, the fullest picture of the monitored medium is obtained. In terms of process control on-line analysis, which incorporates continuous sampling, permits the incorporation of feed-back or feed-forward control loops to optimise process control. This has obvious applications in process engineering, but in the context of environmental analysis its principal application is in the monitoring of remedial processes such as activated sludge technology.

The terminology of sampling just described is often confusing and does not describe precisely the sampling strategy being employed. Perhaps a more consistent set of terms respectively for the three strategies should be non-continuous, semi-continuous and continuous sampling.

3.2 Aims of sampling

Accepting that in real circumstances we must take the properties of an analytical sample to be representative of the whole, then sampling protocol must aim to maximise the representativeness of any sample taken. To achieve this the aims of sampling may be summarised as:

- To select the locations of sampling and the times or frequency of sampling such that requisite information on quality of sample can be calculated or inferred accurately.
- To ensure that the concentrations of analytes of interest are the same as the concentrations in the whole body sampled, at the time and location of sampling. In other words, the act of sampling should not affect in any way the concentration of determinand. A good example of the alteration of the concentration of a determinand due to the act of sampling is the concentration of dissolved oxygen. In sampling water for dissolved oxygen (dO_2) determinations, any physical disturbance of the water during sampling through excessive turbulence or agitation (the slightest occurrence of bubbling may suffice) will result in an erroneous, high measured oxygen concentration.
- To ensure that once the sample has been taken, the concentration of analytes of interest (determinands) does not change from the time of sampling to the time of analysis.

3.3 Sampling point and sampling location

In this regard, sampling point and sampling location differ in that location determines what the sample will represent and sampling point is determined by the requirement for the sample to be representative of the whole. Should it be desirable to determine the effect on the quality of a body of water of discharging into it an effluent, then the sampling locations should represent sites before and after the addition of the effluent. If the water body is a river then these will be suitable sites upstream and downstream. If the determinand was dissolved oxygen, then the sampling point at each location should be representative of the whole water body and therefore be positioned neither at the surface (to avoid excessive turbulence) nor immediately above an anaerobic sediment. Sampling points, therefore, should generally be at mid-depth and mid-stream and on to a prescribed depth. The purpose of prescribing the depth is to achieve consistency of sampling conditions and thus to optimise the final quality of data. In many instances the sampling locations are obvious, whereas the sampling points usually require careful consideration.

3.4 Selection of sampling points and sampling locations

3.4.1 General considerations

Other than the general points regarding suitablity of sampling points and locations discussed in Section 3.3, consideration should be given to:

- *Accessibility.* The ease of access to a site should be ensured when regular samples are to be taken. In some instances it may be an advantage to choose a sampling location or point which is inaccessible, for example where automated and expensive equipment is permanently located. Although the choice of sampling point should ideally take into account accessibility, it should be secondary to choice on the grounds of analytical quality, i.e. the degree to which a sampling point is representative of the body of water (or other environmental medium) should take precedence over personal convenience.

- *Safety and security.* The safety of personnel taking samples must always be a consideration, together with that of the general public. As mentioned above, security may be necessary to protect expensive equipment, but also to protect intruders, albeit uninvited.

- *Avoidance of boundaries.* The concentration of analytes at the banks and beds of rivers and other watercourses, the walls of water pipes etc. are usually not representative of the whole water body. Among the reasons for this is the fact that many biotic elements are concentrated at these locations, particularly plants and micro-organisms, and consequently so are their individual metabolic interactions. In some instances these biotic elements may be the areas of interest, in which case sampling should reflect as closely as possible the water content around them.

3.4.2   Site survey and preliminary sampling programme

Watercourses usually vary more in terms of spatial heterogeneity than do gas streams for example. This arises because stratification often occurs in water bodies, and this only occurs in bodies of gas where laminar flow conditions prevail, or where gas flow is zero, which is rarely the case. In both instances a preliminary survey of the medium of interest should be carried out (a) to determine the spatial heterogeneity and (b) to determine the suitability of any sampling point identified in this way, in terms of quality of data achievable (accuracy, error and variance etc.).

3.4.3   Impact of flow rates

Whether monitoring on a continuous or a semi-continuous basis, information on the flow rate of the medium under analysis may be necessary should mass flows of determinands ultimately be required. There are available many types of flow-metering systems. If continuous flow-rate monitoring is a prerequisite of the analysis, then the choice of sampling points should take into consideration the accuracy with which flow rates are known or can be measured at each point.

3.4.4   Composite sampling

If heterogeneity exists at a required sampling location, then samples must be taken from a number of points at each location. If the assumption can be made that the average of analytical determinations from all of the sampling points within a location is representative of the actual situation, then analytical workload can be reduced by generating composite samples. In other words, samples can be taken from each sampling point and then pooled in equal volumes (or at equal flow rates if sampling is continuous) and analysed as a single sample. Similarly, it is sometimes acceptable for samples, taken at different times, from a single location to be pooled to give a composite value, representing an average value rather than a grab sample. There are commercially available automated samplers for sampling watercourses and which find extensive application in the water treatment industry. Such samplers access a continuous stream of water and a timer is pre-set to capture a given volume of the water flowing through the device and to transfer it to a sample container. The sample container is then retrieved for analysis as often as is deemed necessary. Typically a composite water sampler in a large sewage treatment works operating activated sludge processes would be collected twice daily.

3.5 Time and frequency of sampling

In the case of water quality parameters a single datum is of limited or no value. This may be due to temporal variation in flow and determinand concentration. The aim then is to establish the time and frequency of sampling such that the samples adequately represent the level of determinand during the period of interest. The usual practice, therefore, is to analyse discrete samples over the period of interest. This is not necessary, for obvious reasons, where a technique exists for continuous measurements to be taken or where composite sampling systems operate (see Section 3.4.4).

The following points should be considered when setting sampling times and frequency:

- specifications required by the analysis, e.g. accuracy and precision;
- random and systematic variations of water or effluent quality;
- establishing times of sampling;
- duration of sampling events.

3.5.1 Specifications required by the analysis

The quality of the final information must be defined otherwise a rational approach to designing a sampling regime would be impossible. For general information regarding the quality of analytical results please refer to Chapter 2. However, quality of a water body, for example, may vary randomly, cyclically or both to varying degrees. The degree to which one variation type predominates and the information required from the sampling process determine the sampling times.

3.5.2 Establishing times of sampling

Once the variations prevalent in the quality of effluent have been established, then the choice of times can be set. If the variation is cyclic, then the period should be determined. If the determinand is subject to a consent limit, then it would be sensible to set the sampling times to coincide with a peak in the cycle. If the analyte concentration is to be averaged over a cycle, then the sampling time should not equal the period and preferably be less than half a period.

It is not sufficient to expect the cycle period, or indeed the type of quality variation, to remain constant since this rarely happens. Consequently the whole sampling regime should periodically and systematically be reviewed.

Whether a period of abnormality is the subject under investigation or whether it is insignificant determines the sampling regime. If the former, then usually an additional sampling programme is initiated, with the existing systematic regime still in place. If the latter then the sampling regime should be left in place and reviewed at the earliest suitable opportunity.

3.5.3 Duration of sampling events

Usually, unless the rate of change of effluent or water quality is large, then the rate at which a sample is taken is of little consequence. If the rate of change of quality is great then it must be borne in mind that changes which occur during the sampling event are undetectable. Composite samples are usually used when the rate of change of quality is unimportant.

Identification of the sampling points and times required for a programme may well lead to the situation where there are insufficient resources to allow optimal sampling frequency at all of the sampling points concurrently. Under such circumstances sampling points should be prioritised and sampling events kept to the minimum, which allows a statistically significant result. Alternatively, sampling points may be accessed in rotation.

In summary, two main types of programme are used: where a determinand is measured and remedial action results if water quality fails to meet a predetermined standard, i.e. quality control; where a sampling programme is designed to establish the usual value of a quality parameter over a given period of time and within predetermined levels of accuracy and precision, i.e. quality assessment or characterisation. This is achieved by setting down specific protocols for the acquisition of samples for a given analysis. Such protocols must include sampling point, location, frequency, number of replicates and special requirements dictated by the parameter or analyte concentration to be determined.

3.6   Sample handling considerations

3.6.1   Sample preparation

There is frequently a need to prepare samples for analysis by a treatment which takes place between the sampling event and the analytical procedure. Once portions of water, gas or any other medium for analysis have been obtained and transferred from the sampling points or locations to another location (usually the sample delivery point) for the purpose of analysis, then any sample preparation is carried out at this time.

3.6.2   Sample for analysis

This term denotes all or part of a sample after it has been prepared for analysis. When such preparation is necessary it is often regarded as part of the sampling process. Sample preparation is, however, not always required and in some instances samples are made available for analysis directly on reaching the sample delivery point. Historically in the water industry a 'sample for analysis' refers only to a portion of the gross sample, and the following aspects constitute sample preparation:

- Appropriate identification of samples, such as placing suitable labels on sample containers or collection containers.
- Transfer from collection containers to sample containers.
- The physical and/or chemical treatment of samples to minimise any changes in the concentrations of one or more determinands in the period between the instant of the sample becoming available at the sample delivery point and when the analysis begins. This is the process referred to in sample preservation below.

3.6.3   Sample preservation

The concept of sample preservation is readily apparent when considering water samples, at least as far as common water quality parameters are concerned. When considering gas analysis, in many instances once a sample has been taken it is already in a stable form and ready for instrumental (or titrimetric etc.) analysis. In some cases, however, the gas molecule may be reactive or simply charged, whereupon a proportion of the gas in a sample may adsorb onto the surface of the sampling vessel or react with it.

Specific instances of sample preservation are given in the DoE publication *Methods for the examination of waters and associated materials*. However, general considerations to be made are as follows:

- Adequate stabilisation of samples may often be achieved by the addition of a chemical reagent to them so that the processes causing instability are prevented or reduced. Such processes may be chemical, such as auto-oxidation, physical, such as temperature, or biological, such as biodegradation.
- Stabilising reagents cannot be used when they affect the speciation (relative proportions of each chemical species in the sample) of the sample and this speciation is of interest in the analysis. For example, acidification, a common method of preservation, would be inappropriate in a sample for the determination of the relative proportions of metals in sediments and the surrounding waters, since acidification would almost certainly solubilise many metal salts which are insoluble at neutral pH.
- The concentration of a reagent (or relative volume used) must be sufficiently small that the accuracy and precision of analytical results remain within specification.
- Stabilising reagents may affect the performance of analytical systems (i.e. act as interferences). For example, the addition of acid may prevent the formation of the desired coloured compound (or chromophore) in spectrophotometric procedures.
- Appropriate correction must be made for the dilution of the sample caused by the addition of a reagent in solution. The addition of small volumes of relatively concentrated reagents is, therefore, useful.
- In general terms, it is preferable to add the reagent to the empty sample container before a sample is added. This is not always necessary, but should be adopted until evidence is obtained that this is detrimental in any way to the procedure. The advantage to adding preservatives to sample collection containers is that sometimes the time taken to 'fix' a sample is crucial because of the reactivity of the analyte, and the direct addition of preservative to collection vessels removes systematic and gross error associated with sampling and fixing.
- Relatively large amounts of stabilising reagents are often required and great care is necessary when one or more of their constituents are also to be measured in other samples. The use of reagents such as mercuric chloride and chloroform may well cause contamination problems in laboratories commissioned to measure these materials at low concentration. Indeed this very combination of mercuric chloride and chloroform is commonly used to stabilise samples for analysis of water quality parameters.

3.6.4 Reagents for water sample preservation

Some reagents are suitable for several determinations, for example biocides. Common materials in use as biocides include mercuric chloride, sodium azide, chloroform and toluene. Biocides are necessary in determinations where the analyte concentration may be affected by biological activity (usually microbial), notably determinations of hydrocarbons and compounds containing nitrogen and phosphorus.

3.6.5 Sample transport

The transportation of samples has a number of implications. There are many regulatory considerations regarding safe transport etc., but the following aims of sample transport have analytical procedures in mind:

- to convey the sample from the sample delivery point to the place of analysis as quickly as possible;
- to convey the sample in a condition identical (or as close as possible) to that when the sample was taken.

Regulatory requirements usually include:

- appropriate carriage containers to guard against explosion or fire hazard;
- appropriate placarding to enable emergency services accurately to assess risk when dealing with a vehicle involved in a road traffic accident;
- appropriate documentation to permit accurate logging of the whereabouts of samples which may have different degrees of import, e.g. those required for litigation, those with hazardous or special waste considerations etc.

It should be borne in mind that, for laboratories working under quality accreditation schemes, the exact age, mode of transportation, transportation conditions etc. are important issues and must be recorded to ensure validity of the end result of the analysis under the adopted accreditation scheme.

3.6.6 Sample storage

The three principal methods to be considered for sample storage are:

- ambient storage;
- refrigeration;
- deep freezing.

During storage of environmental samples ambient temperature is rarely used because the number of potential contaminants is so great that to assume no analyte loss due to biological and chemical interactions would be complacent. It should be assumed that an environmental sample should at least be refrigerated to slow down any biological activity to reduce error, unless experience demonstrates otherwise. For example, in some instances it may be detrimental to the sample to refrigerate it, e.g. when the sample is for the counting of nitrifying bacteria, which are temperature sensitive and some degree of kill may be brought about by lowering the temperature unnecessarily.

3.6.6.1 *Refrigeration*

Biological activity is sometimes prevented or reduced by storing samples in the dark and at low temperatures (4 °C is usually the temperature stipulated for arresting biological activity within a sample). Refrigeration also helps in reducing problems which arise from other sources, such as the vapour pressure of volatile organic compounds (VOCs) which may be the analyte of interest.

3.6.6.2 *Deep-freezing*

An extension to refrigeration is to freeze the samples in their sample containers immediately after collection and to store the frozen samples at −20 °C until required for analysis. Deep freezing essentially prevents microbial growth, whereas refrigeration to 4 °C merely reduces the extent to which microbial growth occurs. Typically, solvent-safe refrigerators are required for this purpose.

In some instances it is more convenient to freeze samples in liquid nitrogen. This has the added advantage that freezing is rapid, minimising evaporation of volatile components, and is then sustainable at the same temperature throughout transportation. The disadvantage of using liquid gases as refrigerants is the low-temperature hazard to the operator, associated with the possibility of burns. Vehicles must also be suitably ventilated to prevent the accumulation of nitrogen within them, to the exclusion of oxygen.

Both refrigeration and deep freezing have the advantage that no manipulation of the sample material is required. This reduces the risk of cross-contamination of samples and the introduction of new contaminants. Sometimes, however, refrigeration and/or freezing is required after chemical fixing of a sample.

3.6.7   Sample pre-treatment

For analysis of a sample by instrumental methods, which predominate in pollutant analysis in soils and gases, pre-treatment of samples is determined by the method of analysis. The previous example, of the addition of a biocide to a sample destined for analysis for hydrocarbons to prevent microbial spoilage, may be treated in the following way as an alternative: extraction into a Freon such as 1,1,2,2-tetrachloroethane would transfer the hydrocarbon contained within a soil or aqueous sample into the hydrophobic Freon. In this medium, microbial spoilage would be unlikely to occur, and a biocide would be unnecessary. This approach, however, would constitute sample pre-treatment as the extraction would be necessary prior to analysis by infrared spectrophotometry. Methods of sample pre-treatment are too numerous to catalogue here and usually form part of a standard operating procedure. However, the following examples serve to illustrate the kinds of pre-treatments which samples may need to undergo prior to instrumental analysis:

- sample derivatisation (for volatilisation of analytes to be examined by gas chromatography (GC) or to increase a property such as absorbance, which would in turn increase the sensitivity of analysis by high-performance liquid chromatography (HPLC));
- pre-filtration of sample to remove particulate matter or micro-organisms (such as would be required prior to injection onto an HPLC column, or for soluble metal ion concentration in samples contaminated with micro-organisms);
- solvent extraction of analyte (such as may be dictated by column characteristics employed in chromatographic analysis or to be able to utilise spectral wavelengths at which water would be a major interferent);
- pH correction to enable colorimetric analysis to be carried out, e.g. for protein determinations, or to maintain organic acids of interest in the protonated (free-acid) form;
- adsorption of gas components (e.g. oxides of nitrogen and sulphur, and VOCs) onto solid supports such as activated carbon, prior to thermal desorption and instrumental analysis by GC–mass spectrometry (MS).

The list above is by no means exclusive and new methodologies are continually under development.

3.7   Sampling of environmental media

Environmental media are sampled from a large number of location types: for example, gases may be sampled from stacks or flues, from the workplace environment, from interstitial spaces in soils, from reaction vessels within process plant and dissolved in liquids. Each of these locations requires a different sampling strategy and different equipment. The same principle applies to the other environmental media also. Consequently, the number of strategies available for sampling in the environment is large, there are many types and they are often highly specific to an application. The following sections give examples of sampling of a number of media for environmental analysis.

3.7.1   Sampling of gases and aerosols

3.7.1.1   *General considerations*

It is usually necessary to collect a sample from the atmosphere with separation of the pollutant of interest from other constituents and at the same time bring about a concentration of the analyte prior to analysis. After separation the mass of analyte collected is evaluated by physical or chemical methods and the concentration of the material in the atmosphere prior to collection or concentration is calculated. This dictates an accompanying method for the measurement of the volume of the sample of gas collected. If a flue gas is being sampled, the flow rate of the gas must also be known before emissions to the atmosphere can be determined.

Gas samples are not collected instantaneously but over a predetermined period, varying from seconds to weeks. It should be borne in mind that the longer the sampling event, the more periods of cyclic or random variation in analyte concentration will be excluded from the final analysis. The minimum sampling period is usually determined by pump capacity or by the efficiency of separation process employed. To collect a mass of analyte sufficient for analysis, various combinations of sampling time and flow rate are possible and the principles discussed in Section 3.5 apply.

A system for air pollutant sampling generally comprises the individual components shown schematically in Figure 3.1. The system shown in Figure 3.1 is applied to grab or composite sampling of gases. Continuous sampling of gases may be applied to analytical devices which may be pre-programmed to take measurements at discrete times or to devices which give a continuous real-time signal.

In this system, gas enters via a port, which protects against coarse particle entry (e.g. inverted funnel or coarse filter) and is then passed into a trapping vessel containing adsorbent material which allows subsequent removal. The length of pipe or tubing from the collection point to the trapping–concentration chamber should be the shortest length possible since some pollutants may interact while in this part of the system, giving rise to error. Ozone and the oxides of nitrogen have been observed to undergo a variety of such interfering reactions at residence times greater than 5 s in a sampling inlet tube. The more

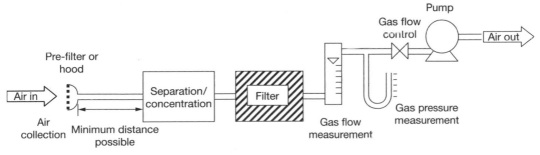

Figure 3.1   Schematic representation of the component parts of a gas sampling system

33

reactive the gas, generally the more prone it will be to this kind of interference. Other types of interference occur as a result of pollutant removal by physical or chemical sorption on the tube walls, or interaction with aerosol deposits within the inlet line or excessive humidity.

3.7.1.2 Techniques for sampling gases

Reactive or absorptive systems

Absorption by bubbling – impingers   The separation and concentration of gaseous pollutants can be done by several techniques, of which absorption in a liquid solution or hygroscopic solid is one of the most common. Absorption is a solubility phenomenon controlled by the equilibrium partial pressure of the dissolved gas over the liquid surface. It can be improved if the absorbed species undergoes an irreversible chemical reaction which produces a non-volatile compound. If this technique is used then the period of time necessary to produce stable compounds is an important factor in the absorption of the analyte of interest. The kinetics governing the mass transfer of a gas to the liquid phase is often the rate-limiting step in the absorption process. This limitation can be minimised by increasing the gas–liquid contact interface. This is usually achieved in impingers by sparging the gas into the trapping liquid via a scintered glass bulb. This breaks up the gas into many small bubbles, thus increasing the surface area of gas in direct contact with liquid. In addition impingers are often used in series, so that gas passing through the first of a series of impingers is further trapped in secondary and tertiary vessels. The simplest gas impinger is the Dreschell bottle. The Dreschell bottle, especially when not fitted with a scintered diffuser, may be relatively inefficient, but its advantage is that it allows high gas flow rates. Where high-efficiency impingers with diffusers are used, there is often an undesirable pressure drop across the vessel. The resulting back-pressure may impart an instability to the vessel, and leaks may occur. In addition, the pressure drop must be accounted for in calculations of analyte concentration in the gas as it passes through the system.

Impregnated filters   Efficient sampling of reactive gases may be achieved by passage through a filter material impregnated with a chemical reagent which, by reaction, removes and concentrates the analyte at the same time. Collection with impregnated filters presents several advantages in comparison with bubbling: the material is less fragile and lighter, occupying less volume, reducing transportation costs and handling difficulties; there are no problems associated with absorbent volume changes due to evaporation or leakage; automation of analysis is usually a viable option; several parameters may often be analysed concurrently. Disadvantages of this system stem from the possibility of interferences arising from the co-collection of particulate material. In addition, it is possible only to add a small amount of reagent to the filter medium, which in consequence may become rapidly exhausted in localised high concentrations of analyte. Water content of the sampled air or gas may also introduce interferences. Figure 3.2(a) shows schematically an impregnated filter device.

The efficiency of this sampling method is determined by the duration of the sampling event and the flow rate of gas through the sampling device.

Denuder systems and Dräger tubes   Denuder systems (Figure 3.2(b)) should be operated where aerosol levels are high, since aerosols may act as interferences on impregnated filters. A denuder system consists of parallel or concentric fine-bore tubes, the walls of

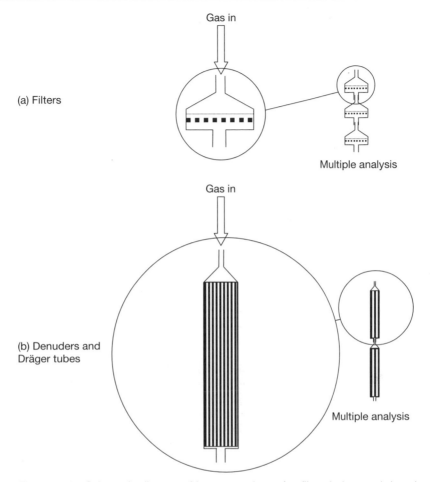

Figure 3.2 Schematic diagram of impregnated reactive filter devices and denuders

which may be impregnated with reactive materials, rather than a filter medium. Gases which carry aerosols pass through such systems ideally under conditions of laminar flow. Under these conditions the aerosol droplets pass through the device and gaseous components react with the impregnated reagents. There is, of course, some degree of interaction of the aerosol with the reactive component of the denuder, but the degree of interference is much reduced in comparison with an impregnated filter, which, by its very design, forces aerosol droplets to come into intimate contact with the filter medium by being forced through it.

Denuder systems function by virtue of the fact that both dissolved and suspended materials in a gas continually move by Brownian diffusivity. The larger aerosol particles interact less so, and consequently, provided that laminar flow is maintained within the denuder tube, the greater motion of dissolved gases allows them to react to a greater extent. This operating principle has the advantage of relatively low interference, but the associated disadvantage is that processing of such systems requires low flow rates of gas to maintain laminar flow conditions, and there may be a considerable pressure drop across the device. Examples of reactive denuder systems which have been used for atmospheric monitoring include aluminium sulphate or magnesium oxide impregnated tubes for the measurement of nitric acid and lead oxide or potassium carbonate for sulphur dioxide sampling.

Adsorptive systems The mechanism of adsorption of gaseous materials onto solid surfaces is complex. It relies on weak intermolecular forces such as electrostatic and van der Waals interactions. Therefore, when a gas or vapour molecule adsorbs onto a surface at a given rate, there is a concurrent desorption rate. These rates reach an equilibrium leaving a fixed concentration of adsorbed gas on the solid surface which is dependent on temperature, relative humidity, vapour pressure of solvents etc. This mechanism can be harnessed to trap environmental pollutants and then the conditions altered to allow subsequent removal of the pollutant for analysis by an instrumental technique such as GC–MS. Usually removal is brought about by a rapid increase in temperature (thermal desorption) under controlled conditions of gas flow and rate of change of temperature. If the analytes being trapped are known, then a GC system may be adequate for the subsequent analysis, but where unknown constituents are present a GC–MS system provides a more reliable method of speciation. Commercially available microprocessor-controlled automated thermal desorption (ATD) devices for linkage to GC or GC–MS systems have provided a reliable method for workplace and natural environmental monitoring for some years. Personal monitors such as Dräger tubes containing adsorbents such as activated carbon, located in small devices which may be attached to clothing, are common in occupational health schemes, and they are processed by ATD with GC–MS.

Remote sensing, drones and balloons Recent developments in the analysis of atmospheric pollution are the use of remote sensing methods. This obviates the need for sampling at all. Remote sensing methods are covered in Chapter 8. More traditional methods of sampling atmospheric gases, however, include the use of drones and balloons. Drone aircraft and gas-filled balloons essentially sample in the same way, using instrumentation which may sample quantities of gas by extraction as described above or alternatively by utilising instrumentation which yields a signal suitable for electronic logging. These, however, are specialist applications, and are outside the scope of this book.

3.7.2 Sampling of water

Water sampling may be carried out by a large number of different devices, depending on the source of the required sample. For sampling fresh water, estuaries and subsurface sea water, simple bottles may be borne on telescopic handles. The majority of sampling carried out by water authorities is done with sampling bottles or stainless steel buckets. Slow-moving streams may be sampled immediately below the surface using an inverted sampling bottle to avoid contamination with surface film, and then turned upright to take in the sample proper. In fast-flowing waters surface samples will be more representative of the whole. Large rivers, estuaries and the sea surface are generally best sampled using hand-held flasks or bottles from a boat, preferably without a powered engine in operation, especially when sampling for hydrocarbon contamination. A schematic diagram of a surface sampling device is shown in Figure 3.3. In this example the bottle is lowered in a cradle or on a telescopic handle to the required depth. Once there, a vent tube is opened and water is permitted to enter the bottle via is tube leading to the bottom of the vessel (which avoids turbulence).

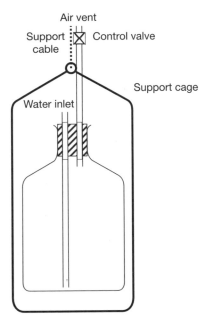

Air vent

Support cable

Control valve

Support cage

Water inlet

Figure 3.3 Surface water sampler

Sampling the surface or subsurface of the sea is fraught with difficulty, since the sea surface microlayer, which is some 50 μm thick, is rich in surfactants which complex with metals and associate with many hydrophobic compounds. These materials may cause interference when sampling as they may preferentially adsorb to the sampling bottle. There are many sampling devices for use in sea water which have different materials of construction from stainless steel to Teflon. Some systems employ dispersant sprays. All systems preferentially sample one or more of the surface film components.

Deep sea water is usually obtained using a hydrowire-borne device. The principle of operation is a weighted wire which conveys an open cylindrical vessel to a predetermined depth and then closes both end caps remotely. This is shown schematically in Figure 3.4. This type of device was originally developed by Van Doorn, but several innovations have been produced. A problem associated with such devices, however, is the leaching of contaminants into water samples from materials of construction, e.g. plasticisers.

Groundwater is almost always sampled via boreholes sunk specifically for the purpose. When water is extracted from such wells, the borehole must be purged of stagnant water. Typically this is carried out by drawing to waste a volume of up to 10 times that contained in the casing of the borehole. It is particularly important that sampling depth and frequency be determined either from a preliminary survey or based on a large number of samples taken over a period long enough to allow steady-state levels of analyte to be drawn off in the effluent.

Sampling for hydrocarbons is a common procedure in environmental protection. Hydrocarbons may arise naturally, but more usually they arise anthropogenically. Sampling is usually performed using stainless steel or Teflon containers. Where continuous sampling occurs it is usual to use pumping systems and tubing constructed from stainless steel. Samples must be collected into borosilicate glass whch has been acid washed or cleaned

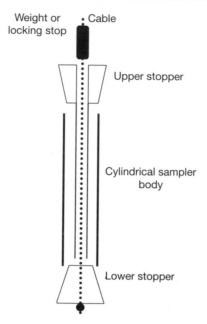

Weight or locking stop

Cable

Upper stopper

Cylindrical sampler body

Lower stopper

Figure 3.4   Schematic representation of deep-water samplers

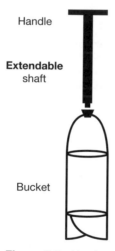

Handle

Extendable shaft

Bucket

Figure 3.5   Hand auger for soil sampling

with a suitable proprietary detergent or cleaner for the specific purpose. Before use the containers are heated to >400 °C to remove VOCs. Sampling for a mixture of hydrocarbons (i.e. volatile and non-volatile) should be carried out in duplicate since different instrumental conditions dictate different analysis times for these materials. A biocide should be added to samples taken for hydrocarbon analysis.

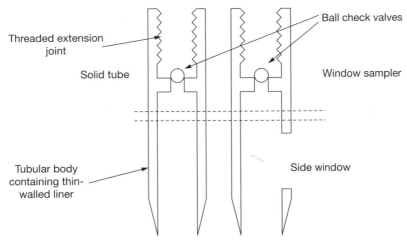

Figure 3.6   Schematic cross-sections through a solid-walled tube and a window sampler (the shafts are shortened for clarity)

3.7.3   Sampling of sediments, sludges and muds

Sediments take two major forms: settled and suspended. Suspended sediments are sampled by filtration. Typically a $0.45\,\mu m$ pore size is required if using a membrane filter.

Sediments which are well compacted may be sampled by scoops or even by hand augers (Figure 3.5) if there is not a significant depth of water overlying the solids. Sediments on the sea bed, however, are usually sampled using a dredge sampler. These devices are trawled using boats and automatically collect a predetermined volume of solids. Alternatively, automated coring devices may be used which contain removable liners.

Sludges in process plant are usually sampled via sampling ports designed for that purpose. Dewatered sludges are cored in the same way as soils (see below).

3.7.4   Sampling of soils

Soils may be sampled in a number of ways depending on the depth to which a sample is required or whether or not a complete or discrete horizon is required. If a complete horizon is needed, then a hinged sampling tube is used. Such samplers are driven into the soil either manually or by pneumatic hammer. Alternatively they may be sunk using rotary drilling devices. On extraction of the sampler, the tube is split and the complete core removed. A variation of this type of device is a 'window sampler'. This is a similar extendable tube, usually constructed of stainless steel with a rectangular window in the side. These devices are particularly appropriate for determining the concentration of hydrocarbons, heavy metals etc. at different depths, where intact soil structure is not important. Tube and window samplers are shown schematically in Figure 3.6.

Sampling of soils at depths of 1 m or less permits the use of hand augers as discussed earlier and shown in Figure 3.5.

3.8 Bibliography

Association of Analytical Chemists 1975 *Official methods of analysis of the Association of Analytical Chemists*

Department of the Environment *Methods for the examination of waters and associated materials*, published in parts, HMSO

Greenberg A E, Clesceri L S and Eaton A D (eds) 1995 *Standard methods for the examination of water and wastewater*, 19th edition, American Public Health Association

Hunt D T E and Wilson A L 1986 *The chemical analysis of water*, 2nd edition, Royal Society of Chemistry, London

Skoog D A, West D M and Holler F J 1997 *Fundamentals of analytical chemistry*, 7th edition, Saunders College Publishing, Philadelphia, PA

4

Classical methods of chemical analysis

4.1 Introduction

A number of analytical chemical methods were available for environmental monitoring before the advent of the modern sophisticated instrumental methods, i.e. chromatography. These are still widely used particularly in field methods when the detection of pollution incidents is most important. Details of the field test kits and methods will be given in a later section. In this chapter the principles and practices of classical chemical analysis will be considered. One of the major advantages of the classical method is the simplicity of the apparatus and the limited skills required to perform the analysis. High precision may require adoption of very careful procedures by skilled operators but, nevertheless, volumetric methods offer a more readily accessible analytical program for the novice than many techniques. Methods based on the use of volumetric reactions and titrations have been extensively used in environmental monitoring and these will be discussed first.

4.2 Volumetric analysis

Volumetric analysis is quantitative chemical analysis involving volume measurements. We determine the exact volume of a solution of accurately known concentration, a standard solution, which is required to react quantitatively with a known volume of a solution of unknown concentration of the substance to be determined. (The unknown material could be sampled by mass if it is in solid form.) The known volume is added from a single calibration device called a pipette and the unknown volume is determined by addition of the other solution from a scaled vessel called a burette. A wide range of pipettes and burettes of varying volumes and precisions are available. The devices chosen will depend on the exact nature of the determination performed. The process of mixing two solutions until the reaction is just completed is called a titration. The point at which the reaction is just complete is called the end point (EP) or equivalence point or stoichiometric (theoretical) EP. The completion or EP of the titration should be detectable by some unmistakable change in the system. This can be either by a visual method in which the colour change in the solution occurs directly or by the addition of an auxiliary reagent, called an indicator, which changes colour only when the EP of the primary reaction has been reached. Alternatively the EP can be determined by some instrumental process of which more will be said later. In the ideal situation the visible EP should coincide with the stoichiometric EP. In practice there is usually a small difference between the two which is termed the titration error. When choosing the system to be used care should be taken to minimise this error.

4.2.1   Reaction requirements

Before a chemical reaction can be used in a volumetric analysis it must fulfil the following conditions:

1. The reaction must be simple and capable of being expressed by a chemical equation. The determinant should react completely with the reagent in stoichiometric proportions.
2. The reaction should be rapid, preferably instantaneous. Occasionally the addition of a catalyst can be used to speed up the rate of reaction.
3. There must be a marked change in some physical or chemical property of the solution at the equivalence point, enabling the end point to be determined by:
    (a) measuring the change in potential between an indicator and reference electrode, a potentiometric titration;
    (b) measuring changes in the electrical conductivity of the solution, a conductimetric titration;
    (c) measuring the change in electrical current between two electrodes dipping into the solution, an amperometric titration;
    (d) measuring changes in optical absorbance at a defined wavelength with a spectrophotometer.
4. If condition (3) cannot be satisfied then an indicator should be available which, by change in physical properties, should sharply define the EP. The normal use of indicators is for a visual change in colour to occur at or immediately after the EP. However, indicators can also be used in instrumental EP determinations when spectrophotometers can be used to monitor the indicator colour change.

Volumetric methods are capable of high accuracy and possess several advantages. They are rapid and simple to perform, using basic, cheap, apparatus, which is easily calibrated and in many cases, with suitable control of experimental conditions, can avoid complex separation procedures. More will be said later about systems and procedures.

4.2.2   Titration methods

Titrations can be performed in a number of different ways, depending on the extent of the principal reaction and the speed at which the equilibrium position is attained.

4.2.2.1   *Direct titration*

In this method the unknown is titrated by direct addition of the standard solution to the unknown. The EP is detected by a number of methods.

4.2.2.2   *Indirect methods*

Here a second material, which reacts with the reagent only after the material to be determined has reacted, is added to the titration flask before the start of the titration.

4.2.2.3   *Back titration methods*

A measured excess of the reagent is added to the solution and the excess of reagent, unused by the determinand, is titrated with a standard solution of a material, which reacts with reagent. The reagent solution can be calibrated by titrating the same reagent volume in a blank determination. This method is sometimes useful when the initial reaction of the determinand is too slow for a direct titration to be practical. The reagent is added and the solution allowed to stand for a period of time under controlled conditions to allow the full reaction to occur before titrating the excess.

4.2.2.4 Displacement methods

An unmeasured excess of a compound formed by a metal ion and a reagent is added into the analyte solution. If the analyte forms a more stable complex than the original cation a displacement reaction occurs and the liberated species can then be titrated with a standard solution in the normal direct method.

4.2.3 Classification of reactions

The reactions employed in volumetric analysis can be classified under two headings:

(a) Those in which no change in valency or oxidation state occurs but depend on the interaction of ions.
(b) Oxidation–reduction reactions which involve a change in valence or oxidation state of both reactants. This is a transfer of electrons between reacting substances.

The first of these two basic groups can be further subdivided into three groups to produce four separate classes altogether.

4.2.3.1 Neutralisation reactions

This type of titration involves the reaction between the hydronium ion, H_3O^+, produced by acids and the hydroxide ion, OH^-, produced by bases, to give salts and water. Acidimetry uses standard acid and unknown base while alkalimetry uses standard base with unknown acid. The acids or bases may be derived indirectly by hydrolysis of the salts of weak acids or bases.

4.2.3.2 Precipitation reactions

These reactions generally depend on the combination of ions, other than H_3O^+ or OH^-, to form an insoluble solid. Also the reaction between a number of organic compounds can produce a solid precipitate. Although this type of reaction occurs frequently in volumetric analysis its wider use is in gravimetric analysis, which will be discussed in another section.

4.2.3.3 Complex formation reactions

Interaction between the analyte and a reagent can result in the formation of a larger molecule, which behaves as a single discrete unit. Interactions between cations and negatively charged or neutral ligands are the most common complex-forming process. Addition of the reagent molecule leads to complex formation.

4.2.3.4 Oxidation–reduction reactions (redox)

All reactions involving a change in oxidation number, i.e. a transfer of electrons between reactants, come into this category even if, subsequently other processes such as precipitation

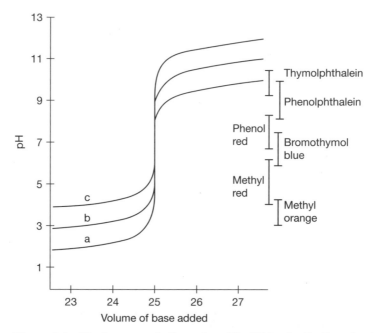

Figure 4.1   Titration curves in the region of the EP for the titration of a strong acid with a strong base at various concentrations: curve a, 1 M; curve b, 0.1 M; curve c, 0.01 M. On the right-hand side are the pH ranges for a number of indicators

occurs. Whenever an oxidation occurs there must be a corresponding reduction. The standard solution can be either an oxidizing or a reducing agent. The chemical equivalent of a material is based on the amount of material involved in a redox reaction in which 1 mol of electrons is exchanged between species.

A number of reactions fall into several categories. For example, the interaction between Fe^{3+} and CN^- leads to formation of a complex ion, $Fe(CN)_6^{3-}$, which can lead to formation of a precipitate if the correct cation is present. Each of the four categories mentioned will now be discussed in detail with the accent on applications of the techniques.

4.2.4   Neutralisation reactions

Titration of acid with base represents reaction of the hydronium ion, H_3O^+, produced from the acidic material with hydroxide ion, OH^-, produced from the base. The reaction can be followed by monitoring the H_3O^+ ion concentration. If both acid and base are strong electrolytes then the H_3O^+ ion concentration, measured as pH $= -\log[H_3O^+]$, will follow the curve shown in Figure 4.1. The stoichiometric equivalence or EP occurs at pH 7. This is independent of the acid or base concentration as indicated in the figure by the data for the titrations of differing concentrations of hydrochloric acid with the corresponding concentration sodium hydroxide solution. Although the initial and final solution pH will depend on the acid concentration, the end point always occurs at pH 7 for all strong acid and strong base reactions. However, if the acid or base is a weak electrolyte, incomplete ionisation

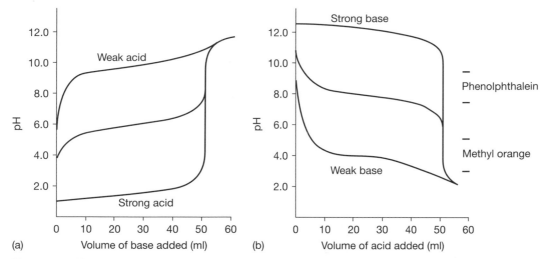

Figure 4.2 Effect of acid and base dissociation constants on the solution pH during acid–base neutralisation reaction. Titration curves are for 50 cm³ of 0.1 M solutions. (a) Strong base (0.10 M NaOH) against various strength acids and (b) strong acid (0.10 M HCl) against various strength bases. Note the pH at the EP is a function of the dissociation constant but the volume of titrant is not. Again pH ranges for two common indicators are shown on the right-hand side

will occur and then hydrolysis of the salt will occur producing an EP at a pH other than 7, as shown in Figure 4.2. The exact pH at the EP will depend on the extent of dissociation of the weak acid or base and can be calculated from the measured ionisation constants of the acid or base. Since most common acids and bases are colourless, in both the free acid and salt forms, an auxiliary agent, an indicator, which changes colour at the EP, must be used to give a visual EP.

4.2.4.1 *Indicators*

Most indicators are weak organic acids or bases which change colour when moving from the free acid form to the salt form. The indicators are weak electrolytes and hence their behaviour in water can be represented by the equations

$$H_2O + HIn \rightleftharpoons H_3O^+ + In^-$$

or

$$InOH \rightleftharpoons OH^- + In^+$$

The un-ionised indicator has a different colour from the ionised, In^+ or In^-, forms (Figure 4.3). Both of these reactions are equilibria and therefore we can write an equilibrium constant in the approximate form

$$K = \frac{[H_3O^+][In^-]}{[H_2O][HIn]}$$

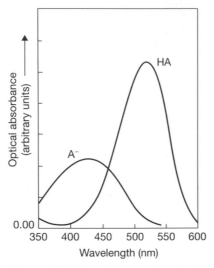

Figure 4.3   Absorption spectra of methyl red in the anionic form A⁻ and in the un-ionised form HA

Since only low concentrations of indicators are used the water concentration remains unaffected by the indicator and the equilibrium constant becomes

$$K' = \frac{[H_3O^+][In^-]}{[HIn]}$$

The equilibrium lies very much to the un-ionised side so that, when the indicator is added to an acid solution, the excess of H_3O^+ in the solution suppresses the indicator ionisation totally and the concentration of In⁻ is very small. When the solution is alkaline the concentration of H_3O^+ is reduced and the indicator ionises to maintain the equilibrium and hence the In⁻ concentration increases. At a definite pH HIn and In⁻ will be present in equal concentrations and any further addition of OH⁻ will increase In⁻ concentration and hence complete the solution colour change. The change in colour will occur over a limited pH range and this should correspond to the pH at the EP of the principal reaction, if the indicator has been correctly chosen. The volumes of solution required to bring about the change in solution pH and take the indicator through its colour change should be less than $0.1\,cm^3$: two additional drops of reagent solution at most. Similar considerations apply to the weak base indicators. Some common acid–base indicators and the pH at which they change colour are given in standard textbooks. On the right of the titration curve (Figure 4.1) are shown the pH ranges for four possible indicators for the titrations. Neutralization curves for strong acid–strong base, strong acid–weak base and weak acid–strong base are also shown in the diagram. In the strong acid–strong base then at high concentration, 1 M, any of the indicators will suffice but, at the 0.01 M level, only bromothymol blue and phenol red will suffice. In the strong acid–weak base situation because the EP occurs at pH < 7 only methyl red and methyl orange of the four shown are suitable indicators. When titrating a weak acid with a strong base, the EP occurs at pH > 7 and hence of the indicators shown, only phenol- and thymolphthaleins are suitable. Ethanoic acid, CH_3CO_2H, a product of the oxidation of ethanol and present in vinegar, is a weak acid with pK_a of 4.75. Aqueous solutions, even at 0.1 M concentrations, have pH values which are within or close to the colour change range of methyl orange and hence, if this indicator is chosen, a false EP will occur after about 1 cm^3 addition of strong base solution. Titrations of weak acid with weak bases are not suitable for indicator EP determination since the pH change is very slow over a wide range of added volumes.

4.2.4.2 Polyprotic acids

When a molecule contains two or more acidic protons then the shape of the titration curve will depend on the relative magnitude of the acid dissociation constants. The strongest proton is usually titrated first and the EP appears as a relatively sharp curve but for second and subsequent protons the EP may be less clearly defined depending on the acid dissociation constants. Although sulphuric acid is a diprotic acid it only shows a single EP because the first acid dissociation constant is so high that the solution of the acid only contains HSO_4^- and SO_4^{2-} ($K_{a2}(HSO_4^-) = 10^{-2}$) and the EP seen coincides with the HSO_4^- titration. Sulphurous acid, H_2SO_3, produced by some bacteria extracting oxygen from SO_4^{2-} under anaerobic conditions, is a weak acid ($K_{a1} = 10^{-2}$ and $K_{a2} = 10^{-7}$); hence both ionisation stages are observed during titrations with sodium hydroxide. The titration is imprecise since sulphurous acid is an equilibrium with dissolved SO_2 gas, which can be lost from the solution to the atmosphere. For acids with more than two protons, the titration curve for the third and subsequent protons may be indistinct. As an example the dissociation of phosphoric acid is a three-stage process which corresponds to the reactions:

$$H_3PO_4^{2-} + H_2O \rightleftharpoons H_3O^+ + H_2PO_4^-$$

$$H_2PO_4^- + H_2O \rightleftharpoons H_3O^+ + HPO_4^{2-}$$

$$HPO_4^{2-} + H_2O \rightleftharpoons H_3O^+ + PO_4^{3-}$$

The acid dissociation constants, K_a, for the three stages are 7.5×10^{-2}, 6.23×10^{-8} and 2.2×10^{-13} respectively and this means that the titration curve for the third proton cannot be seen under normal titration conditions. Figure 4.4 shows the pH as a function of added base and the sensitive ranges for indicators suitable for the first and second acid protons. The position of the EP for the third proton can only be predicted from the positions of the first two EPs.

For very weak acids the second proton may only show acidity in extremely basic media. In the determination of total nitrogen in organic samples, the nitrogen present is converted to ammonium salts, by reflux with sulphuric acid in the presence of a catalyst. Distillation of the solution, after it has been made alkaline with sodium hydroxide solution, evolves ammonia, which is trapped in boric acid solution:

$$H_3BO_3 + NH_3 \rightarrow NH_4^+ + H_2BO_3^-$$

The dihydrogen borate ion produced can then be titrated with a standard acid as indicated in the equation

$$H_2BO_3^- + H_3O^+ \rightarrow H_3BO_3 + H_2O$$

The acid dissociation constants for boric acid are $K_{a1} = 7 \times 10^{-10}$, $K_{a2} = 1 \times 10^{-13}$ and $K_{a3} = 1 \times 10^{-15}$; hence the ammonia is incapable of neutralising more than a single proton of the boric acid.

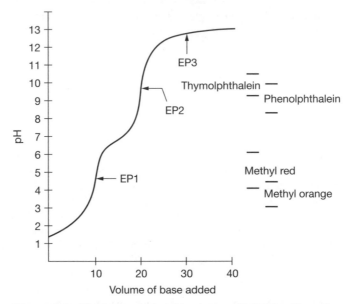

Figure 4.4   Titration curve for phosphoric acid H_3PO_4 with sodium hydroxide solution. The first two EPs are visible but the third proton EP can only be predicted from knowledge of the first two EPs

4.2.4.3   *Water alkalinity (acidity)*

Neutralisation reactions of an acid with a base are used to determine the alkalinity (or acidity) of water using a number of different methods for determining the EP of the reaction. Carbon dioxide dissolves in water to form the very weak carbonic acid:

$$CO_2 + H_2O \rightleftharpoons H_2CO_3$$

The solubility of carbon dioxide in water to form H_2CO_3 is very low; $K = [H_2CO_3]/[CO_2]$ $= 2 \times 10^{-3}$. The dissociation of the carbonic acid into hydronium ions and anions is given by the two following equations:

$$H_2CO_3 + H_2O \rightleftharpoons H_3O^+ + HCO_3^-$$

$$HCO_3^- + H_2O \rightleftharpoons H_3O^+ + CO_3^{2-}$$

The acid dissociation constants for the carbonic acid are $K_{a1} = 5 \times 10^{-7}$ and $K_{a2} = 5 \times 10^{-11}$. Because of the weakness of the carbonic acid the pH of distilled water, in normal equilibrium with absorbed carbon dioxide, is about 4.5. In the presence of bases such as OH^- the hydronium ions are removed and the acid dissociates to give increasing concentrations of the bicarbonate and carbonate ions. Alkalinity of water can thus be due to the presence of OH^-, HCO_3^- and CO_3^{2-}. Calculations of the equilibrium constant for the reaction

$$HCO_3^- + OH^- \rightleftharpoons H_2O + CO_3^{2-}$$

show that the concentration of HCO_3^- in equilibrium with OH^- is negligible and hence the solution can only contain OH^- and CO_3^{2-} or HCO_3^- with CO_3^{2-}. The presence of HCO_3^- can arise from the reaction of the carbonic acid in the water with insoluble calcium carbonate according to the equation

$$CaCO_3 + H_2CO_3 \rightleftharpoons Ca(HCO_3)_2$$

This is the mechanism whereby natural soft rain water becomes hard following contact with limestone rocks for extended periods of time. The solution can thus contain differing concentrations of OH^-, HCO_3^- and CO_3^{2-} depending on the origin of the water. Titration with an acid can provide values for the concentrations of the three sources of alkalinity. Notice that HCO_3^- can act as a base and react with acid. If some phenolphthalein indicator is added to a water sample and it turns pink then OH^- and/or CO_3^{2-} are present. The combined quantity of these two ions can be determined by titrating the solution with standard sulphuric acid solution until the pink colour is just discharged. The pH of the solution is now 8.2 and the solution only contains HCO_3^- ions. Addition of a mixed bromocresol green–methyl red indicator allows the titration to continue down to pH 4.5 when all the HCO_3^- will have been neutralised and converted to H_2CO_3. The difference between the titres to pH 8.2 and pH 4.5 is due to OH^-. The results are usually expressed as ppm $CaCO_3$ equivalent. Similarly the titre from pH 8.2 to pH 4.5 is a measure of the temporary hardness due to $Ca(HCO_3)_2$, which is water soluble.

4.2.4.4 Instrumental end point determinations

Alternatively the alkalinity can be determined using a pH electrode to indicate the EP of the two reactions. The plot of pH against volume added should exhibit two inflections as shown in Figure 4.5. The first at pH 8.5 is due to the neutralisation of OH^- and CO_3^{2-} and the second at 4.5 is due to reaction of HCO_3^- to form H_2CO_3.

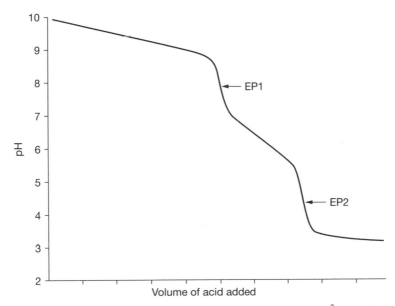

Figure 4.5 Typical pH titration curve for water sample ~200 cm³ containing mixture of hydroxide and carbonate ions with 0.1 M hydrochloric acid. Indicated EP 1 shows removal of OH^- and conversion of CO_3^2 to HCO_3^-. Second EP shows conversion of HCO_3^- to H_2CO_3

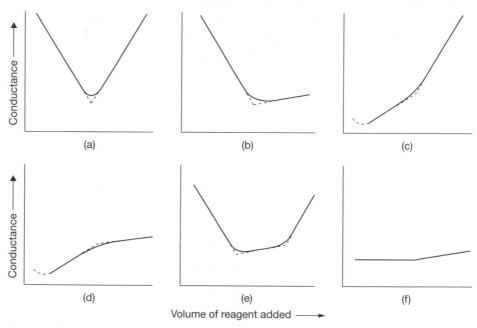

Figure 4.6 Conductimetric titration curves for (a) strong acid–strong base, (b) strong acid–weak base, (c) weak acid–strong base, (d) weak acid–weak base, (e) mixture of strong and weak acids with strong base and (f) precipitation of AgCl from water sample

Acid and base neutralisation reactions can be followed by the changes in solution conductivity following the neutralisation reaction. For an acid in solution the ionisation process is

$$HA + H_2O \rightleftharpoons H_3O^+ + A^-$$

The anion A^- has a much lower conductivity than H_3O^+ and hence the conductance of the solution is largely due to the H_3O^+. Strong acids are almost completely dissociated, > 95%, in solution and their solutions will show high conductance. Addition of a base adds OH^- to the solution according to the scheme

$$BOH \rightleftharpoons B^+ + OH^-$$

Since the OH^- will neutralise the H_3O^+, the less mobile B^+ ions are replacing H_3O^+ ions and the solution conductivity will fall, until the EP is reached, when the excess of OH^- will cause the conductivity to increase again (Figure 4.6(a)). If the acid is strong but the base is weak then the rise in conductance, after the EP, will be slight (Figure 4.6(b)). If the original acid was a weak acid then the fall in conductance on addition of the base will be less, owing to the reduced H_3O^+ concentration, and it may increase depending on the cation conductance (Figure 4.6(c)). After the EP there will be an increase, when the excess of OH^- from the strong base occurs. Weak acids and weak bases are difficult to titrate since the hydrolysis of the salts produced masks the EP (Figure 4.6(d)).

One advantage of the conductivity titration over the indicator method is the ease of determining each component in a mixture of strong and weak acids. The solution conductivity follows the course indicated in Figure 4.6(e). The strong acid is neutralised first followed by the weak acid. Mixtures of a number of strong and weak acids are very difficult for any of the methods so far described and normally only the total acid content is determined.

An additional advantage of conductimetric titrations over both indicator and pH electrode methods is the limited manual dexterity required by the operator. Exact volume additions to just reach the EP are required by both of the other methods. In conductimetric titrations the solution conductivity is measured as a function of the added volume of reagent and the EP is detected from the extrapolation of straight line regions of plots of conductivity against volume added. The volumes added must be accurately measured but do not have to be specified values and, provided that sufficient data points are available to define the linear portions on either side of the EP, the method can achieve high accuracy.

4.2.5 Precipitation reactions

Because of the importance of halides in chemistry and particularly water treatment, the most common precipitation processes in volumetric analysis involve the determination of halides in water by the use of silver nitrate reagent. When silver nitrate is added to a solution containing Cl^- or Br^- or I^- then, as soon as the concentrations of Ag^+ and X^-, where X^- indicates any halide, exceed the solubility product of AgX, given by the reaction

$$AgX(s) \rightleftharpoons Ag^+(aq) + X^-(aq)$$

$$K_{sp} = [Ag^+][X^-]$$

a precipitate of AgX is formed. The EP occurs when no more silver halide is precipitated. Detection of the EP by visual observation of complete precipitation is impossible so either a silver ion electrode is used to indicate the excess of Ag^+ ions or an indicator is used. Two types of indicator are available: those which form a coloured precipitate, i.e. chromate ions, or those which form a coloured soluble complex, i.e. thiocyanate ions. Because the solubility product of silver chromate is higher than that of the silver halides the silver chromate only precipitates after the halide precipitation is complete. The silver thiocyanate adsorbs onto the surface of the precipitated halide, when there is an excess of silver ions, to produce a red colour on the precipitate. In both cases the coloured compound is not formed until the silver halide precipitation is complete. Mixtures of Cl^- and I^- can be determined by performing the titration first in ammonia solution and then repeating the titration in acid solution. In ammoniacal solution the silver ion concentration is reduced below that required to exceed K_{sp} for AgCl by formation of the $Ag(NH_3)_2^+$ complex ion, i.e.

$$Ag^+ + 2NH_3 \rightleftharpoons [Ag(NH_3)_2]^+$$

The silver iodide solubility product is so low that the ammonia cannot lower the silver ion concentration below that required for silver iodide precipitation and therefore it continues to precipitate. The difference between the titres in ammoniacal and acid solutions is equivalent to the chloride content.

The precipitation titration of silver chloride can also be followed by solution conductance measurements since as the chloride ion is precipitated initially it is replaced by an equal concentration of the similar-conductance nitrate ion. At the EP then an excess of silver and nitrate ions is being added and the solution conductance will rise as indicated in Figure 4.6(f).

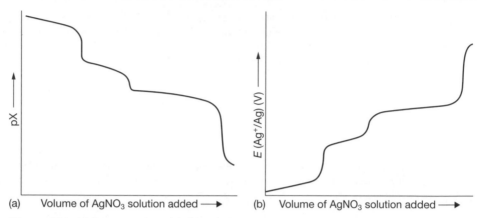

(a)   Volume of AgNO$_3$ solution added ⟶   (b)   Volume of AgNO$_3$ solution added ⟶

Figure 4.7   (a) Concentration of Ag$^+$ ions, pX, against added silver nitrate for a solution containing a mixture of I$^-$, Br$^-$, Cl$^-$ ions and (b) the potential of the silver electrode immersed in the solution

If the silver electrode is used to follow the titration then all three halide ions Cl$^-$, Br$^-$ and I$^-$ can be determined since the level of free silver ions in solution rises with the completion of each precipitation stage. The negative logarithm of the silver ion concentration, pAg$^+$, for a precipitation titration of a mixture of all three halides with silver nitrate solution will show three steps which are proportional to the solubility product of the corresponding silver halide. A plot of the silver ion concentration as a function of added silver nitrate for a mixture of all three halides, which corresponds to the potential of a silver ion electrode immersed in the solution, is shown in Figure 4.7. Both conductimetric and potentiometric titrations with silver nitrate have been used to determine the chloride ion content of water.

4.2.6   Complex formation reactions

Mixing of solutions of two ions can lead to the formation of a different species. Examples of these involve the reactions of cyanide ions with either iron or silver ions:

$$6CN^- + Fe^{3+} \rightleftharpoons Fe(CN)_6^{3-}$$

The hexacyanoferrate is a stable entity which, under normal conditions, behaves differently from either CN$^-$ and Fe^{3+}. The silver complex is less stable and is easily decomposed in solution by addition of a competing ligand, i.e.

$$2CN^- + Ag^+ \rightleftharpoons [Ag(CN)_2]^-$$

$$[Ag(CN)_2]^- + 2NH_3 \rightleftharpoons Ag(NH_3)_2^+ + 2CN^-$$

Cyanide can be determined by titration with silver nitrate but the reaction is difficult since initially silver cyanide precipitates and then re-dissolves, forming the complex dicyanoargentate ion, [Ag(CN)$_2$]$^-$, identified above. If silver ions are in excess then silver dicyanoargentate is precipitated. The silver cyanide precipitated by the local excess of silver during the addition is very slow to re-dissolve. Addition of ammonia to the titration improves the rate of dissolution of the silver cyanide.

Some metals form complexes with excess of cyanide in alkaline solution and the excess cyanide can be determined by the silver nitrate titration. Nickel, cobalt, mercury, zinc, cadmium and platinum metals form complexes of the type $[Ni(CN)_4]^{2-}$, which can be used to determine the metal ion. Fortunately there are other better methods than cyanide complex formation for the determination of these elements.

Ligands so far discussed have only one functional group which can combine through donor bonding so that the process, indicated by the simple equation for the hexacyanoferrate, in reality involves a series of intermediate stages and thus the EP is drawn out and unsuitable for analytical purposes.

Most cations have preferred co-ordination numbers of 2, 4 or 6; as a result molecules which contain multidonor groupings, usually organic molecules, can form complexes via a reduced number of intermediate species and hence exhibit sharper EPs. The organic complexing molecules have another advantage over monofunctional donors, owing to the wide range of metal ions they react with and their ability to be selective under the correct conditions. The most widely used molecule, ethylenediaminetetraacetic acid (EDTA), has the structure

$$HOOC-CH_2 \qquad\qquad CH_2-COOH$$
$$\diagdown\qquad\qquad\diagup$$
$$N-CH_2-CH_2-N$$
$$\diagup\qquad\qquad\diagdown$$
$$HOOC-CH_2 \qquad\qquad CH_2-COOH$$

There are six potential electron donor groups in this molecule: lone pairs on the two nitrogen atoms and four 'double-bonded' oxygens in the carboxyl groups. So it is capable of completely filling the co-ordination sphere of many cations; also, the carboxyl groups can lose ionisable hydrogen species forming anions with up to 4 negative charges, which then assist the complex formation with the positively charged ion. The free acid is insoluble in water but reacts with two equivalents of sodium hydroxide to produce the di-anion, which is water soluble. Because of the presence of four acid groups on the EDTA the stability of complexes of EDTA with metal ions is very dependent on the solution pH and relative electron acceptor properties of the metal ion.

The ready formation of complexes between EDTA and metal ions can be used for the determination of many cations. Unfortunately the reaction is not specific but by careful choice of experimental conditions some selectivity can be introduced. Many elements can be determined by EDTA titrations using different indicators under the different pH conditions. Table 4.1 lists some of the elements determined by EDTA titrations and the solution pH required for successful completion of a number of metal ion analyses.

4.2.6.1 *Hardness of water*

A standard method for the determination of calcium in water is the complexiometric titration with EDTA. Hardness in water can be due to the presence of calcium and magnesium salts. Both of these elements form undesirable solid precipitate scums with soaps and detergents as well as forming deposits in boilers following water evaporation. Both elements form complexes with EDTA in the stoichiometric rapid reactions as indicated in the equation

$$M^{2+} + EDTA^{2-} \rightleftharpoons MEDTA$$

One EDTA molecule can completely occupy the co-ordination sphere of the metal ions and the equilibrium favours the complex formation by a large amount; hence the reaction

Table 4.1 *Metals determined by EDTA titrations*

Metal	Buffer pH	
Barium	12	
Bismuth	1	
Cadmium	5	
Calcium	12	
Cobalt	6	
Copper	6 or 10	Ammonia to keep Cu in solution
Iron	2–3	
Lead	6	
Magnesium	10	
Manganese	10	
Mercury	6	
Nickel	7–9	
Strontium	12	
Thorium	2	
Tin(II)	6	
Zinc	6 or 10	
Indirect titrations		
Calcium	10	Substitution method
Gold	10	
Palladium	10	
Silver	7–8	Back titration with reagent excess
Aluminium		

is suitable for quantitative analysis. A sample of water is treated with a buffer to control the solution pH and then titrated with a standard EDTA solution using an Eriochrome Black T indicator. This indicator is a complex between an organic ligand and Mg^{2+} ions, which is a weaker complex than the Mg^{2+} complex with EDTA. When all the calcium and magnesium in the water sample have been complexed with EDTA at the EP, the excess EDTA decomposes the weak, red, magnesium indicator complex releasing the free indicator molecule, which is a different colour, blue, from the complexed ligand. The initial titrations at pH 10 give results for both magnesium and calcium. If the pH is adjusted to 12 then $Mg(OH)_2$ is precipitated and only the calcium is titrated. Permanent hardness due to salts other than $Ca(HCO_3)_2$ can be determined by boiling the solution to remove the temporary hardness. The soluble $Ca(HCO_3)_2$ is converted to $CaCO_3$, which is precipitated, and, after filtering of the solid, the filtrate can be titrated with EDTA solution in the usual way. The filtering stage can be omitted, but the titration must be performed rapidly since the EDTA will re-dissolve the precipitated $CaCO_3$ over a limited period of time. Dry $CaCO_3$ solid is used as a standard to calibrate the EDTA solution prior to the determination. Results can be compared with the values obtained by the acid–base titration. For optimum results careful control of the solution pH is required. As with other reactions buffers are used to control the pH and thus reduce interferences.

4.2.7   Redox reactions

In an oxidation–reduction reaction electrons are transferred from the material being oxidised to the material being reduced. The oxidant thus accepts electrons and the reducing agent donates electron.. In any redox reaction there are conjugate redox couples:

$$oxidised(A) + reduced(B) \rightleftharpoons reduced(A) + oxidised(B)$$

The position of the equilibrium depends on the relative strengths of the two redox pairs oxidised(A)–reduced(A) and reduced(B)–oxidised(B).

Each redox reaction is a combination of two half-reactions and a study of these enables the equilibrium position for the complete reaction to be predicted. This method is very useful when trying to construct balanced equations for a reaction. In these reactions species which are not oxidised or reduced serve only as carriers for the active species. A typical redox reaction used in analytical chemistry is that between ferrous (II) ions and manganate (VIII) ions, which can be ascertained by the addition of the two half-reactions

$$5e^- + MnO_4^- + 8H_3O^+ \rightleftharpoons Mn^{2+} + 12H_2O$$

$$5Fe^{2+} \rightleftharpoons 5Fe^{3+} + 5e^-$$

giving the overall equation:

$$8H_3O^+ + 5Fe^{2+} + MnO_4^- \rightleftharpoons 5Fe^{3+} + Mn^{2+} + 12H_2O$$

Overall reaction equilibrium lies sufficiently to the right, owing to the MnO_4^- being a stronger electron acceptor than the Fe^{3+} ion, for the reaction to be used quantitatively. The reaction is self-indicating since the MnO_4^- ion has an intense purple colour, while the Fe^{3+} ion is a pale yellow colour and the Mn^{2+} ion a pale green colour. Thus when excess of MnO_4^- is present the solution develops a deep purple colour.

4.2.7.1   *Chemical oxygen demand*

When studying the chemical oxygen demand (COD) of water the redox reaction of MnO_4^- is again used but this time to oxidise chemicals present in water samples. The equation given above for the MnO_4^- reaction is still valid but it is impossible to write a balanced equation for the oxidation of the organic material present in the water because of the ill-defined nature of the oxidisable species. A measured sample of the water is treated with a measured excess of standard MnO_4^- solution, in a sealed container, for 2 h at 40 °C. The residual MnO_4^- is determined by reaction with excess iodide, in acid solution, liberating iodine, which is then titrated with thiosulphate in a further redox reaction:

$$16H_3O^+ + 2MnO_4^- + 10I^- \rightleftharpoons 2Mn^{2+} + 24H_2O + 5I_2$$

$$I_2 + S_2O_3^{2-} \rightleftharpoons 2I^- + S_4O_6^{2-}$$

A reagent blank should be determined by using a sample of freshly distilled water. The volume of standard MnO_4^- used can be related to the COD of the water by the above equations. If the concentration of the MnO_4^- solution is 0.0025 M and that of the thiosulphate standard is 0.0125 M then the difference, in cm^3, in volumes of thiosulphate used for the blank and the sample titrations is equivalent to ppm of COD of the water sample.

Unfortunately the oxidising power of MnO_4^- is so high that it oxidises species which are not considered part of the normal COD processes, e.g. chloride to chlorine. Hence this method has been superseded by a similar digestion process using acidified potassium dichromate as the oxidant. The digestion has to be performed under reflux for 15 min to achieve satisfactory oxidation. The excess of dichromate is determined directly by titration with standard ammonium ferrous sulphate solution using diphenylamine in sulphuric acid as the indicator. Although the dichromate $Cr_2O_7^{2-}$ is not as powerful an oxidant as the MnO_4^{-}, $E_0(MnO_4^{-}/Mn^{2+}) = 1.52$ V and $E_0(Cr_2O_7^{2-}/Cr^{3+}) = 1.33$ V, it has two advantages. Chloride ion is not oxidised by dichromate, $E_0(Cl_2/Cl^-) = 1.36$ V. Also, potassium dichromate can be used as a primary standard, i.e. sampled by weight when dry, because it is easily prepared in a pure form with good stability and reproducible composition. Solid potassium permanganate, $KMnO_4$, on the other hand, has variable composition and is unstable. Solutions of the $Cr_2O_7^{2-}$ are stable indefinitely under closed vessel conditions, but the MnO_4^{-} solution is difficult to prepare from the solid and decomposes with time. The decomposition is catalysed by exposure to UV illumination, and therefore fresh solutions of MnO_4^{-} must be prepared and standardised regularly. Standardisation is usually performed with sodium oxalate solid, a suitable primary standard, which is easily oxidised by the MnO_4^{-} ion to carbon dioxide:

$$5 \begin{array}{c} COO^- \\ | \\ COO^- \end{array} + 2MnO_4^- + 16H_3O^+ \rightleftharpoons 2Mn^{2+} + 5CO_2 + 24H_2O$$

More recently a spectrophotometric method based on dichromate has been developed and this will be discussed in a later section.

4.2.7.2 *Residual chlorine determination*

The thiosulphate titration can also be used for the determination of residual chlorine in water present as either dissolved chlorine gas or as hypochlorite ion ClO^-. A sample of potassium iodide is added to a measured water sample. After addition of some acid any residual chlorine or hypochlorite, OCl^-, liberates iodine from the iodide according to the equation

$$ClO^- + 2I^- + 2H_3O^+ \rightleftharpoons Cl^- + 3H_2O + I_2$$

The liberated iodine is titrated with standard thiosulphate solution using an aqueous solution of a starch indicator to sharpen the EP determination. The starch is turned blue by free iodine but not by iodide. Hence the EP occurs when the blue colour disappears.

4.2.7.3 *Dissolved oxygen in water*

The oxygen content of water samples can be determined by performing a redox titration. The water sample, normally $250 \, cm^3$, is treated with $2 \, cm^3$ of a 50 wt% solution of manganese sulphate followed by $2 \, cm^3$ of an alkali–azide–iodide solution (50%, 15% and 0.2% respectively) in a stoppered bottle. Care should be taken to eliminate air from the space above the water sample by completely filling the sample bottle before commencing the determination, otherwise the oxygen in the entrapped atmosphere will give false results. The white precipitate of manganese hydroxide initially produced rapidly turns brown owing to the oxidation, in alkaline solution, of the manganese from the +II to the +III oxidation state by the dissolved oxygen according to the equation

$$4Mn^{2+} + O_2 + 8OH^- \rightleftharpoons 4MnO(OH) + 2H_2O$$

After being shaken several times to allow complete reaction the solution is allowed to settle. When the solution is acidified, the oxidised manganese state will then oxidise the iodide to liberate iodine according to the equation

$$2MnO(OH) + 6H_3O^+ + 2I^- \rightleftharpoons 2Mn^{2+} + 9H_2O + I_2$$

The liberated iodine can be titrated with standard thiosulphate using the starch indicator as described earlier. A range of oxygen concentrations can be accommodated either by titrating the entire contents of the sample bottle, assuming the volume used is known accurately, or by withdrawing a sample of the acidified solution and titrating it separately. The volume of thiosulphate used for the titration can be related to the dissolved oxygen content through the above equations. If the concentration of the thiosulphate is 0.0125 M then each cm^3 of the titre is equivalent to 0.1 mg of dissolved oxygen. Nitrite ions, which are also capable of oxidising the iodide ion to iodine, interfere with the determination and hence azide species are added to the reagent to destroy any nitrite. If the water sample contains appreciable copper or lead content then an alternative to the azide must be used since these ions form insoluble explosive compounds with azide ions.

Although modifications of the above method have been used to determine the dissolved oxygen level in 5 day biochemical oxygen demand determinations, they are prone to numerous interferences and have largely been superseded by the instrumental dissolved oxygen meter measurements. The dissolved oxygen meter can be calibrated with distilled water samples whose dissolved oxygen levels have been determined by the titrimetric method detailed above.

4.2.7.4   *Instrumental end point for redox reactions*

Redox reactions can be followed by methods based on indicators, which combine with one of the components or are oxidised by the reagent, but also by potentiometric measurement. Electrode potentials are concentration dependent and thus, when an excess of oxidant is present, the potential of an electrode responding to the oxidant will increase dramatically. Again taking the Fe^{2+}/MnO_4^- reaction as an example the potential of an inert platinum electrode, with respect to a standard calomel electrode, will increase as the MnO_4^- solution is added to the Fe^{2+} ion solution and the ratio of Fe^{3+} to Fe^{2+} increases.

When two oxidisable species are present in solution then the one with the lowest electrode potential is titrated first and, if the two electrode potentials differ by 0.2 V, then a clear distinction of two EPs is seen. Figure 4.8 shows the typical curve for a mixture of Ti^{3+} and Fe^{2+} with MnO_4^-. The first EP is due to the more powerful reducing species, Ti^{3+}, and the second to the Fe^{2+} titration.

4.2.7.5   *Automatic titrations*

A number of titrimetric methods involving instrumental EP determination have been developed by the water industry and standards agencies in order to allow rapid automatic titrations to be used for the determination of a number of key parameters for the large number of routine analyses. These have to be performed on a regular basis with a high sample throughput, high confidence and accuracy and rapid turn-round. Potentiometric, including pH, conductimetric and colorimetric titrations, using spectrophotometric EP determination, are all easily controlled by microprocessor or minicomputer and the results can be calculated by the machine with a precision specified in the method control package. The

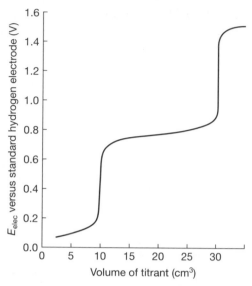

Figure 4.8   Potentiometric titration of $25\,cm^3$ of a mixture which is $0.1\,M$ in Ti^{3+} and $0.2\,M$ in Fe^{2+} with $0.05\,M$ MnO_4^- solution. The first EP is for the Ti^{3+}

machine is almost ideally suited to calculate the two best-fit straight lines to the data from the conductimetric titration system and also the intercept EP. In a number of systems the sample identification is bar coded on the sample bottle and these details can be read automatically by the machine before the analysis starts and the details married up with the determination results obtained by the machine. The output of such determinations is already in a suitable form for a number of data-processing packages, which can form part of an environmental monitoring and management package. These determinations can be used to automatically control the addition of reagents for treatment of water supplies in order to maintain specification and provide an alarm signal, if any of the determinations exceeds set parameters.

4.3   Gravimetric analysis

4.3.1   Principles

Gravimetric analysis depends on the measurement of mass of analyte or of compounds of known composition that contain the analyte. Two alternative techniques can be applied to gravimetric analysis:

1.   The analyte, or a species chemically related to the analyte, is precipitated as a sparingly soluble compound that either has a well-characterised composition or can be converted to a product of known composition by a simple suitable heat treatment.
2.   The analyte, or its decomposition product, can be volatilised at a suitable temperature and the volatile product collected and weighed or the product weight determined indirectly from the loss in weight of the sample.

4.3.2   Requirements for gravimetric methods

The factors which determine a successful analysis for the more common precipitation procedures will be discussed before the experimental details are examined:

1. The precipitate must be so insoluble that no appreciable loss of material occurs during the collection and washing of the precipitate. The quantity in solution at the end of the precipitation should be less than the minimum quantity detectable by the analytical balance, usually 0.1 mg.
2. The physical nature of the precipitate must be such that it is readily separated from the solution by mechanical methods, e.g. filtration, and can be washed free of impurities.
3. The precipitate must be either of definite known chemical composition or convertible to a pure substance of definite chemical composition.

4.3.3   Solubility product

The solubility of a precipitate AB is due to the dissociation of the solid, producing ions which are surrounded by molecules of the solvent. The process is controlled by the solubility product which, assuming that the un-ionised AB is insoluble, is given by the relationship

$$AB \rightleftharpoons A^+ + B^-$$

$$K_{sp} = [A^+][B^-]$$

This is a simple equilibrium expression with the terms for the water and the undissociated solid AB assumed to be constant. If the A^+ and B^- species are derived from the pure AB solid then the solution will contain equal molar concentrations of A^+ and B^-. If A is the species to be determined and an excess of B^- is added, the concentration of analyte A^+ is reduced to maintain the solubility product. The solubility data for silver chloride in the presence of KCl demonstrate the effect ($K_{sp}(AgCl) = 1.1 \times 10^{-10}$):

Cl$^-$ concentration	Ag$^+$ concentration
1.04×10^{-5}	1.04×10^{-5}
$1.0 \ \times 10^{-3}$	$1.1 \ \times 10^{-7}$
$3.0 \ \times 10^{-1}$	$3.7 \ \times 10^{-8}$

Thus to precipitate all of the silver add an excess of chloride. Note that the addition of too much chloride leads to dissolution of the silver chloride by formation of the soluble $AgCl_2^-$ complex. The precipitate is less soluble in cold water and hence filtration at ambient temperatures reduces solubility losses. Washing the precipitate with small quantities of cold solvent reduces the quantities of analyte lost while removing most of the contaminants. It is preferable to wash with a large number of small-volume washes rather than a small number of large-volume washes.

4.3.4   Characteristics of precipitates

4.3.4.1   *Particle size and colloid formation*

The physical state of the precipitate is conditioned by the circumstances under which it is produced. Under certain conditions, the precipitate particles can be of such a small particle size that they pass through conventional filters and do not settle to the bottom, but remain suspended in solution even on standing. During the precipitation process, when the initial precipitate is formed, then, if the surface absorbs ions which repel the incoming ions, further precipitation occurs at a new nucleation site, instead of increasing the dimension of existing particles, and a colloid suspension is formed. The colloid can be made to coagulate by heating the suspension with added electrolyte, which reduces the charge on the particles by reducing the number of ions absorbed on the surface. Also some of the colloidal material re-dissolves at the higher temperature and is re-precipitated, on existing nucleation sites, on cooling.

Other conditions which improve precipitate crystallinity are dilute solutions, slow addition of the precipitating agent with good stirring and finally an excess of the reagent. Sometimes the solution acidity is important in order to control interference from hydrous oxides and hydroxides. As an example, calcium oxalate precipitates as good crystals from acid solution and precipitation is completed by slow addition of ammonia. If the oxalate is precipitated in alkaline solution, a gelatinous precipitate is formed. Similarly, aluminium hydroxide forms a colloidal precipitate when ammonia is added to a solution of aluminium ions, but, if the hydroxide is formed by hydrolysis of urea, the precipitate is highly crystalline and easily filterable:

$$(NH_2)_2C{=}O + 3H_2O \rightarrow CO + 2NH_3 + 2OH^-$$

The reaction takes ~2 h at 95 °C, i.e. almost boiling.

4.3.4.2 *Purity*

Besides the control of particle size for filtration purposes, the precipitate must be of high purity, otherwise impurities will be dried and weighed at the same time as the required determinand. The factors affecting the purity are co-precipitation, occlusion, mixed crystal formation and surface adsorption.

Surface adsorption is most important for precipitates with large surface areas, particularly colloids. Coagulation of the colloid particles does not significantly decrease the amount of adsorption, since large internal surfaces are still exposed to solvent. Dissolution and re-precipitation is the most effective cure for surface adsorption.

Mixed crystal formation occurs when some of the ions of the precipitated solid are replaced by ions of another element or group. The two ions must have the same charge and their sizes differ by <10%. Also the two salts must be isomorphous, i.e. belong to the same crystal structure class. Examples of mixed crystal formation occur with barium and lead sulphates. When mixed crystal formation is a problem, the only solution is separation of the interfering ions.

During crystal growth, if the rate of deposition is high, then the counter ions to the determinant cannot escape from the charge surface region before the crystal grows round it, and it becomes occluded. Similarly, solution can be mechanically trapped by the crystals in a pocket formed by their growth.

Both of these sources of impurity can be reduced by using low precipitation rates, adequate digestion over a period of time, particularly at higher temperatures, and dissolution followed by slow re-precipitation.

4.3.5 Precipitating agents

A variety of reagents can be used to produce compounds which are insoluble in common solvents. They fall naturally into several categories.

Table 4.2 *Precipitation of metal hydroxides and hydrous oxide*

pH	Metal ions precipitated
3	Fe^{3+}, Sn^{2+}, Zr^{4+}
4	Th^{4+}
5	Al^{3+}
6	Cu^{2+}, Cr^{3+}, Zn^{2+}
7	Fe^{2+}
8	Ni^{2+}, Cd^{2+}, Co^{2+}
9	Ag^{+}, Hg^{2+}
11	Mg^{2+}

4.3.5.1 *Inorganic precipitating agents*

These reagents typically form sparingly soluble solids, whose lattice energy exceeds the enthalpy of solution of the ions. Generally they involve interaction between large anions and cations to form neutral species. Sometimes hydroxides or hydrous oxides, the precipitated species, are converted to the oxide, by ignition, before weighing. Some of the species precipitated as hydroxide species are given in Table 4.2 as a function of solution pH. The ignition temperature for conversion to the oxide varies considerably and sometimes the hydroxide can be measured quantitatively. Drying temperatures can affect precipitate weights for a number of different solids. For some, e.g. AgCl, the water is lost at 120 °C. Others, such as Al_2O_3, retain the water strongly and require heating to high temperatures, 1000 °C, before the last traces of water are removed. An additional problem sometimes occurs with complex anion precipitates. Oxalates can be decomposed to oxide via a series of intermediate stages, each of which requires careful control of experimental conditions, if it is to be used as the determined species. The precipitate can often be used to determine either cation or anion concentrations, e.g. if excess of a barium salt solution is added to a sulphate ion solution the barium sulphate precipitate can be used to determine sulphate levels. Similarly, excess of sulphate can be used to determine barium ion concentration. Sulphides of many metals are very insoluble (see Table 4.3) but these are difficult to collect because many readily form colloids and hence they are used in a separation method rather than as an analytical reagent.

4.3.5.2 *Reducing agents*

Some reagents are capable of converting the analyte to its elemental form, which can be filtered off, dried and weighed. The coinage and precious metals are typical of this type of reaction. Their oxides can be reduced by reagents such as SO_2, H_2, H_2S, CO, C and hydroxylamine H_2NOH to the metal, which can be separated, washing and weighed. For silver the oxide is thermally unstable and decomposes directly to the element. Alternatively the metal can be produced at the cathode in an electrochemical cell provided that chemicals which react with the metal, typically halogens, are absent from the solution.

Table 4.3   *Solubility products for metal sulphides*

Compound	K_{sp}
Ag_2S	1.6×10^{-49}
As_2S_3	2.5×10^{-28}
CdS	3.6×10^{-29}
CoS	1.9×10^{-27}
CuS	8.5×10^{-45}
FeS	1.5×10^{-19}
HgS	4.0×10^{-53}
MnS	1.4×10^{-15}
NiS	1.4×10^{-24}
PbS	4.2×10^{-26}
ZnS	1.0×10^{-20}

4.3.5.3   *Organic precipitants*

A complete range of organic reagents have been developed for gravimetric inorganic species determination. Generally, these reagents are more specific in their reactions than the
inorganic precipitants. Two types of reagents are involved: those forming sparingly soluble
non-ionic co-ordination compounds and those involved in essentially ionic bonding. Those
forming co-ordination complexes contain at least two functional groups, so arranged in the
molecule that the electron pair donor bonds form five- or six-membered rings containing the
metal ion. These compounds being non-polar are insoluble in water and hence are easily
separated and dried. Examples of the chelating reagents forming five- or six-membered
rings are 8-hydroxyquinoline and dimethylglyoxines, both of which contain acidic hydrogens,
capable of being lost when forming the chelates. This loss of one or more protons, during
the complex formation, means that stability and solubility of the complex are pH dependent. Thus, the selectivity of the precipitation process can be improved by control of the
solution pH. Because of the relatively large reagent molecular mass the limit of detection
is reduced.

The second reagent type is exemplified by the tetraphenylboron ion, $B(C_6H_5)_4^-$, where
C_6H_5 is the phenyl group. This reagent reacts with the heavier alkali and alkaline earth
cations to form insoluble salts (large cation, large anion). The composition of the precipitate
is stoichiometric, based on simple charge ratios.

Several of the organic-type reagents are capable of being used for the determination of
organic molecules or, more particularly, an organic functional group. Many of these reactions are between non-ionic molecules, but all of them contain strong dipoles, which
provide the initial bonding processes to start the reaction. A number of standard texts are
available covering this extensive range of chemical reactions.

4.3.6   Indirect methods

The processes discussed so far have depended on mass measurement of precipitates after
drying, but there are some methods which depend on volatilisation techniques: either by
measurement of the residue remaining after some characteristic molecules have been

evolved or by determining the mass of the evolved species. The former suffers from the potential pitfall of uncertainty about the nature of the material volatilised, while the latter may suffer from a lack of sensitivity.

Typical evolved materials are water and carbon dioxide. Many inorganic materials eliminate water on ignition. Collection of this water in a suitable pre-weighed quantity of desiccant followed by re-weighing of the desiccant allows the water evolved to be determined by difference. These indirect methods assume that only water is lost, but a number of materials undergo decomposition irrespective of the presence of water and hence erroneous results can be obtained. If in doubt, the evolved gasses are analysed to identify any non-water products. The moisture content of cereal grains is determined by weight loss under conditions which do not result in any decomposition.

Carbon dioxide can be evolved and absorbed in a suitable absorbent, Ascarite II or Carbasorb, in which the reaction is

$$2NaOH + CO_2 \rightarrow Na_2CO_3 + H_2O$$

In order to retain sensitivity, the absorbent must contain a desiccant to trap the water produced, particularly since the heat of the reaction may increase the water vapour pressure.

The classical determination of carbon and hydrogen in organic compounds is by combustion of the material in excess of oxygen and determination of the mass of carbon dioxide and water evolved by the use of absorbent tubes. Modern systems use instrumental detection methods based on either a series of katharometer detectors, each followed by an absorption tube, or chromatography.

4.3.7   Experimental techniques – solid precipitation and collection

After slow addition of the reagent, the precipitate is allowed to stand, preferably warmed above ambient, for a period of time, to allow the solid to coagulate and reduce adsorption. When the solution has cooled to ambient, the solution is filtered either through a filter paper or a weighed sintered glass crucible. In all cases, the solid must be completely transferred from the precipitation vessel to the filter device. The solid is well washed with small quantities of solvent in which the solid is insoluble, usually water, and then heated to dryness. The solid is cooled to room temperature in a desiccator, before weighing. The heating and cooling processes are repeated until the solid reaches constant weight. When a cellulose filter paper is used, because the paper can contain variable amounts of water, the drying must be performed at a high enough temperature to burn the paper, i.e. 600 °C, leaving no paper-derived residue. Papers cannot be used for filtering solids which will be decomposed to unidentified products by the heating at 600 °C.

4.3.8   Calculation of results

The mass of solid precipitate having been successfully determined, the next step is to convert this to a mass of determinand and hence to composition of the original material. The following example will show the procedure.

A sample of 1.1324 g of a dried iron ore was dissolved in acid and the hydrate oxide of Fe(III) precipitated by addition of ammonia. After filtration and washing, the solid was ignited at 600 °C to produce 0.5394 g of pure Fe_2O_3. The steps in the calculation are each mole of Fe_2O_3 contains 2 mol of Fe

159.69 (formula weight) g of Fe_2O_3 contain 2×55.847 (formula weight) g Fe

1 g of Fe_2O_3 contains $\dfrac{2 \times 55.847}{159.69}$ g of Fe

0.5394 g of Fe_2O_3 contain $\dfrac{0.5394 \times 2 \times 55.847}{159.69}$ g of Fe

This quantity of iron was contained in 1.1324 g of ore:

% Fe in ore is $\dfrac{0.5394 \times 2 \times 55.847 \times 100}{159.69 \times 1.1324}$

% Fe in ore was 33.32

Generally the calculation can be simplified into one of the type

$$A = \frac{\text{weight of product}}{\text{weight of sample}} \times GF \times 100$$

where GF is the gravimetric factor defined as the mass of determinand per gram of determined material given by the expression

$$GF = \frac{\text{formula weight of substance sought} \times \text{atom ratio}}{\text{formula weight of substance weighed}}$$

The atom ratio is the ratio of the number of A atoms in the weighed substance to the number in the determined substance.

4.3.9   Evaluation of gravimetric analysis

4.3.9.1   *Specificity*

This is generally poor, since most precipitants react with groups of ions or compounds. Hence interference is a problem with mixed samples. Prior separation of interfering species may be the only solution.

4.3.9.2   *Sensitivity*

The detection limit is imposed by the solubility of the compound. Most micro-balances can weigh to less than 0.1 mg but at this level transfer losses can be significant and compound solubility usually exceeds this limit. Collection devices have a mass of several grams, hence 1 mg represents a high degree of calibration precision.

4.3.9.3   *Precision*

Provided that the sensitivity limits detailed above are adhered to, gravimetric methods are capable of high precision, <0.1%.

4.3.9.4   *Equipment*

Apart from the analytical balance, the rest of the equipment is cheap and easy to use. Simple procedures requiring attention to detail are used throughout.

4.3.9.5   *Time*

Comparatively long periods of time elapse between start and finish of the analysis, but a number of limited-time activities are interspersed with longer periods of inactivity as indicated in the scheme below.

Action	Time taken
Precipitate	10 min
Digestion	1 h
Filter	15 min
Dry	>1 h
Weigh	5 min

4.3.9.6   *Calibration*

No calibration procedures are normally required once the stoichiometry of the reaction product has been established and the solid acts as its own primary standard. By adoption of a cyclic procedure an improved throughput of samples can be achieved, but the process cannot currently be readily adapted to automatic procedures.

4.3.10   Applications to environmental measurements

Currently five water quality parameters are measured regularly by gravimetric methods and these are only because of the lack of a suitable alternative. Total, suspended and dissolved, solids, oils and fats, and sulphate are regularly determined gravimetrically, although for the last of these nephelometric methods based on light scattering by the solid barium sulphate particles are also available.

4.3.10.1   *Suspended solids*

For the suspended solids determination, a sample of the water is taken and filtered through a sintered disc or a demountable filter paper, which is then dried to constant weight at 110 °C. The gain in weight of the sintered disc, or paper, is due to the suspended solids. If this mass is divided by the volume of water sample then the suspended solids concentration is obtained. The demountable filter paper is favoured because of the large volume above the filter paper while a small-dimension paper is retained. Any solid caught on the glass walls can be easily washed onto the paper. The difficulty with the determination is in producing a representative homogeneous sample since most of the suspended solids are of variable density and size. The sample must be adequately agitated immediately prior to sampling. For some suspended solid data the filter is immersed in the water course and a measured volume of liquid drawn up through the filter, which is then removed, dried and weighed.

4.3.10.2   *Dissolved solids*

For the dissolved solids the filtrate from the suspended solids determination is added to a pre-weighed glass beaker and the water removed by evaporation followed by drying to constant weight at 110 °C. If the water is low in suspended solids then a separate larger-volume sample may be needed.

4.3.10.3 *Total solids*

Total solids are normally determined by the sum of suspended and soluble solids content but as a periodic check water samples are added to a weighed beaker and the water evaporated to dryness at 120 °C. The weight difference of the beaker is the total solids content.

4.3.10.4 *Sulphate*

Sulphate is determined gravimetrically by precipitation of barium sulphate, although work is in progress to develop a suitable determination based on light scattering by the precipitated barium sulphate. This will be discussed in the section on small-scale test kits. Similar scattering experiments may prove useful for the suspended solids giving a value for the volume percentage of suspended matter without any mass implications.

4.3.10.5 *Oils and fats*

Oils and fats in waste waters are determined by gravimetry but the method involves no chemical reactions. The waste material is extracted with petroleum spirit at 40–60 °C and after separation of the petroleum phase the solvent is removed and the residue weighed. Water samples are shaken with the petroleum spirit and then allowed to separate in a separating funnel. The lighter petroleum layer is drawn off and the solvent removed by distillation under vacuum. Solid samples are extracted with the petroleum spirit in a standard Soxhlet extraction thimble under reflux. On completion of the extraction evaporation of the solvent again leaves the fat residue which can be weighed. The determination yields a total soluble fats value, without the extensive chemical reactions required for individual fat type content determinations.

4.3.10.6 *Acid–solvent-soluble material*

Solid samples are occasionally analysed by gravimetric methods to determine the amount of material soluble or insoluble in a number of common solvents. This is important when studying leaching of materials from soils. A weighed sample of the solid is treated with an excess of the solvent, i.e. water, acids, alkali and/or organic solvents such as methanol, and the mass remaining is determined. The solution can be evaporated to determine the soluble amount. Results are quoted as $x\%$ solvent soluble or insoluble.

4.4   Spectrophotometric analysis

4.4.1   Introduction

Molecular spectroscopies throughout the spectral range are widely used for the identification and quantification of an extensive list of inorganic and organic materials present in the environment. Some of the spectroscopies, notably infrared and microwave, have very limited applications in environmental monitoring owing to their reduced sensitivities. They may be used in chromatographic systems as a selective detector and discussion of this aspect is deferred to Chapter 7 on combination methods. Infrared absorption spectroscopy has been used in workplace airborne monitoring where the infrared absorbance of vapours and gases, over several metres path length, can identify and quantify the level of airborne contaminants, ensuring that the atmosphere is below the Occupational Exposure Level (OEL) for potential pollutants. Ultraviolet (UV)–visible and X-ray spectroscopies have the required sensitivities and specificities. X-ray spectroscopy has wide usage in environmental monitoring as an element detector because of the minimum interference between elements and also minimum sample preparation for natural samples. Drying may be the only requirement for full quantitative analysis. X-ray spectroscopy has limited use in molecular determination, while X-ray diffraction has some capabilities for molecular–compound fingerprinting in limited circumstances. Only major components, >3% concentration, in limited-composition mixtures can be detected. Since many species present in low concentrations produce absorption spectra in the UV–visible spectral region, UV–visible spectroscopy has wide applications in both quantitative and qualitative analysis, hence its widespread use for detection in chromatography. Chemical reactions can be used to improve the sensitivity and selectivity of UV–visible spectrometric determinations. Fluorescence spectroscopy is also applicable to environmental monitoring particularly because of its high sensitivity and occasional specificity. Many molecules of biochemical interest can be readily studied in liquid or gaseous forms by fluorescence spectroscopy and also from surface layers of solids suspended in solutions. The latter is particularly useful for studies of materials absorbed on solid substrates.

Two processes associated with the transfer of energy between atoms or bonds can occur depending on whether the experiment is looking at energy absorbed by the electron species during promotion of the electron from a ground to an excited state, i.e. producing absorption spectra, or whether the experiment is measuring the energy emitted when the excited species returns to the ground state, producing emission spectra. The absorption of radiation by an atom or molecule is extensively used in environmental monitoring particularly in field test kits where simplicity is of the essence and sophisticated electronics can be avoided. Emission spectra require energetic sources for best results and will be discussed after the more widely used absorption methods. Both absorption and emission techniques rely on the application of the Beer–Lambert law governing the energy exchange process and the proportionality of the effect to the concentration of the active species in the measurement apparatus. This process has been discussed fully in another section.

4.4.2 Absorption spectroscopy

The origins of the UV–visible absorption in molecules are electronic. Absorption occurs when the energy of the incident radiation corresponds to the energy difference between the ground electronic state and the lowest excited electronic state. Unfortunately molecules are in rotational and vibrational motion and each electronic state has associated with it a number of rotational and vibrational states. Excitation of the electron invariably produces changes in the vibrational modes of the molecule and hence the absorption bands are associated with a range of energy changes, producing very broad bands. In the gaseous phase the vibrational fine structure is sometimes apparent (Figure 4.9). In solutions the spectral resolution is degraded by collisions with solvent molecules. This is particularly important when the solvent can chemically interact with the absorbing species as shown in Figure 4.9. The non-polar inert solvent cyclohexane still allows some of the vibrational fine structure, seen in the vapour phase, to occur whereas polar water interferes with the vibrations by exerting a chemical interaction with the molecule. When the species concerned has a number of states with similar energy levels then multiple absorption bands can be observed. The MnO_4^- ion is a good example of this effect; see Figure 4.10. Both inorganic and organic molecules exhibit UV–visible absorptions but often reagents are added to increase the sensitivity or to adjust the wavelength of the absorption into a more suitable spectral region or increase the specificity for the analyte.

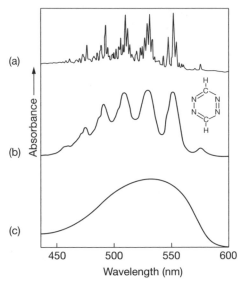

Figure 4.9 Absorption spectra for 1,2,4,5-tetrazine in (a) vapour phase, (b) non-polar (hexane) solution and (c) polar (water) solution (Reprinted with permission from Mason S F 1959 *J. Chem. Soc.* 1265, Royal Society of Chemistry)

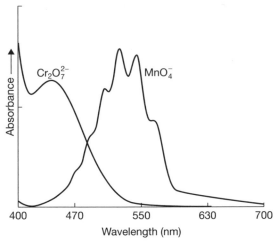

Figure 4.10 UV–visible absorption spectra of aqueous $Cr_2O_7^{2-}$ and MnO_4^- ions showing absorption maxima at 550 nm and 440 nm enabling both components to be measured

4.4.3 Beer–Lambert law

The intensity of the light beam passing through an absorbing sample, I, is related to the incident beam intensity, I_0, by the relationship

$$\log (I/I_0) = \varepsilon c l$$

where ε is the molar absorbance or extinction coefficient, c is the molar concentration of the absorbing species in moles dm^{-3} and l is the path length through the absorbing medium, usually measured in centimetres. The normal path length for many instruments is in the range 1 mm to 5 cm. ε is dependent on the quantum efficiency of the excitation process and the stability of the excited state. The Beer–Lambert law tacitly assumes a monochromatic light source and absorption process. Since this is rarely the case, and ε can only be calculated approximately in most matrices, calibration of the system is required. Discussion of the applications of UV–visible spectroscopy to environmental diagnostics naturally falls into two classes, organic and inorganic compounds, and these will be discussed separately.

4.4.4   Organic compounds

Organic molecules absorb UV–visible radiation by interactions either with electrons which are shared between two or more atoms forming direct covalent bonds between the atoms or with unshared outer electron pairs. The latter are usually associated with the atoms in groups V, VI and VII of the periodic table, i.e. those with p orbital electrons, nitrogen, oxygen and halogens etc. The molecular absorption wavelengths depend on the binding energy of the electrons involved and the magnitude of the excitation energy compared with the lowest excited state. Quantum mechanical molecular orbital theory shows that, the more strongly bound the electron, the higher the energy of the excited state and hence the greater the energy of the incident photons required to produce the excitation. Excitation usually involves an electron from the highest-energy occupied molecular orbital (HOMO) being promoted to the lowest-energy unoccupied molecular orbital (LUMO).

The binding energies of electrons in carbon-to-carbon and carbon-to-hydrogen bonds are so high that the energies required for the excitation process correspond to wavelengths <180 nm. Atmospheric water absorption and the absorption edge of silicon dioxide cause experimental difficulties at these wavelengths. Vacuum spectrometers using special optics based on lithium or calcium fluorides are required. Compounds containing unsaturated double or triple bonds exhibit usable UV–visible absorptions due to the presence of the more weakly bound electrons in π orbitals (HOMO). Excitation to the π^* antibonding orbitals (LUMO) corresponds to the UV–visible spectral region. The process is indicated schematically in Figure 4.11. Any organic and inorganic molecules which incorporate these unsaturated linkages as a functional group will exhibit suitable absorptions. Suitable

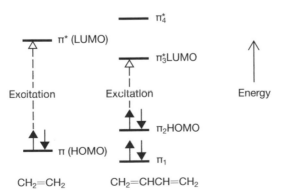

Figure 4.11   Schematic absorption process for molecules containing electrons in π orbitals excited to empty π^* antibonding molecular orbitals

Table 4.4   *UV–visible absorption bands for some organic compounds*

Chromophore	Example	λ_{max}	ε_{max}	Transition
Alkyl	R–CH$_3$	122	–	$\sigma \rightarrow \sigma^*$
Alkene	R–CH=CH$_2$	177	13 000	$\pi \rightarrow \pi^*$
Conjugated	R–CH=CH–CH=CH$_2$	217	21 000	$\pi \rightarrow \pi^*$
Aromatic				
Benzene	C$_6$H$_6$	204	7 900	$\pi \rightarrow \pi^*$
		256	200	
Alkyne	R–C≡C–R′	178	10 000	$\pi \rightarrow \pi^*$
		196	2 000	
		225	160	
Carbonyl	R–C=O	186	1 000	$\pi \rightarrow \pi^*$
	R′	280	20	$n \rightarrow \pi^*$
	R′=H	180	20 000	$\pi \rightarrow \pi^*$
Carboxyl	R′=OH	204	41	
	R′=NH$_2$	214	60	
Nitro	R–NO$_2$	280	20	$n \rightarrow \pi^*$
	R–NO	300	100	$n \rightarrow \pi^*$
	R–ONO$_2$	270	12	$n \rightarrow \pi^*$
Nitroso	R–NO	665	20	$n \rightarrow \pi^*$
Azo	R–N=N	347	10	$n \rightarrow \pi^*$
Halogen	CH$_3$Cl	173	2 000	$n \rightarrow \pi^*$

groups, called chromophores, are detailed in Table 4.4. Also given in the table are the wavelengths of their maximum absorption, λ_{max}, and the molar extinction coefficient, ε. The higher the value for the latter the more sensitive the absorption is and the lower the limit of detection for that compound. Enhanced absorption is shown by molecules which contain multiple unsaturated double bonds separated by single bonds, i.e. conjugated systems.

Sulphur, bromine and iodine atoms present in organic compounds can produce usable UV–visible absorption since their outer, non-bonded, electron pairs are more loosely bound and hence are more easily excited than bonded electrons. Also, the presence of these elements affects the wavelengths and sensitivity of the chromophores in Table 4.4 by interactions of their electrons with the bonded electrons. The substitution of hydrogen by electron-active groups in the benzene ring can also affect very strongly the position of the absorbance and also the molar extinction coefficients for the aromatic ring. Many phenols, especially in alkaline solution when they are present in the anion phenolate form, exhibit absorbance around 270 nm with a high molar extinction coefficient, $\varepsilon = 2000$, as opposed to $\varepsilon = 200$ for benzene, and thus can be directly determined by direct optical absorption measurements on the solution.

When two or more substituents are present in the aromatic ring then the relative position of the substituents, as well as their electronic properties, is very important. Typical ring absorbance can vary by an order of magnitude depending on the position of the substituents. The data in Table 4.5 show the effect for amino and nitro substituent positions on the intensity of the absorption spectra.

Table 4.5 *Absorption maxima for some disubstituted benzenes*

Substituents	Positions	λ_{max} (ε_{mol})
$-NH_2$, $-NO_2$	1, 2	227 (16 000), 275 (3000), 405 (6000)
	1, 3	235 (16 000), 373 (1500)
	1, 4	229 (5000), 375 (16 000)

When several aromatic rings are fused together (polyaromatic hydrocarbons) or when each ring is separated by a single carbon-to-carbon bond then the molar extinction coefficient is also increased. Notice that the extinction coefficients are from around 10^4 (terphenyl) to 10^5 compared with 10^3 for benzene and also the increasing absorbance at longer wavelengths, 300–500 nm. Addition of hetero-atoms in the ring structure also increases the absorption sensitivity.

Since many organic molecules exhibit UV absorbance in the 200–300 nm region simple measurement of absorbance in this region is a good test for the presence of organic material in waters but is not definitive as to the molecules responsible for the absorbance. Hence the widespread use of UV absorbance as a detector in high-performance liquid chromatography (HPLC) systems when the UV–visible spectrometer provides the sensitivity and the HPLC provides the selectivity.

4.4.5   Inorganic compounds

The absorbing species present in inorganic compounds can be divided into a number of different groups. Compounds of the main group elements are very similar to organic compounds. Electrons in single covalent bonds have UV–visible absorption spectra in the <180 nm region. Unsaturated double-bonded systems containing π-bonded electrons absorb in the UV–visible range. Compound ions, such as nitrate, sulphate, phosphate and halates, exhibit absorptions in the 200–220 nm spectral region.

4.4.5.1   *Direct nitrate measurements*

Direct measurement of absorbance at 220 nm can be used for the determination of nitrate ions in water in the 10–1000 ppm range. Care has to be taken to ensure that no other absorbing species, particularly organic materials, are present. This is usually done by also measuring the absorbance at 275 nm. If the absorbance at 275 is >10% of the absorbance at 220 nm then interferences are present and the direct measurement technique is invalid. Limits of detection are inferior to those for the corresponding organic chromophores.

4.4.5.2   *Transition metal spectra*

Almost all transition elements, those in the middle of the periodic table, are coloured in one or all of their oxidation states as either solvated ions or compounds and complexes. This is due to the presence of incomplete inner electron shells. Absorption in the visible and UV arises from transitions of electrons from partially filled d orbitals to unfilled d orbitals. Normally this would be disallowed by quantum mechanical theory, spin forbidden, because

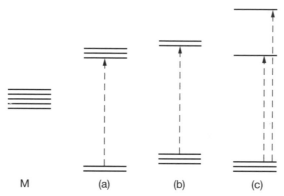

Figure 4.12   Splitting of d orbitals on a transition metal cation in the presence of (a) tetrahedral, (b) octahedral and (c) square planar arrangements of identical ligands

the d orbitals are all of equal energy and are centrosymmetric. The symmetry of the orbitals involved must be different for allowed transitions. The transitions are allowed by the removal of symmetry during some of the thermal vibrations of the ligands and the central species. In the presence of ligands around the transition metal ion the energy levels of these d orbitals are no longer equivalent. The number of d electrons in the transition metal species affects the ligand geometry around the metal ion which can vary between tetrahedral, square planar and octahedral. The d orbital splitting pattern and energy levels are very dependent on the ligand geometry. Some orbitals closest to the ligands are raised in energy and others lowered in energy. Energy level diagrams are shown in Figure 4.12 for the three ligand geometries, octahedral, square planar and tetrahedral. The electrons occupy the lowest energy levels possible and for partially filled d levels the excitation from the highest occupied d orbital to the lowest unoccupied d orbital, as indicated in the figure, corresponds to the visible region of the spectrum. The energy separation of the d orbitals is dependent on the donor properties of the ligands surrounding the transition metal ion and the number of d electrons on the ion. The ligands can be arranged in order of their donor properties, a spectrochemical series, which indicates the size of the energy split in the d orbitals. Since the energy levels associated with the transitions are a function of ligand geometry so are the wavelengths of the absorptions. The absorption associated with tetrahedral complexes is more intense than for octahedral complexes because the tetrahedral arrangement naturally reduces the symmetry of the d orbitals and does not rely on thermal vibrations to remove the centre of symmetry. Therefore some d-to-d transitions become spin allowed. The octahedral $[Co(H_2O)_6]^{2+}$ ion is a pale red in solution with a molar absorbance coefficient of about 4 but, on addition of excess of chloride ions, the solution becomes an intense blue colour, with a molar absorbance of 200, owing to the formation of the tetrahedral $[CoCl_4]^{2-}$ complex. Ligands capable of forming tetrahedral complexes with transition metal ions can be used to increase the sensitivity of direct spectroscopic methods for the metal ion. Also, mixtures of ligands around the central metal ion enhance the d-to-d transition intensity. This is typified by the series of complexes formed by the ion Co^{3+} with a mixture of chloride ions and ammonia. Complexes containing both chloride and ammonia, e.g. $[Co(NH_3)_3Cl_3]$, exhibit more intense colours than either the $[CoCl_6]^{3-}$ ion or the $[Co(NH_3)_6]^{3+}$ ion. For more detailed discussion of the quantum mechanics of the d–d transitions the reader is referred to some of the standard quantum mechanics texts.

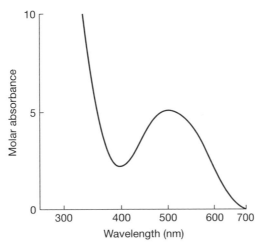

Figure 4.13 Absorption spectrum of Ti^{3+} ion in aqueous solution showing the d–d transition at 500 nm

These d-to-d transitions do not produce the strongest absorptions. Figure 4.13 shows the absorption spectrum for the Ti^{3+} ion in aqueous solution. Absorptions arising from the excitation of electrons from d to p orbitals, because they are spin allowed, are the most intense, but these occur in the UV region, many below 200 nm. The short-wavelength side of the figure shows the edge of the d-to-p transition as well as the other absorption mechanism, charge transfer transitions.

4.4.5.3   Charge transfer spectra

When the ligand is an electron donor, bonded to an electron acceptor molecule or ion, then an additional electron excitation called a charge transfer can occur. An electron is excited from an orbital essentially on the ligand to an orbital essentially associated with the acceptor species and hence a net transfer of electrical charge between donor and acceptor. The molar extinction coefficient ε_{max} for charge transfer absorptions is typically >10 000, thus producing a high sensitivity for analysis. A typical charge transfer process occurs with the thiocyanate, SCN^-, complex of iron(III) ions, Fe^{3+}. In this complex the Fe^{3+} is surrounded tetrahedrally by four SCN^- ligands with the S towards the iron. Absorption of a photon of ~450 nm radiation excites an electron from the sulphur-centred molecular orbital to an empty orbital on the Fe^{3+} ion, temporarily reducing the oxidation state of the iron from Fe^{3+} to Fe^{2+} and oxidising the anion to the SCN radical: an internal oxidation–reduction has occurred. Because the absorption occurs in the blue region of the spectrum the solution appears red. Absorption by this complex is used as a qualitative and quantitative test for Fe^{3+}. Reduction of the Fe^{3+} to Fe^{2+} by the addition of Sn^{2+} destroys the complex and causes the colour to disappear.

In most charge transfer processes the metal ion serves as the electron acceptor and the ligand as the electron donor. For some metal ions in low oxidation states, e.g. Cu(I), with certain ligands the roles are reversed and the metal ion acts as the donor. The ligands are usually neutral species in this type of charge transfer because of the need to accept electrons. A good example is the complex of Fe^{2+} iron with 1,10-phenanthroline. Here the ligand is the acceptor and the Fe^{2+} the donor.

The best ligands for accepting the electron from the metal ion in a charge transfer process are those ligands which possess empty π or π^* orbitals which can accommodate the extra electron. Typical ligands are SCN^-, NCO^-, CN^- and pyridine C_5H_5N. Similarly the best electron donor ligands are those containing electrons in π bonding orbitals or non-bonded lone pairs. Some analytical methods for transition metal determinations involve measuring the UV–visible absorbance of complexes with these ligands.

Inner transition elements contain unfilled f orbitals which are usually two levels below the bonding orbitals and are thus shielded from the effects of molecular vibration–rotation on the electron energy levels. Ligands around the metal ion split the f orbitals into different energy levels and hence transitions can occur between occupied f orbitals and higher-energy, unoccupied, f-type orbitals. Because there is no vibration contribution to the orbital energy separations the absorption spectra are sharp and well resolved.

Many of the determination of inorganic metal species involve addition of a suitable ligand to either produce a charge transfer band or move the normal d–d-type transition into a suitable region for observation. For main group elements, with normal transitions in the far-UV region, analysis involves measuring the intensity of the absorption spectrum of a molecular chromophore, usually an organic reagent, shifted by reaction with the main group element. Some typical reactions, which demonstrate these processes, are given below.

4.4.5.4   *Nitrates and nitrites*

Although nitrates can be determined directly from their UV–visible absorption the optimum determination of nitrites and nitrates at the ppm level is achieved by use of colour developing reagents. Nitrites in acid solution will diazotise 4-aminobenzenesulphonamide, sulphanilamide, which can then react to produce a diazo dye either by self-condensation or by reaction with N-(1-naphthyl)-ethylenediaminedihydrochloride. The dye absorbs at 550 nm with a sensitivity of 0.1 µg. Nitrate can be determined by a similar procedure if it is first reduced to nitrite. An alternative determination for nitrate involves the formation of a compound with sulphosalicylic acid, which absorbs at 415 nm.

4.4.5.5   *Orthophosphate*

Orthophosphate PO_4^{3-} reacts with molybdate ion in acid solution to form phosphomolybdic acid. Partial reduction of this complex with hydrazine compounds produces a strong blue colour. The intensity of the absorption is proportional to the orthophosphate content in the original solution. Unfortunately the optimum wavelength and the extinction coefficient are both dependent on the solution acidity during the reduction process which leads to the colour development.

4.4.5.6   *Anionic detergents*

Many anionic detergents are based on the sodium salts of alkyl sulphonic acids. Low concentrations of these detergents in water can be determined by reaction of the anionic sulphate species with the cation of methylene blue dye to form a neutral species which can be extracted into an organic solvent, e.g. trichloromethane. The colour intensity in the organic layer is proportional to the anion concentration in the original water sample. Since the dye reagent is a salt it and the anion detergent are both insoluble in the organic phase and do not interfere with the determination. Similar processes can be applied to the determination of quaternary ammonium cation salts by the reaction with the anionic indicator bromophenol blue. Normally used as an acid–base indicator, in weakly acid solution of pH > 4.5 the blue anion will react with the quaternary ammonium cation to form a neutral, trichloromethane-soluble, complex which can be determined spectrophotometrically.

4.4.5.7 *Fluoride determination*

The displacement of one component in a complex by another species can also be used to measure the new species if there is a change in the absorbance spectrum accompanying the substitution reaction. Determination of fluoride ions by the reaction of the fluoride ions with the lanthanum complex of alizarin fluorine blue is possible owing to the change in the absorption spectrum when the fluoride is incorporated into the complex.

4.4.6 Solvent and materials choice

Because all materials show UV–visible absorption the solvents available for presentation of the sample are limited. No measurement can be made in the region of the solvent's own absorption and hence choice of solvent may restrict the methods available for the analysis. The solvent cut-off wavelengths for some common solvents are given in Table 4.6. Notice that propanone, an excellent solvent for many organic materials, has a cut-off which inter-feres with the determination of most organic functional groups Pure solvent cannot be used as a background correction in the region of the absorbance edge, because of the spectral artefacts produced by the subtraction of two very large quantities which are marginally different owing to interaction with the solute. Many of the solid materials used for cells absorb UV radiation and, if measurements are to be made in the 200–340 nm region, fused silica cells must be used. Normal glass absorbs strongly below 360 nm and many plastics, as expected from an earlier section, also strongly absorb UV light and hence cannot be used for UV spectroscopy measurements.

Solvents which are immiscible with water can be used to extract the coloured species from water and hence reduce the interference by other components dissolved in water. A number of heavy metal contaminants can be determined in sequence by using a colour developing reagent and performing a series of extractions at different pH conditions. The reagent contains a number of acid groups which can be replaced by metal ions to produce a neutral species which is insoluble in water but soluble in non-polar solvents. Because of

Table 4.6   *Absorption cut-off wavelengths of some common solvents*

Solvent	Cut-off (nm)	Solvent	Cut-off (nm)
Water	190	Dichloromethane	233
Hexane	199	Trichloromethane	247
Diethyl ether	205	Tetrachloromethane	257
Ethanol	207	Benzene	280
Methanol	210	Pyridine	306
Cyclohexane	212	Propanone	331

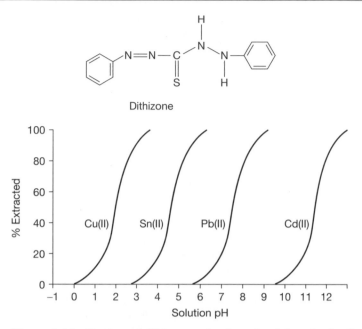

Dithizone

Figure 4.14 Structure of dithizone molecule and variation of extraction efficiency for some divalent metals as a function of solution pH

the presence of the acid groups the formation constant, stability, of the metal ion complex is dependent on the solution pH and the particular metal ion. Dithizone is an organic molecule with two hydrogen ion functional groups (Figure 4.14). When these hydrogen ions are removed the absorption spectrum of the molecule is changed, reducing the interference from the neutral reagent species. If this molecule is added to a solution of metal ions then zero-charge complexes can be formed, which can then be extracted into the non-polar solvent. The optical absorbance of the dithizone reagent in the complex with the metal is different from the uncomplexed reagent and hence the metal concentration can be determined from the absorbance. The formation for a number of the metal ion complexes as a function of solution acidity is also shown in Figure 4.14. A large number of reagents of this type are available and a number of reviews have been published.

4.4.7 Methods for spectrophotometric determinations

Methods available for the spectrophotometric determination of analytes can be divided into those based on colour measurement and those based on spectral measurements.

4.4.7.1 *Duplication method*

A measured volume of the sample is placed in a tube and the reagents are added to develop the colour or absorption spectrum. In a similar tube the same reagents are mixed and a standard solution is added until the colour or spectrum matches the unknown. From the volume of standard added the concentration of the analyte is calculated.

4.4.7.2 *Balancing method*

A fixed volume of the sample solution is placed in a tube and a variable quantity of a standard solution is added to a second identical tube until the colour intensity in the two tubes is identical, as judged by the human eye viewing the split eyepiece. Assuming Beer's law is applicable, the products of analyte concentration and depth of solution in the two tubes are identical and hence the unknown concentration can be calculated.

4.4.7.3 *Standard series method*

A measured volume of the test solution is contained in a glass tube and, after development of the colour by addition of a specified quantity of reagent, the colour is compared with a series of identical tubes containing different quantities of the standard in a search for the closest match.

4.4.8   Instruments for spectroscopic measurements

The three methods described above can all be performed using the human eye as the comparator, but this is prone to error and for improved precision a photometer can be used. This is essentially a photoelectric cell which measures the light absorbed by the sample and unlike the human eye does not measure colour. For optimum performance the samples are examined with a selected spectral region radiation source and not a white light source. The spectral region can be selected by means of an optical filter, which transmits only a selected region of the emission from the white light source.

The highest accuracy and specificity are achieved when a spectrophotometer is used to measure the optical absorbance as a function of wavelength, i.e. to record the absorption spectrum. Measurements are then performed using a selected narrow spectral band. In the simplest form the absorption from the sample is compared with that of a series of standard solutions using a single light path but for highest precision a dual-beam instrument, in which the solution absorption is compared with a background reagent blank, is preferred; see Chapter 5.

A variation sometimes used is flow injection analysis. The sample is injected into a solvent stream, mixed with the reagent in a flow system and the resulting solution passed through a measurement cell in the spectrophotometer. Concentration of reagents and contact time can be varied over wide ranges and adjusted to optimise the sensitivity and reduce the interferences. This method is particularly important for measuring transient species when the experimental manipulation time is important.

4.4.9   Advantages of spectrophotometric analysis

Spectroscopic analysis has a number of advantages, detailed below, which are responsible for its wide application in environmental diagnostics:

(a)   It is widely applicable to the determination of a large number of inorganic organic and biochemical species either by direct measurements or by application of a chemical conversion reaction prior to measurement.

(b)   Simple procedures with a high sensitivity are available, particularly when charge transfer absorption spectra are used. Limits of detection in the region of 10^{-7} M are easily obtained by careful selection of experimental conditions.

(c)   Selectivity of the determination can be improved by careful selection of the observation wavelength or, where overlapping absorptions occur, by measurements at several wavelengths. As an example of this dichromate, $Cr_2O_7^{2-}$, and permanganate, MnO_4^-, can be determined in a mixture of the two by measuring absorbance at different wavelengths. At 550 nm the absorbance is largely due to MnO_4^- but at 430 nm the absorbance is largely due to $Cr_2O_7^{2-}$ (Figure 4.10). If the measurement is performed at the two wavelengths both species can be determined. Additional chemical reactions can also be used to eliminate interferences. As an example the addition of fluoride ions to a solution of ferric ions forms a stable complex ion, $[FeF_6]^-$, which prevents the iron ions from interfering with the formation of oxine complexes of other metals during spectrophotometric determinations.

4.4.10   Atomic absorption spectroscopy

Atoms in the ground state are capable of absorbing light photons of the correct energy. Light emitted by an excited atom returning to the ground state possesses the correct energy to excite an identical ground state atom. This is the basis of atomic absorption spectroscopy (AAS), discussed more fully in Section 5.4.2. Light of the correct wavelength is generated by passing an electrical discharge to a cathode of the element to be determined. Passage of this light output through a flame containing atoms of the element to be determined produces a reduction in the beam intensity at the detector. The reduction in beam intensity follows the Beer–Lambert law and is proportional to the concentration of the element concerned. Solutions from environmental samples can be drawn into the flame and elements determined by use of the correct light source. Since there are higher concentrations of ground state atoms than excited state atoms in the flame the sensitivity of AAS is much greater than that of simple flame emission spectroscopy discussed in a later section. As a comparison the normal working range for Ca by flame emission is 0.5–50 ppm whereas for AAS the concentration range is 0.5–10 ppb: three orders of magnitude less. If the graphite furnace is used increased sensitivity can be obtained; see Section 5.4.2.3. A schematic diagram of a typical assembly is shown in Figure 5.18. Because the emission line from an atom is very sharp, 0.1 nm, it is very specific for the element being determined. Interferences occur either from accidental coincidence of absorption from another element, avoided by selection of another wavelength, or when atoms of another reactive species present in the flame form a compound with the determinant element. This interference can be reduced by addition of a third material which will preferentially react with the interfering element. Careful control of the flame conditions is extremely important and this is discussed elsewhere.

4.4.11   Additional criteria for spectroscopic analysis

Before an analyte can be determined by spectrophotometric or spectroscopic methods the following criteria must be met:

1.   suitable sensitivity for the analyte in question;
2.   there should be a direct proportionality between the intensity of the absorption and the concentration of analyte;
3.   because spectroscopic methods are not absolute a calibration curve from a suitable standard will be required;
4.   the absorbing species should be stable for sufficient time for measurements to be performed, even for those reactions for which the colour development is not instantaneous;

Table 4.7 *Proportion of atoms in upper state as a function of 'flame' temperature*

Element	Wavelength (nm)	Temperature		
		2000 °C	3000 °C	4000 °C
Na	589	9×10^{-6}	6×10^{-4}	4×10^{-3}
Ca	422	1×10^{-7}	4×10^{-5}	6×10^{-4}
Zn	213	7×10^{-15}	6×10^{-10}	2×10^{-7}

5. the absorption spectra must be reproducible under specified conditions;
6. interferences either must be absent or can be corrected for by suitable means;
7. absence of precipitates from the solution if absorption measurements are performed.

4.5   Emission spectroscopy

If the outer valence shell electrons of an atom are excited by some process, involving the absorption of energy, then the excited atom can return to the ground electronic state by the emission of radiation. The energy of the emitted radiation will correspond to the energy separation between the ground and the excited electronic states. This energy difference between the two electronic states is a function of atomic structure and is characteristic of the atoms present. The number of excited atoms present at any one time is a function of this energy difference and, since thermal vibrations and collisions are methods of transferring energy between atoms, the effective temperature the atom experiences. The proportion of excited atoms, $\hat{A}$, is given by the Boltzmann distribution function

$$\hat{A} = \exp(-E/kT)$$

where E is the energy difference between the electronic states, k is the Boltzmann constant and T is the absolute temperature. The smaller the energy difference between the two electronic states the greater the proportion of excited atoms and also the higher the temperature the greater the proportion of excited atoms present. The energy of the radiation liberated is equal to the energy difference between the excited and ground electronic states. The intensity of the radiation is proportional to the number of excited atoms present. The higher the temperature the greater the emitted intensity. Table 4.7 shows values for $\hat{A}$ for some elements. Measurement of the intensity of the emitted characteristic radiation is a method of determining the concentration of the emitting species. The process is not absolute and requires calibration using known standard concentrations of the species concerned.

Two methods of providing the excitation energy to the atoms are a hydrocarbon–oxygen flame and an electric discharge. Since these two sources are very different in their effects and require different instrumental properties they will be discussed separately.

4.5.1   Flame photometry

If a sample of an easily excited atom is injected into a non-luminous flame then in the hot regions of the flame the electrons in the atoms are excited and in the cool region of

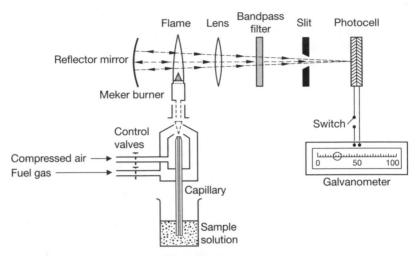

Figure 4.15 Elements of a flame photometer

the flame the excited atoms emit radiation as they return to the ground electronic state. Measurement of the emitted radiation intensity by a photocell can be used to determine the analyte. Because the energy provided by the flame is limited only those elements with low excitation energies can be determined in this manner. This means that essentially the alkali and alkaline earth elements with a few transition metals are reliably detected by this method. A typical instrumental arrangement is shown in Figure 4.15. The sample, preferably in solution, is drawn into a capillary tube by the passage of the air for the flame over the end of the tube and is thus directly injected into the flame. This means that the only sample preparation required is filtering to prevent solid particles from blocking the pick-up capillary. Elements studied by this technique are given in Table 4.8 together with their characteristic emission wavelengths and the limits of detection by flame emission and typical ranges examined. At too high a concentration of emitting species the light output is no longer linear and errors can be introduced into the determination. Some selectivity

Table 4.8 *Elements routinely determined by flame photometric methods: observation wavelength and sensitivity ranges*

Element	Wavelength (nm)	Sensitivity (ppm)	Limit of detection (ppm)
Li	671	0–15	0.005
Na	589	0–5	0.0003
K	766	0–10	0.003
Rb	780	0–20	0.004
Cs	852	0–20	0.004
Mg	285	0–20	0.008
Ca	423	0–50	0.005
Sr	461	0–30	0.001
Ba	422	0–40	0.004

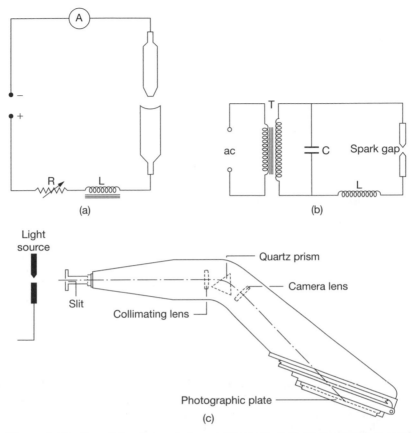

Figure 4.16   Essentials of an emission spectrograph. Both (a) dc and (b) ac spark sources are used and the optical arrangement with photographic detector is shown in (c)

can be introduced by using a filter between the flame and the detector. Improved resolution can be obtained by using a monochromator in place of the filter. Analysis of alkali and alkaline earth cations in water samples is routinely performed by flame photometry.

4.5.2   Spark emission spectroscopy

If the excitation source is changed from a simple flame to a dc or ac spark, Figures 4.16(a) and 4.16(b), then the instrument is capable of detecting every element above helium. This is particularly advantageous since some of the non-metals such as boron, carbon and phosphorus are difficult to detect by other methods. Most of the non-metal detection methods rely on identifying a particular chemical compound whereas the spark emission mode can detect these elements regardless of the chemical nature of the material. The number of excited states available to the electrons in the atoms is greater and as a result a complete spectrum of lines is emitted rather than a limited number of fundamental frequencies present in the flame emission mode. The sensitivity and reliability of the measurements are greater. Accidental coincidences of spectral wavelengths can be resolved by considering multiple wavelengths for identification. The system is capable of detecting most of the elements present in a sample above the ppm range and in many cases to <1 ppm. The emitted radiation is normally detected with either a photographic film or a Vidicon image analyser after a prism monochromator–analyser. A basic schematic of a typical optical arrangement for an emission spectrograph is given in Figure 4.16(c). The instrument is not quantitative and careful calibration is required for precise determinations of trace analytes present in some matrices. A major advantage of this system is the lack of sample preparation. Solid samples or evaporated samples can be readily analysed in the dc cup electrode without any major preparative work although for best results prior removal of volatile species is necessary. The ac spark system is best suited for metal and alloy samples.

4.6   Rapid small-scale test kits

4.6.1   Introduction

Rapid, accurate and simple methods for the field determination of indicator species enable potential pollution incidents to be identified and limit their impact by rapid decisions and response. Field determinations can eliminate the need for sample storage and transportation prior to laboratory analysis. Mobile analysis can be used as a screening tool to eliminate non-critical samples, thus reducing the number of samples to be examined by more sophisticated and expensive laboratory analytical instruments. Many of the kits available only test for specific contaminants or a generic series, e.g. a number of chlorinated phenols in any one test reaction, and hence different tests have to be performed for each analyte. Careful checks must be made for interferences and these eliminated if possible.

Because of the experimental difficulties associated with field determinations, most of the rapid test kits use packages containing pre-measured quantities of reagents and require the minimum of operator manipulation and skill. Most of the test kits are available for liquid phase, but some do operate in the solid and gas phases. The following sections concentrate on the liquid phase.

4.6.2   Chemical screening

Test kits available cover the complete range of classical chemical analytical methods with the exception of gravimetric methods where light scattering methods are used to estimate mass of solids present. Field test kits naturally divide into colorimetric and volumetric analysis methods, which are the most widely used techniques. Sampling methods and analysis procedure vary with supplier but are usually detailed in the literature provided with the kit.

4.6.2.1   *Sampling methods*

The simplest analytical methods involve *in situ* measurements by dipping an instrumental probe or a simple test paper treated with reagent into the material to be screened. The former is typified by the dissolved oxygen probe and the dissolved solid conductivity cell and the latter is typified by normal pH test paper, which can be used to monitor acidity levels in soils, liquids and gas streams. In this case the result is determined by visual comparison and matching of colours with a calibration chart: the only operator skill

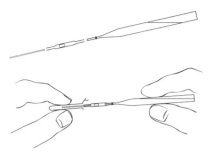

Figure 4.17   Test kit sampling using a capillary filling and dilution system

required. Since the result is a measure of concentration then the sample volume is non-critical.

When the sample volume is critical to the determination then a number of different methods are used in the kits. Simple filling of a container to the specified mark or to overflowing, preferably through a side arm, is preferred since this reduces handling problems. Filling of a glass capillary with the liquid under test is also used, Figure 4.17, and can be extended to provide a method of sample dilution in some kits which draw dilution water through the capillary into a reagent vial. More precise volumes can be obtained by using either a positive displacement pipette or an evacuated vial which takes in the required volume when the vacuum seal is broken under the surface of the liquid. Both of these systems can be adapted to introduce variable quantities of sample.

4.6.2.2   *Reagent addition*

Test papers impregnated with the reagents, which the operator contacts with the sample prior to measurement, offer the simplest method of reagent control. Displacement pipettes can be used to add reagent and are ideally suited for repeat addition to a large number of samples if multi-tip pipettes are used. A dropping bottle is a cheaper alternative but suffers from the disadvantage that the operator may have to count the number of drops added, which can be a source of error if either a large number of drops are required or a large number of samples are to be analysed. Dropping bottles are ideal for the addition of conditioning chemicals to the sample prior to the determinant reagent. Drop volume varies with experimental conditions such as temperature and solution composition but manufacturers usually allow for these variables in their test kit design.

Measurement of reagent quantity by volume, e.g. a level scoopful, is prone to a number of errors and is only used when exact quantities are not required. Addition of excess of pre-treatment chemicals to the sample either to produce the analyte in the correct chemical species or to reduce interference by other species present is amenable to this method. Use of pre-formed reagent tablets for the addition overcomes these difficulties, with the added advantage that a number of reagents and conditioning chemicals can be pre-mixed and added as a single operation, ensuring a more uniform test solution. The reagents can be sealed into easily cleaned sachets to prolong their shelf life and also to prevent accidental solution contamination. If evacuated vials are used for sampling then all the reagents for the determination can be contained in the vial and the only operation required is to break open the vial under the sample surface. Again the use of a preconditioning tablet may be necessary to eliminate interferences.

4.6.3 Colour development tests

Kits are formulated so that the observed colour change is directly proportional to analyte concentration in the sample. Simple dipping of a test paper in the sample solution and making a visual comparison with a standard chart, after suitable reaction time has elapsed, is used for the normal pH test paper. More sophisticated versions examine the test paper in a photometer, which measures either the total light reflected from the surface or the intensity at a pre-set wavelength selected by the filter. Alternatively a colour comparator can be used in which the sample vial is placed inside a housing and the colour compared with a calibration strip or disc containing a series of standard colour elements. In some test kits the sample is either drawn into an evacuated vial or used to fill a tube and after adequate mixing to remove any air bubbles the vial is compared with a series of standard tubes, when the closest colour match with the standard is looked for. This is essentially a direct reading process but instrumental comparators, which enable the tubes to be matched electrically, are available. Again a colour chart can be used to make the comparison. These comparison methods are essentially of limited scope and accuracy but suitable for preliminary incident investigation.

Electronic measurements of solution optical properties are more accurate and provide the opportunity for a number of determinations from the same sample. In the simplest form the absorption of the light from a white light source by the solution is measured by a photocell detector. Although not specific the system is rugged and compact for field operations. Some systems use a light-emitting diode as the source, which provides some wavelength selectivity, without loss of durability. Simple photodiodes are used as the detectors.

Spectrophotometers enable measurements to be performed in selected wavelength regions of the spectrum using either filters to select broadband spectral regions or, for smaller bandwidths, monochromators. The basic principles of both types of system were discussed earlier. Filters offer a more robust system for field use since there are no critical moving parts in the optical system. Diode array detectors as developed for HPLC systems measuring the throughput from a permanently fixed dispersion prism or grating can produce a limited-resolution, robust, compromise spectrometer. The sample is treated with a measured quantity of reagent to produce the coloured solution whose optical absorption is then measured on the spectrophotometer. Simple normal laboratory plastic cuvettes can be used for measurements in the visible region. Machines which can accept a range of cell sizes from 1 mm to 4 cm offer a wider determination range without altering the solution conditions.

Evacuated reagent ampoule systems can be measured directly in the instrument but suffer from optical aberrations produced by the circular tube and show only limited flexibility. Similarly kits using screw-capped bottle reaction vessels measure *in situ* in the instrument. This system can be used for the direct determination of contaminants on solids, provided that the solid can settle to the bottom of the tube before reading, whereas the vacuum filling system requires transfer of the analyte to a suitable liquid phase before optical measurement. This can be an advantage since some separation of analyte from interferences can be undertaken at the same time. COD of a sample can be determined by measuring the absorbance due to chromium(III) ion produced by the chemical reaction of a quantity of dichromate in a screw-capped sealed vessel. Optical absorbance can be converted to ppm COD using a calibration graph, a conversion sometimes done directly by the machine. All of the instruments are normally checked on a regular basis.

The tests available range from the simple addition of the sample to prepared quantities of the reagent(s) followed by optical measurements, ideal for field measurements, to kits involving a sequence of careful additions of a series of chemicals, over a period of time, in order to eliminate interferences, convert the analyte to the correct form and produce the optimum measurement solution. These kits are more suitable for preliminary screening in the laboratory environment.

4.6.4 Titrimetric methods

The simplest titrimetric method available in kit form is the addition of tablets to a measured volume of solution until a prescribed indicator reaction occurs. By counting the number of tablets required the analyte concentration can be calculated from a calibration chart. This method has the disadvantage of limited resolution unless the number of tablets used is high but will readily give an indication of contamination levels. Improvements in resolution can be achieved if a prescribed number of tablets are placed in a calibrated volumetric container and the water sample is added until the prescribed colour change occurs when the measure of added sample volume yields the concentration from either a direct reading scale on the container or a calibration graph. Instead of tablets a reagent solution contained in a dropping bottle can be used with again the addition of the reagent to a fixed sample volume or addition of variable sample quantities to a fixed number of reagent drops being the alternatives. The latter technique does enable a wider range of sample levels to be accommodated with acceptable resolution and is easier from the operator viewpoint.

If an evacuated vial is used to contain the reagents then, by addition of a valve to the inlet tube, the quantity of liquid sample drawn into the vial can be controlled and again the volume of liquid sample required to produce the desired colour change against a fixed volume of reagent enables the concentration to be determined by reading the liquid level against a calibration scale, printed on the rear of the ampoule. The operator must posses some manual dexterity in order to control the valve effectively and produce the correct result.

Some kits provide the operator with calibrated burettes or pipettes of the type used in conventional laboratory-based volumetric systems but these are not as convenient for field usage as the kits described so far. Some indication of the test kits available are given in many chemicals suppliers' catalogues where details of methods available for the testing of water for a number of common analytes are given.

4.6.5 Airborne monitoring – Dräger tubes

Chemical reaction test kits are available for airborne monitoring. A predetermined volume of air is drawn in through a glass or plastic tube containing a reagent. Any analyte present causes a chemical reaction in the tube, producing a colour change in the reagent. The intensity of colour or the length of the colour change region measured against a calibration scale is proportional to the analyte concentration. Both systems are based on visual estimation of the readings. For short-term measurements, the seals on the ends of the tube are broken open and the tube is inserted in the inlet port with the airflow direction indicating arrow pointing to a manually operated bellows pump. Each operation cycle of the bellows draws 100 cm^3 of air through the sample tube. The colour develops in the direction of the air flow as the scale shows. The volume of air sampled is controlled by the number of operations of the pump and varies for each analyte depending on the sensitivity of the reaction used for the determination. The manufacturers' calibrations supplied specify the number of operations required. Detector tubes are specific for either single materials or groups of chemicals, e.g. oxidising agents, hydrocarbons. The tube is constant but the reagent contents and mode of operation are different. A number of tube configurations are available with the measurements being undertaken on a short- or long-term basis in either the active or the passive mode. In the basic tube the reagent is provided in a single uniform layer and the colour development along the tube is the measure of concentration. In addition to the normal reagent indicator layer, some tubes contain additional pre-layers which can be used to condition the sample prior to the measurement, e.g. by moisture removal, trapping of interfering compounds and conversion of the analyte into a suitable form for the determination. Typical of the last of these is the detection of hydrogen cyanide by the following reaction occurring in the pre-layer:

$$2HCN + HgCl_2 \rightarrow Hg(CN)_2 + 2HCl$$

The hydrogen chloride produced is detected by an acid indicator reagent layer. Sometimes the preconditioning layer is contained in a separate tube connected to the front of the detector tube by a flexible tube. This arrangement is preferred for heavily contaminated samples since the pre-tube capacity is greater than that of a simple single layer. Additional tubes may be connected behind the detector tube to absorb any toxic material produced by the chemical reactions occurring in the indicator tube.

Incompatibility between the reagents used in the detection reaction can be overcome by confining the reagents in a separate glass tube inside a flexible region of the main tube, which is broken open immediately prior to the measurement. Some of the more complex detection reaction chemistry involves multiple tubes inside the main tube, which have to be broken and reagents dispersed in the correct sequence. In the more complex determinations a series of tubes are broken open in sequence over a period of time.

Simultaneous determination of a number of analytes can be achieved by two techniques. If there are no compatibility problems, then the detector tubes can be connected in series with the same air sample being drawn through each tube in sequence in a similar manner to the preconditioning tube operation. If reaction products from the detection chemical reactions are incompatible then a number of tubes can be connected in parallel in a special adapter fitted to the normal pump. Suitable adjustments must be made to the number of pump strokes required and also the resistance of the tubes to the air flow must be carefully balanced, in order to maintain an even distribution of the sample air between the tubes. A pack of five tubes for the monitoring of toxic inorganic combustion products is available.

For long-term monitoring, e.g. the working day, different techniques are employed. Continuous pumping with a battery-powered pump operating at ~1 litre h^{-1} and tubes containing either higher concentrations of reagents or less sensitive reagents are used. A passive tube which relies on the diffusion of the analyte into the colouring region is also available. However, diffusion tubes and sampling tubes filled with absorbing material, e.g. charcoal or molecular sieve, are more widely used for extended sample collection at remote sites followed by later laboratory examination.

Reaction tubes are also available for the monitoring of volatile material in water and soil samples. In both of these techniques the volatiles are entrained in a turbulent flow of air drawn through the water or over the soil into a standard measurement tube.

4.7   Bibliography

Denney R C and Sinclair R 1997 *Applications of ultraviolet and visible spectrometry*, Analytical Chemistry by Open Learning (ACOL), Wiley, New York
Mendham J 1990 *Chemical analysis of water*, RSC Monographs, London

Mendham J, Denney R C, Barnes J D and Thomas M J K (eds) 2000 *Vogels textbook of quantitative chemical analysis*, 6th edn, Prentice Hall, Pearson Education, Harlow

Patnaik P 1997 *Handbook of environmental analysis: chemical pollutions in air, water, soil and solid wastes*, Lewis Publishers, Boca Raton, FL

Phillips J P, Bates D K, Feuer H and Thyagarajan B S (eds) 1995 *Organic electronic spectral data*, Vol 31, Wiley, New York

Simons W W (ed) 1995 *Sadtler handbook of ultraviolet spectra*, Heyden, London

Skoog D A, West D M and Holler F J 1996 *Fundamentals of analytical chemistry*, 7th edn, Harcourt Publishers, Sidcup

Thomas M J K 1996 *Ultraviolet and visible spectroscopy*, Wiley, New York

5

Instrumental analysis – optical and spectroscopic techniques

5.1 Fundamentals of optics

5.1.1 Nature of light

Thinking men from all ages, including Isaac Newton in the 17th century, pontificated about the nature of light. According to Newton, light is a stream of corpuscles emanating from its source. Later, Thomas Young in his famous double-slit experiment showed that two beams of light interfere with each other to produce light and dark patches and propounded the wave theory of light.

An understanding of the nature of the 'light waves' came from a completely different approach. Intensive research on the connection between electricity, magnetism and light gave birth to the electromagnetic (EM) wave theory of light. Although this theory explained many properties of light, including the interference phenomenon, it failed to explain the photoelectric effect, i.e. the ejection of electrons from solid surfaces by light. Albert Einstein could explain this phenomenon by considering light as a stream of particles called 'photons' having different energies corresponding to its colours. Max Planck later discovered that these photons can have only discrete amounts of energy. The wave–particle duality was resolved by associating the energy of the photons with the frequency of the EM wave. Light is, therefore, an EM wave or a stream of photons, depending on the type of interaction to be explained.

5.1.1.1 *Electromagnetic wave concept*

According to this concept, every point in a beam of light is associated with sinusoidally varying electric and magnetic fields[1] – one is alternately created by the other. These electric and magnetic fields are perpendicular to each other and also to the direction of their motion, i.e. the direction of the propagation of the beam (Figure 5.1). The contours in space or in time of the maximum values of the electric (or magnetic) field vectors constitute the light wave. The rate at which this vector quantity changes from zero to maxima in opposition directions and back to zero (one complete cycle) is called the vibration rate or wave frequency. The difference in the frequencies in different wave packets accounts for

---

[1] The term 'field', coined by Michael Faraday, describes the state of tension in a region which would not be there if the agent creating this is removed. The direction and magnitude of this quantity is represented by an arrow.

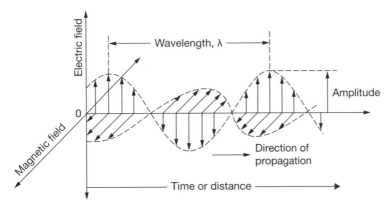

Figure 5.1 Concept of EM wave

the difference in the colour of the light. Later, it transpired that light is only a small portion of the vast EM spectrum which extends from low-frequency radio waves up to extremely high-frequency γ-radiation corresponding to a wavelength range from kilometres down to a millionth of a micrometre.

Parameters associated with EM waves are as follows:

- **Amplitude** (A) – the maximum length of the electric vector, expressed in **volts**.
- **Period** (p) – the time, in **seconds**, for the successive maxima of the waves to cross a fixed point in space.
- **Frequency** (v) – the number of peaks (crest of the wave) crossing a fixed point in space **per second** (s^{-1}). This is the same as the rate of oscillation of the electric field vector and is commonly expressed in the unit of **hertz** ($v = 1/p$).
- **Wavelength** (λ) – the linear distance between two successive peaks in a wave train. The commonly used unit of wavelength depends on the spectral region. For infrared and visible (light) wavelengths, the unit is **nanometres** ($1\,nm = 10^{-9}\,m$) or **micrometres** ($1\,\mu m = 10^{-6}\,m$). For X-rays or γ-rays, the unit is the angstrom ($1\,\text{Å} = 10^{-10}\,m$).
- **Velocity** (c) – the speed in a particular direction in **metres per second**. In a uniform medium, the EM radiation propagates with a constant velocity v, so that $v = v/\lambda$. The velocity is different in different media, being maximum in empty space and only very slightly different in air. The velocity of light or any EM radiation in air is a constant quantity and is given as:

$$c = v\,\lambda \approx 3 \times 10^{8}\,m\,s^{-1} \tag{5.1}$$

- **Power** (P) – the energy reaching an area A_t, per second, expressed in units of **watts**. The power density (P/A_t) is related to the electric field vector:

$$P/A_t = \varepsilon_0\,c\,E^{2} \tag{5.2}$$

where ε_0 is the dielectric permittivity of the medium and E is the magnitude of the electric field vector. The bar means the time-average values.

5.1.1.2 *Quantum concept*

The observation that light can strike off electrons from solid surfaces led to the foundation of the quantum theory of light. This effect and the phenomena associated with the absorption and emission of light by atoms and molecules cannot be theoretically explained by the wave model of light. The electrons associated with atoms cannot acquire sufficient energy from the energy of the light waves distributed over the entire wave front to react instantly to account for the observed effects. The energy needs to be concentrated in a point known as 'photon'. A compromise of wave–particle duality demands a wave aspect of these photons. Therefore, the photons are considered to be 'packets of waves'. Each packet contains waves of a discrete frequency and the energy of the photon or wave-packet is related to the frequency by

$$E\ (J) = h\nu = hc/\lambda \tag{5.3}$$

where the Planck constant $h = 6.625 \times 10^{-34}\,J\,s$.

It can be experimentally demonstrated that particles such as electrons, protons and light atoms exhibit 'wave-like' properties by producing interference fringes (pattern) in a double-slit experiment and a density distribution pattern in space after crossing the edge of an obstacle. It was therefore concluded that, like photons, the particles are also associated with a wave aspect. All particles must, therefore, have a corresponding wavelength. To describe the interactions of particles and photons, between themselves and between each other, each was attributed a wavefunction (ψ) which describes the periodic variations of a phantom quantity, called amplitude, in space and time. The mathematical formalism based on this wavefunction can explain all the interaction phenomena without invoking any model (wave or particle) for either light or particles. This formalism is known as quantum electrodynamics (QED).

5.1.2   Geometrical optics

For all practical purposes the propagation of light can be approximately described by tracing rays as straight lines originating from the source. This enables the design of optical instruments for directing light beams and focusing (imaging) light sources using the criterion that light travels at different speeds in different media. The design of optical instruments is essentially based on the laws of reflection, refraction and dispersion of light in highly polished surfaces (mirrors) and transparent media.

5.1.2.1   *Reflection*

The law of reflection states that the angle of reflection is equal to the angle of incidence, i.e. $\theta_r = \theta_i$. The angles are measured from the normal to the surface. Hero of Alexandria deduced this law from the concept that light rays take the shortest possible path, POQ (Figure 5.2) to travel from a point P to another point Q, after reflection from the glass–air interface (mirror).

5.1.2.2   *Refraction*

For a ray of light to travel from a point P in one medium (e.g. air) to a point Q′ in a denser medium (e.g. glass) as shown in Figure 5.2, the ray does not take the shortest path PQ′ as dictated by Euclidian geometry. It turns out that the propagation of light is governed by Fermat's principle of 'least time'. It states that, among all possible paths between two points, a ray of light will choose the one that takes the least possible time to travel. Since the light travels more slowly in a denser medium, it must optimise the paths in the air and in the glass so that the total time taken is the least of all possible paths. It can be easily

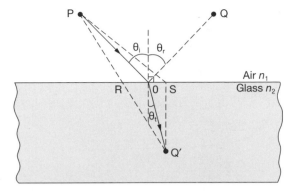

Figure 5.2 Reflection and refraction of light

shown that out of all possible paths (e.g. PRQ′, PSQ′, etc.), the path POQ′ takes the ray minimum time. To achieve this, the light ray has to bend towards the normal, and this phenomenon is known as refraction of light.

The amount of bending, i.e. the angle of refraction θ_t, depends on the speed of light in the medium relative to that in the vacuum. This property of the medium is characterised by a parameter called the refractive index, n. For glass, $n_g = c/v_g$, where v_g is the speed of light in the glass and c is that in vacuum (or in air). The refraction of light is given by Snell's law as:

$$n_1 \sin \theta_i = n_2 \sin \theta_t \tag{5.4}$$

where θ_i and θ_t are the angles of incidence and refraction respectively and n_1 and n_2 are the refractive indices of the media 1 and 2 respectively.

The performance analysis and design criteria of optical components and systems by ray tracing rely almost exclusively on the above relation. The refractive index varies widely for different media and the values for the most important materials can be found in textbooks and the literature.

5.1.2.3 *Dispersion*

The refractive index depends not only on the type of medium but also on the wavelength of the light (or any EM radiation). Therefore, rays of different colours (i.e. wavelengths) bend to different degrees when passing from one medium to another. This phenomenon is known as dispersion and is the cause of many natural and man-made optical effects such as the rainbow, the red sunset and the separation of colours in a beam of white light by a prism or a grating. The last of these is the basis of all spectroscopic equipment.

Dispersion curves of dielectric materials are represented by plots of the refractive index against the wavelength. Such curves for some common optical materials are shown in Figure 5.3.

Cauchy's empirical equation expressed as

$$n = A + B/\lambda^2 + C/\lambda^4 \tag{5.5}$$

represents the dispersion curve of a medium and the coefficients A, B and C are experimentally evaluated for that medium.

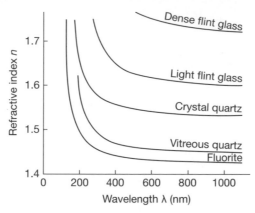

Figure 5.3   Dispersion characteristics of some optical materials

It should be noted that, at some wavelengths, the refractive index may become abnormally high and is not accounted for by Cauchy's equation. The effect is known as anomalous dispersion and is due to a resonance effect as a result of coincidence between the frequency of the light waves and the natural frequency of vibration of the particles in the medium which are bound to each other by elastic forces. This gives rise to selective absorption and reflection.

5.1.2.4   *Imaging*

The process of bringing light emitted or reflected from an object (source) into focus by using the refractive or reflective properties of curved surfaces is called imaging. Because of inadvertent imperfections in curved surfaces (e.g. lenses), all rays from an object are never brought to a perfect focus. Geometric optics based on approximate analysis of rays close to the optical axis of the lens or curved mirror defines the field known as 'Gaussian optics'.

An optical element, i.e. a lens or a curved mirror, is characterised by its focal length, f, i.e. the distance from the centre of the element to the point where a parallel bundle of rays converges to a focus after reflection or refraction, and by its effective diameter, d, known as the aperture. The focal length is governed by the curvature of the surface. The relations between the object distance, u, image distance, v, the focal length, f, and the image size, designated by the magnification parameter, m, are as follows.

For the object outside the focal point of a convex lens or mirror:

$$\frac{1}{u} + \frac{1}{v} = \frac{1}{f} \tag{5.6}$$

and

$$m = -u/v \tag{5.7}$$

(the minus sign indicates an inverted image).

The radius of curvature, R, of an element is related to the focal length as $f = R/2$. For a convex mirror and concave lens, f is negative by convention, so that $f = -R/2$.

For a combination of two or more thin lenses the above formulae will apply. In such an arrangement the image of the first lens becomes the object of the second lens.

For some elements, e.g. a double concave lens, light rays do not converge to an image point; instead, they appear to diverge from an image point behind the optical surface. This apparent image is known as a virtual image and this, unlike the real image, cannot be projected onto a screen.

5.1.3 Physical optics

This topic deals with the physical properties of light governed by phenomena related to the interactions of light with itself. These properties are classified as interference, diffraction and polarisation.

5.1.3.1 *Interference*

The intensity distribution over the cross-sectional areas of two beams of light are not affected by each other when they cross. However, at the region of crossing, the intensity distribution shows a periodic pattern of low and high values and appears as light and dark fringes as is shown in Figure 5.4. The effect, known as interference of light, is most pronounced for monochromatic light (light having a unique colour) and was first demonstrated in the celebrated Young double-slit experiment. The phenomenon is explained by considering light to be propagating EM waves having peaks and troughs and designating the relative positions of these peaks (or troughs) with a parameter called the phase or the phase angle. This, in effect, determines the relative position of the peaks or troughs both in space and in time. The interference phenomenon is explained with reference to Figure 5.4 as follows.

A parallel beam of monochromatic light has straight parallel wavefronts. In a single train of waves, there exists a constant relationship between phases at each point in the train. The beam within the time span of the train, or within the corresponding space, is said to be coherent. It should be noted that there may exist some degree of coherence between successive wave trains. When two coherent beams are superimposed on each other, their electric field vectors either combine additively to produce brightness, i.e. constructive interference, or nullify their intensities to give rise to darkness, i.e. destructive interference, depending on their relative phase angles (0° – same phase, 180° – opposite phase) at the region of crossing.

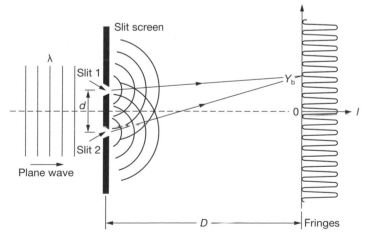

Figure 5.4   Interference of light from two slits

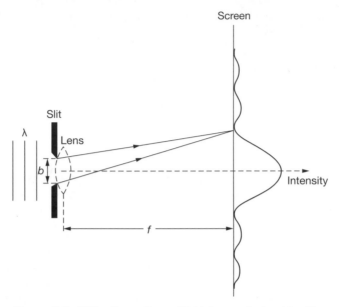

Figure 5.5 Diffraction pattern of light from a single wide slit

For slits having widths ideally narrower than the wavelength of the light, the emerging radiation fields will have spherically expanding wave fronts. The slits, according to Huygens' principle, act as secondary point sources of coherent radiation. By ray tracing and using simple geometry, it can be shown that the position of the mth bright fringe on a plane at a distance, D ($\gg d$), away from the plane containing the slits is given by

$$y_b = m\lambda D/d \qquad (5.8)$$

and the position of the dark fringe is given by

$$y_d = (m + \tfrac{1}{2})\lambda D/d \qquad (5.9)$$

where λ is the wavelength of the light and d is the separation between the slits. The distance between successive fringes (bright or dark) on the screen is given by $S = \lambda D/d$.

By measuring the distances D, d and the average separation distance between two successive fringes on a screen, the wavelength of monochromatic light can be calculated from the above equations.

5.1.3.2 *Diffraction*

The intensity distribution among the fringes in the double-slit experiment depends on the size of the slits. For slit width $d < \lambda$, the intensity distribution is uniformly periodic. However, for wider slits the distribution is different and convoluted by the intensity distribution which would be observed for individual wide slits providing a fringe system called diffraction patterns.

For a parallel beam of monochromatic light passing through a single slit of width b and brought to a focus by a lens of focal length f ($\gg b$), placed very close to the slit (Fraunhofer diffraction), as shown in Figure 5.5, the linear distance between successive minima, S, is given by

$$S = f\lambda/b \tag{5.10}$$

The width of the intensity pattern is, therefore, directly proportional to the wavelength and inversely proportional to the slit width.

Diffraction at the edges of lenses plays an important part in defining the resolving power of a telescope (or a lens). According to Rayleigh's criterion, two objects (sources) are resolved, i.e. distinguished, if the central maximum of the fringe pattern of one falls on the first dark ring of the other. The angular resolution, i.e. the minimum angular separation that can be resolved, is expressed as

$$\theta = 1.22\lambda/D$$

and the diffraction-limited size (i.e. radius, r) of the image at the focal point of a lens is given as

$$r = 1.22\lambda f/D \tag{5.11}$$

where D is the diameter and f is the focal length of the lens.

5.1.3.3 *Diffraction grating*

For multiple slits, placed very close together, interference and diffraction patterns are convoluted. The distributions of intensities in the first, second, etc. diffraction patterns are called the first-order, second-order diffraction patterns, etc. Evidently, the intensities in the higher-order patterns will be progressively lower. For any number of slits of equal widths and separations, the principal maxima will appear approximately at an angular distance (θ) given by

$$d\sin\theta = p\lambda$$

where p is the order of diffraction starting with 1.

A two-dimensional array of high-density grooves or slits on a reflecting surface or a transparency, called a diffraction grating, can be used to separate wavelengths in a particular order ($p = 1$, 2, etc.) according to the above equation. The density of the slits (or transparent lines in an opaque plate) determines the slit separation and hence the dispersion (or resolution) of the diffracted beam. The number of slits (N) illuminated dictates the sharpness and intensity of the diffracted order, and the resolving power of the grating is given by

$$RP = \lambda/\Delta\lambda = pN \tag{5.12}$$

where $\Delta\lambda$ is the minimum wavelength difference that the grating can resolve at a particular order, p.

Gratings are most commonly used in reflection mode in modern spectrometers to disperse a light beam according to its wavelength components or in a monochromator to select a particular wavelength from a mixture of wavelengths (white light).

5.1.3.4 *Polarisation*

This deals with the direction of vibration of the electric field of an EM wave. In light, the electric field vibrates at a rate of about 10^{14} Hz, normally in all possible directions in the plane perpendicular to the direction of propagation of the wave. It is therefore impossible to trace the temporal and directional changes of the electric vectors. If it were possible, then, viewed along the direction of the beam, the tips of the electric vector would have described a circle. Such a beam is called circularly polarised light. However, because of the random fluctuations in amplitude between successive oscillations, the tips would generally describe an ellipse and the beam then becomes randomly polarised, which is sometimes referred to as an unpolarised beam. When a beam of unpolarised light is passed through polaroids, i.e. transparent materials with specific crystal orientations, the transmitted beam will have electric vectors vibrating only in one specific direction. The transmitted light is therefore polarised in a particular direction.

Polarised light can also be produced by reflection from the surfaces of dielectric media such as glass. At specific angles of reflection, depending on the refractive indices of the media, the reflected beam will be linearly polarised. This angle is called Brewster's angle (ϕ) and is related to the refractive index by $n = \tan \phi$; for glass $\phi \approx 57°$.

5.1.4   Laser radiation

The term 'laser' is an acronym for 'light amplification by stimulated emission of radiation' and the terminology is universally accepted to mean the source of radiation. It is a process by which light beams with unique properties are generated. The uniqueness of laser radiation lies in its high monochromaticity, high degree of coherence, very low beam divergence and, most importantly, in its extremely high radiance. Both pulsed and continuous-wave (CW) lasers operating at any wavelength between far infrared (~100 μm) and ultraviolet (~0.1 μm) are now commercially available. These lasers could operate at pulse rates as high as a few kilohertz and could have pulse widths as short as a few picoseconds ($\sim 10^{-12}$ s). Because of their unique properties, lasers are finding important applications in many branches of science and technology including the field of environmental diagnostics.

5.1.4.1   *Properties of laser radiation*

Monochromaticity   This is defined by the spectral spread known as linewidth, $\Delta v'$ (or $\Delta \lambda$), and is normally expressed in cm^{-1}. If the maximum and minimum wavelengths of the radiation field are λ_{max} and λ_{min} respectively around a central wavelength (line centre), λ_0, then the linewidth (LW) is expressed as

$$\text{LW (cm}^{-1}) = \Delta v' = 1/\lambda_{min} - 1/\lambda_{max}$$

Note that λ is expressed in cm (1 cm = 10^4 μm = 10^7 nm). Sometimes linewidths are expressed in frequency units, i.e. LW (Hz) = $c \times \Delta v'$ where c is the speed of light (2.998×10^{10} cm s^{-1}) and $\Delta v'$ is the linewidth in cm^{-1}. Linewidths expressed in wavelength units, e.g. 'cm', are meaningful only for specified line centre wavelengths (λ_0), e.g. $\Delta \lambda \equiv \lambda_{max} - \lambda_{min} \sim \Delta v' \times \lambda_0^2$, assuming $\lambda_0^2 = \lambda_{max} \lambda_{min}$. The spectral linewidth of thermal monochromatic light, e.g. light from a low-pressure sodium lamp (sodium D lines at about 589 nm) could be between 0.1 nm and 0.5 nm (depending on pressure), whereas the linewidth of a laser light could be as small as 10^{-4} nm.

Coherence   Light is emitted from each point of a thermal source as successive trains of EM waves lasting approximately for 10^{-8} s. These trains are emitted randomly and there is no constant phase relation between successive trains, although each train is coherent over a distance of a few metres. However, for a broadband source, the radiation from each point in the source at a point of observation will have no definite phase relations. If the source is small, e.g. a pinhole in front of a broadband source, the radiation emanating from this point source will have some degree of coherence over a distance of a few metres, although the available power outside the pinhole will be very low. In contrast, a single-mode laser could be coherent over a distances in excess of 1 km or so. The coherence length (l_c) and the linewidth of a radiation field are related by $l_c = \lambda_0^2/\Delta\lambda$.

Beam divergence Most lasers produce almost parallel beams. Conventional sources, by contrast, emit radiation nearly isotropically over a solid angle of 4π sr. The angular beam divergence of lasers varies with the type, design and the mode of operation and could be as low as 10^{-5} rad.

The beam divergence of a laser beam operating at only one single specific wavelength (single-mode laser) is diffraction limited and is given by $\phi = \lambda/\pi r$, where r is the radius of the beam at the exit widow of the laser. An He–Ne laser ($\lambda = 0.69\,\mu$m) of radius $r = 1$ mm will have a beam divergence of ~2×10^{-4} rad. At a distance of, say, 1 km, the beam radius will be only about 200 mm. Such small beam divergences cannot be achieved with thermal radiation sources.

Radiance Lasers can produce beams with unparalleled brightness. Consider a 1 mW He–Ne laser commonly encountered in most laboratories. A typical beam divergence of 10^{-4} rad corresponds to a solid angle of approximately 3×10^{-8} sr. For a beam radius of 1 mm at the exit of the window of the laser (area $3.14 \times 10^{-6}\,$m^2), the radiance of the laser source $L = 10^{10}\,$W m^{-2} sr^{-1}. This can be compared with the radiance of the Sun at its surface of merely $1.3 \times 10^6\,$W m^{-2} sr^{-1}. Note that the Sun's radiance is distributed over a wide band of wavelength whereas the radiance of the laser beam is confined to only a very narrow linewidth. This property of the laser allows it to be focused to a diffraction-limited spot (equation (5.11)) and enables extremely high power density to be achieved. High-power lasers (pulsed) having radiance in excess of $10^{20}\,$W m^{-2} sr^{-1} are now commercially available.

5.1.5   Detection and measurement of radiation

Radiation detectors are mainly classified into thermal detectors and photon detectors. In thermal detectors, either the temperature rise of a target due to the absorption of radiation is detected by a thermocouple or the changes in the resistance of an active element due to a temperature rise (bolometer) are detected. These parameters, i.e. the changes in the temperature or current, are then correlated with the incident power of the radiation.

Photon detectors are based on the absorption of photons and either the consequent emission of electrons such as in vacuum photodiodes and photomultipliers or the consequent creation of electron–hole pairs such as in solid state photodiodes. The current thus generated by the flow of such electrons is detected in an external circuit. Since the current is proportional to the incident optical power this provides a measure of the power of the beam. Usually the current is amplified in an external circuit. In photomultipliers, the amplification is primarily achieved internally through many stages of electron multiplication. Electron gain in a typical photomultiplier could be as high as 10^6, depending on the accelerating voltage applied across the dynodes.

The responsivity of photodetectors, R, is defined by the ratio of the output current, i_0, to the input radiation power, P_i, i.e., R (A W^{-1}) $\equiv i_0/P_i$. The performances of photodetectors are characterised by the following parameters.

5.1.5.1 *Frequency response*

A common technique for improving the signal-to-noise ratio (S/N) in radiation detection is to modulate the beam (e.g. by pulsing the source or chopping a continuous beam) and electronically averaging such repetitive pulses. An important characteristic of a detector therefore is its frequency response to radiation modulated at a frequency, f. In most practical circumstances this can be characterised by the modulation frequency f_c corresponding to the time constant, τ_r, i.e. the time taken for the electrical signal to fall to one-half of its maximum value, and is given as

$$f_c = (2\pi\tau_r)^{-1}$$

5.1.5.2 *Electrical bandwidth*

If an electrical system (e.g. an amplifier) uniformly responds to modulation frequencies in the range between f_1 and f_2, the electrical bandwidth, $\Delta f = f_2 - f_1$, for an exponential decay time, τ_e, is given as

$$\Delta f = (4\tau_e)^{-1}$$

and if the system integrates an electrical output over a sampling time t_s, then

$$\Delta f = (2t_s)^{-1}$$

5.1.5.3 *Noise equivalent power*

The random fluctuation in the output of a photodetector constitutes noise. This can often mask weak optical signals and thus set a limit to the minimum detectable signal. The noise equivalent power (NEP) is defined as the minimum input power that produces an output signal with $S/N = 1$ for an electrical bandwidth $\Delta f = 1\,\text{Hz}$.

The quality of a photodetector may also be defined by a figure of merit called the specific detectivity, D^*:

$$D^* = \sqrt{A\Delta f} \times \text{NEP}$$

where A is the effective area of the detector.

5.1.5.4 *Radiometric quantities*

The radiometric quantities are defined as follows:

- Power (P) – the amount of energy Q that is generated, transmitted or received per unit time, t:

$$P\ (\text{J s}^{-1} = \text{W}) = Q/t$$

- Exitance (M) – the amount of power that 'exits' (emerges) from a unit area of the source:

$$M\ (\text{W m}^{-2}) = P/A_s$$

- Intensity (I) – the power emitted by a point source per unit solid angle, ω:

$$I\ (\text{W sr}^{-1}) = P/\omega$$

Since a point source radiates over a solid angle of 4π sr,

$$I\ (\text{W sr}^{-1}) = P/4\pi$$

- Radiance (*L*) – the power per unit projected area of the surface of the source, per unit solid angle:

$$L \ (\mathrm{W\,m^{-2}\,sr^{-1}}) = P/A_s \cos\theta$$

- Irradiance (*E*) – the power incident per unit area of the target, from a point source:

$$E \ (\mathrm{W\,m^{-2}}) = P/A_t$$

- Fluence (*F*) – the energy incident per unit area of a surface per radiation pulse ($\mathrm{J\,m^{-2}}$).

5.2   Fundamentals of energy levels

5.2.1   Atomic orbitals

5.2.1.1   *Bohr's model*

Absorption and emission of EM radiation by an atom take place through its orbital electrons. The theory governing the motion of these electrons around the nuclei was introduced by Niels Bohr. According to this, each electron can be represented by a wave function, ψ, having the mathematical significance that the probability of finding the electron in an element of volume, dv, somewhere in a sphere around the nucleus is $|\psi|^2$ dv. This region can be considered to be an electron cloud (even if only one electron is considered). The highest probability of finding the electron is at the region of highest density of the electron cloud and is normally represented (as a cross-section of a sphere) by a circle called the atomic orbit.

According to Bohr's theory the electrons can move only in specific orbits with radii given by multiples of an integer number, n (= 1, 2, 3 etc.):

$$r = n^2 h^2 / 4\pi^2 m e^2 z \tag{5.13}$$

where h is the Planck constant, m is the mass of the nucleus, e is the charge of the electron and z is the number of protons in the nucleus and the integer n is called the principal quantum number. The energy, E, of the electron in each orbit defined by n is given as

$$E = -2\pi^2 m e^4 z^2 / n^2 h^2 \tag{5.14}$$

The negative sign arises from the fact that the potential energy of the electron is zero when it is at infinite distance away from the nucleus, i.e. free from its influence. It is important to note that, although the magnitude of E is lower for higher values of n (larger orbits), the energy state is represented as opposite, i.e. higher for higher n values.

The electrons can change their orbits by either absorption or emission of energy. Since the orbits are quantised, i.e. their radii and consequently their energy values change only by an integral multiple (n), the absorption and emission energy values must also be quantised. According to Planck's theory, if an electron jumps from an orbit of energy E_2 (quantum number n_2) to an orbit of lower energy E_1 (quantum number $n_1 < n_2$) the energy given out, i.e. radiated, ΔE will have an associated frequency, ν, given as:

$$\Delta E = E_2 - E_1 = \Delta n\, h\nu, \text{ where } \Delta n = n_2 - n_1$$

From equation (5.14), ν (Hz) = $(2\pi^2 m e^4 z^2 / h^3)(1/n_1^2 - 1/n_2^2)$. Expressed in wavenumber $\nu' = \nu/c = 1/\lambda$,

$$\nu' \ (\mathrm{cm^{-1}}) = R z^2 (1/n_1^2 - 1/n_2^2) \tag{5.15}$$

where $R \equiv 2\pi^2 m e^4 / c h^3$ is called the Rydberg constant.

Possible discrete values of the absorption or emission energies of some simple atoms can be calculated from equation (5.15). In practice, the optical radiation from energised (excited) atoms is analysed by a wavelength-selective instrument known as a spectroscope or spectrometer. The radiation is passed to a dispersive element through a narrow slit at the entrance of the instrument. On a photographic plate the image of this slit (source) appears as a discrete bright line. The position of this line corresponds to a specific wavelength.

For hydrogen atoms the transitions between various energy states corresponding to different quantum numbers will give rise to emission or absorption lines at the following wavelength bands (equation (5.15)):

n_1 (=1) $\leftrightarrow$ n_2 (= 2, 3, 4 etc.)   Lyman series (ultraviolet region)

n_1 (=2) $\leftrightarrow$ n_2 (= 3, 4, 5 etc.)   Balmer series (visible region)

n_1 (=3) $\leftrightarrow$ n_2 (= 4, 5, 6 etc.)   Paschen series (infrared region)

5.2.1.2 *Extension of Bohr's model*

To account for atomic spectroscopic results, two modifications of Bohr's model were necessary, which are as follows.

The nucleus of an atom is never at rest. Its motion is accounted for by replacing the electron mass, m in equation (5.14), by an effective (reduced) mass, $\mu = mM/(m + M)$, where M is the mass of the nucleus. Now the frequency of a transition will be different for different isotopes of the same atom, allowing measurement of isotopic abundance in a sample using high-resolution atomic absorption spectroscopy (AAS).

The second modification concerns the concept of circular orbits for the electrons. Although this concept worked well for the hydrogen atom, it did not apply to atoms with higher atomic numbers. A concept was needed of elliptical orbits described jointly by the principal quantum number n and a subsidiary or azimuthal quantum number l (= 0, 1, 2, ..., $n - 1$). Together, these two parameters describe orbits of different ellipticities, and the circular orbit is a special case when $n = l$. The spectroscopic notation for the subsidiary energy levels is s, p, d and f, corresponding to $l = 0$, 1, 2 and 3 respectively.

5.2.1.3 *Energy levels*

The transition of electrons from one orbit to another is represented by the energy level diagram. In this, horizontal lines at appropriate positions in the energy scale (energy increasing with n) represent the energy of the states and are arranged in a vertical column in order of increasing n value. The energy levels for different l values corresponding to a specific n value are represented by horizontal lines in a horizontal array for the sake of better presentation. The horizontal axis has no significance. An example of some of the energy levels and possible transitions for a lithium atom is shown in Figure 5.6.

The number of transitions, shown by broken arrows, is limited by a 'selection rule', which states that only transitions for which l changes by ± 1 are allowed. Thus the transitions 2p ($n = 2$, $l = 1$) $\leftrightarrow$ 3p ($n = 3$, $l = 1$) are not allowed, as $\Delta l = 0$. Similarly, the transitions 2s $\leftrightarrow$ 3d are not allowed, as $\Delta l = 2$.

5.2.1.4 *Electron spin*

Detailed spectral analysis shows that the emission lines of some atoms are actually made up of two very closely spaced discrete lines, i.e. these are doublets. This is explained by attributing

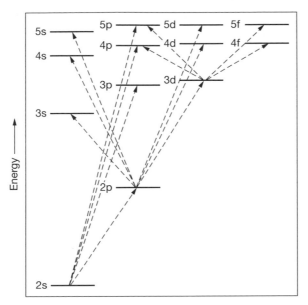

Figure 5.6 Energy level diagram and some possible transitions for the lithium atom

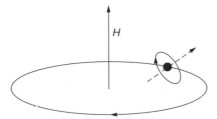

Figure 5.7 Orbital and spin motion of an electron

a spinning motion to the orbital electrons. An orbiting electron gives rise to a magnetic field, H, in the direction perpendicular to the plane of the orbit as shown in Figure 5.7.

A spinning electron acts like a small magnet, and its magnetic field either reinforces the field H or opposes this, depending on the direction of the spin – clockwise or anticlockwise. Therefore the electronic state, uniquely defined by the principal quantum number, n, and the so-called azimuthal quantum number, l, is split into two very close lines. A spin quantum number, m_s, having values of either $+\frac{1}{2}$ or $-\frac{1}{2}$ is, therefore, required to account for the fine structure in the energy levels due to electron spin.

For the complete description of the energy levels, a fourth quantum number, m, defining the orientation of the plane of the electron orbit is needed. The orientation is not arbitrary and in a strong magnetic field changes in such a way that the corresponding changes in energy are quantised. This quantum number is known as the magnetic quantum number and can have values of $-l, \ldots, 0, \ldots, +l$.

It is important to note that no two electrons in an atom can have the same set of four quantum numbers. This restriction requires that no more than two electrons can exist in an orbital and these two electrons must have opposite spin states. Under this circumstance the spins are said to be paired (spin up and spin down) and the atoms do not have a net magnetic field due to electron spin.

5.2.2 Molecular orbitals

5.2.2.1 *Basic criteria*

The concept of molecular orbitals is formed from the mathematical procedure of linear combination of atomic orbitals (LCAO). The concept leads to an understanding of the electronic configurations and their corresponding energies in a molecule. The LCAO model is based on the following criteria:

(a) The nuclei of atoms in a molecule are in their equilibrium position.
(b) Electrons are associated with all the nuclei.
(c) The motion of the electron is described by a wavefunction ψ, where $|\psi|^2 dv$ is the probability of finding the electron within the volume dv inside the molecular dimension.
(d) The lowest-energy molecular orbitals are filled first with the available electrons of the atoms (Aufbau principle).
(e) Each orbital can accommodate a maximum of two electrons with opposite spins (Pauli principle).

5.2.2.2 *Linear combination*

For simplicity we consider a diatomic molecule, e.g. H_2, with the H atoms positionally identified as A and B. With their atomic orbitals described by the wavefunctions ϕ_A and ϕ_B, the molecular orbital due to their linear combination can be expressed as

$$\psi = N(C_A\phi_A + C_B\phi_B)$$

where N is a normalising constant and C_A and C_B are numerical coefficients. Wave mechanical analysis shows that there can be two linear combinations, i.e.

$$\psi_+ = NK\,(\phi_A + \phi_B) \quad \text{giving } \psi_+^2 = N^2K^2(\phi_A^2 + \phi_B^2 + 2\phi_A\phi_B)$$

and

$$\psi_- = NK\,(\phi_A - \phi_B) \quad \text{giving } \psi_-^2 = N^2K^2(\phi_A^2 + \phi_B^2 - 2\phi_A\phi_B)$$

The physical significance of these functions is that ψ^2 is proportional to electron charge density, so that $|\psi_+|^2$ represents a charge cloud density which is more than the electron densities contributed by the individual atoms and is primarily confined in the region between the nuclei. This, in essence, provides a shield between the positively charged nuclei and allows bonding between the atoms. The energy associated with the molecular orbital compared with that of the orbitals of the separated atoms is therefore reduced.

The electron density at the internuclear region represented by the function $|\psi_-|^2$ is lower than that expected for the added individual atomic orbitals. This corresponds to a state of higher energy (a strong repulsion between the nuclei) and is called an antibonding orbital.

5.2.2.3 *Sigma (σ) and pi (π) bonding*

The bonding molecular orbitals formed by the linear combination (overlap) of 1s electrons, 2s electrons, are called σ1s, σ2s type orbitals and for the antibonding case they are known as $\sigma*1s$, $\sigma*2s$ etc. The characteristic feature of these orbitals is that they are symmetrical about the molecular axis.

The combination of p-type orbitals is rather complicated. For $n = 2$ this orbital can have only one azimuthal quantum number, i.e. $l = 1$ (p orbital), and hence has three components perpendicular to each other corresponding to $m = 1, 0, -1$ and represented by p_x, p_y and p_z. The p orbitals are considered to be dumbbell shaped with opposite mathematical signs for the two halves.

When two $2p_x$ orbitals combine they form molecular orbitals similar to σs and $\sigma*$s orbitals and are designated as $\sigma2p$ and $\sigma*2p$. However, for $2p_y$ and $2p_z$ orbitals, both the positive and the negative lobes of the orbitals overlap. This gives rise to what are known as bonding pi (π_y2p) and antibonding pi (π_y^*2p) orbitals. It is to be noted that the energies of π_y2p and π_z2p orbitals are the same and the states are 'degenerate'. The same is true of π_y^*2p and π_z^*2p. From spectroscopic observation, the arrangement of the molecular states in order of increasing energies is as follows:

$$\sigma1s < \sigma*1s < \sigma2s < \sigma*2s < \sigma2p < \pi_y2p = \pi_z2p < \pi_y^*2p = \pi_z^*2p < \sigma*2p$$

5.2.2.4 *Potential energy curves*

Molecular energy states, described so far, are based on the idea of an equilibrium position for the atoms. The atomic configuration in a molecule, however, is analogous to 'balls connected by springs'. When the spring and ball system is compressed or stretched, its potential energy increases. Any vibratory motion, i.e. stretch, bend, etc., will alter internuclear distance and therefore cause a periodic variation in the potential energy of the molecule. A general form of such a variation representing the potential curve, i.e. energy state of a molecule, is shown in Figure 5.8. In this the horizontal axis represents the internuclear distance during the vibration.

A molecule will have many such potential energy curves representing the ground state and many possible electronic excited states. The minima in the curves correspond to the most stable configuration for that particular state in which the atoms concerned are at the equilibrium internuclear positions. If the atoms are stretched too far, they dissociate and hence a levelling off exists in the potential curves at higher internuclear distances. If a molecule is excited to an antibonding state, it will, most likely, return to a state of a more stable, lower-energy electronic configuration.

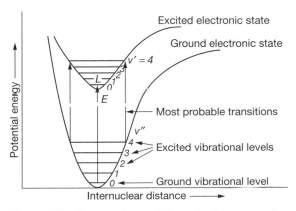

Figure 5.8 Representation of the potential energy of a molecule

5.2.2.5 *Vibrational energy levels*

Horizontal lines drawn at the minima of potential energy curves will not be a true representation of the electronic energy states of a molecule. This is because the atoms are always vibrating and the energy associated with this must be added to the electronic energy of the molecules.

Vibrations of atoms cannot be considered in isolation as each atom in a molecule is bonded with one or more other atoms. All vibrations in relation to a centre of charge for adjacent atoms are classified into two basic types, i.e. stretching and bending. In the former, the internuclear distance changes and, in the latter, the angle between bonds changes. The stretching vibration could be symmetric or asymmetric and the bending vibration could take the form of scissoring, rocking, wagging or twisting.

Similarly to electronic states, the vibrational levels are also quantised, i.e. can have only discrete energy values given by

$$E = \left(v + \frac{1}{2}\right)\frac{h}{2\pi}\sqrt{k/\mu} \tag{5.16}$$

where h is Planck's constant, v is the vibrational quantum number, i.e. any positive integer value including zero, k is the force constant pertaining to the particular bond and μ is the reduced mass. For two relevant atoms of masses m_1 and m_2, $\mu = m_1m_2/(m_1 + m_2)$.

It has been found from experimental measurements that, for most single bonds, $k \sim 5 \times 10^2\,\mathrm{N\,m^{-1}}$ and $\sim 1.3 \times 10^3\,\mathrm{N\,m^{-1}}$ for double bonds.[1] By definition, a newton $\mathrm{N} = \mathrm{kg\,m\,s^{-2}}$. Therefore, $\sqrt{k/\mu}$ has units of $\mathrm{s^{-1}}$ if μ is in kg. For $h = 6.626 \times 10^{-34}\,\mathrm{J\,s}$, the energy is expressed in joules. For a particular mode of vibration, the energy levels are equally spaced for $v = 0, 1, 2$, etc. The state with $v = 0$ is the lowest, i.e. most stable, state pertaining to an electronic state (ground or excited states). The number of vibrational quantum numbers that an electronic state can accommodate depends on the level of the dissociation energy. At higher values of v, some anharmonicity takes place whereby the excited vibrational levels become progressively closer, as shown in Figure 5.8.

5.2.2.6 *Vibrational transitions*

In the infrared wavelengths the transitions are confined to the vibrational levels of the ground electronic state. For such transitions the selection rules are $\Delta v = \pm 1, \pm 2, \pm 3$ etc. In vibrational transitions between levels belonging to two electronic states, no such selection rule applies.

It is known that the highest transition probability in infrared absorption corresponds to a change $\Delta v = \pm 1$. Also, the Boltzmann distribution law predicts that nearly 99% of the molecules will be at the ground vibrational state at normal room temperature. Therefore, the most probable transitions are likely to be between the ground ($v = 1$) vibrational state and the next higher vibrational state ($\Delta v = \pm 1$). Therefore the energy of the radiation emitted or absorbed in infrared spectroscopy is given as (see equation (5.16))

$$E = h/2\pi \times \sqrt{k/\mu}$$

In spectroscopy, the energy is usually expressed as a wavenumber, v' ($= 1/\lambda$), so that

---

[1] The carbon atom has four electrons in the outer shell (2s, $2p_x$, $2p_y$, $2p_z$). It forms single bonds with four adjacent atoms and the compound thus formed is called saturated. If there are only three neighbouring atoms the extra electron forms an extra bond giving rise to the double bond (π bond) and unsaturation.

$$\nu'(\text{cm}^{-1}) = 1/2\pi c \times \sqrt{k/\mu} \tag{5.17}$$

where c is in cm s^{-1} and k and μ are, as before, in MKS units.

At higher values of ν, the above selection rules do not apply rigorously. As a result, transitions involving $\Delta\nu = \pm 2$ or ± 3 may be observed. These give rise to overtone absorption lines at energies approximately 2 or 3 times that of the fundamental line. In addition, two different vibrations in a molecule can interact with each other to give rise to transitions with energies that are some combinations of the individual energies, i.e. sum or difference energies. Absorbances (i.e. the line strengths) corresponding to overtones and combination lines are much smaller than that of the fundamental absorption lines. It is to be noted that infrared absorption takes place only if the radiation field causes a change in the dipole moment of the molecule, i.e. a change in the centre of the charge. Therefore, homonuclear diatomic molecules do not have any infrared absorption spectra.

5.2.2.7   *Vibrational modes*

Complex organic molecules contain many types of atoms and bonds. Therefore many different types of vibrations are possible which give rise to many absorption lines spread over a broad infrared band (spectrum). The number of possible vibrations is calculated from the degrees of freedom in a Cartesian coordinate system. If the molecule contains N atoms, there are $3N-6$ possible vibrations (normal modes). For a linear molecule, there are $3N-5$ 'normal modes' of vibration because of the restriction on its rotation about the bond axis.

For each normal mode, there exists a potential energy curve with the corresponding vibrational levels and the same selection rule applies for any transition. In practice not all absorption lines, governed by the selection rule alone, are observed. This is because of the fundamental restriction that a transition between two vibrational states will be possible only if there is a temporary displacement of the electrical centre of gravity during the vibration, i.e. a change in the electrical dipole moment. For example, CO_2 being a linear molecule has, in theory, $3N-5$, i.e. four, fundamental modes. During symmetric stretching vibration at frequency ν_1 the centre of charge remains unchanged and there will not be any infrared absorption to excite this mode.

The absorption lines corresponding to the asymmetric stretching vibrations (ν_3) and the bending vibrations (ν_2) are observed in the infrared absorption spectra. The remaining fundamental vibration corresponds to the frequency ν_2 of the bending vibration and is degenerate.

5.3   Molecular spectroscopy

5.3.1   Basic concepts

Molecular spectroscopy encompasses absorption, emission, fluorescence and scattering processes. In the UV–visible wavelengths electronic states are involved in transition processes, whereas vibrational states are excited or de-excited in the infrared wavelengths. The energy level scheme pertaining to UV–visible and infrared absorption and the subsequent de-excitation and emission of radiation are shown in Figure 5.9.

At normal room temperatures, the majority of molecules in a sample occupy the lowest vibrational state of the ground electronic level, $E_g(v'' = 0)$. This energy level is taken as the reference level for the energy values of other states. The population of molecules, N_n, in a vibrational state, E_n, with respect to the population, N_m, at the energy state E_m ($E_m < E_n$) at a temperature T (in kelvins) is given by the Boltzmann distribution law:

5   Instrumental analysis – optical and spectroscopic

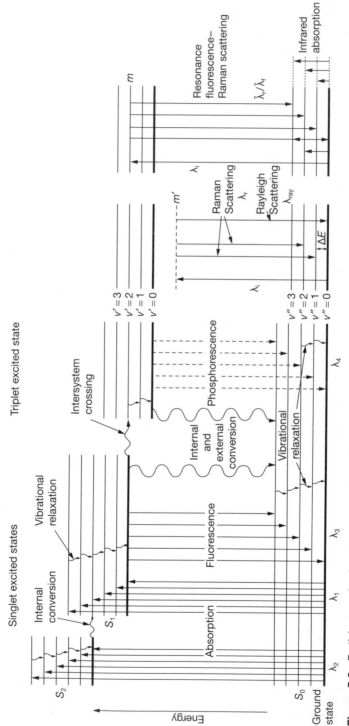

Figure 5.9   Partial energy level and some transition schemes for a photoluminescent system

$$N_n/N_m \propto \exp\left(\Delta E/kT\right) \tag{5.18}$$

where $\Delta E = E_n - E_m$ and $k = 1.38 \times 10^{-23}\,\mathrm{J\,K^{-1}}$ is Boltzmann's constant. From this the energy difference between successive vibrational levels, ΔE, in wavenumber units is estimated to be of the order of $10^3\,\mathrm{cm^{-1}}$. For a temperature of 300 K the ratio of populations at $v'' = 1$ and $v'' = 0$, from the above equation, is ~ 0.008.

In other words, the population of the $v'' = 1$ state is less than 1% of that of the ground vibrational state ($v'' = 0$). Thus, to a very good approximation, all transitions in UV–visible spectroscopy at room temperature may be considered to originate from the ground vibrational state.

For a molecule in an excited vibrational state, the energy retention time (lifetime) is extremely small. Normally a molecule excited to a vibrational level of the excited electronic state loses its energy almost instantaneously ($<10^{-12}$ s) in steps down the vibrational energy ladder without emitting any radiation (vibrational relaxation). At the lowest vibrational state of the excited electronic level ($v' = 0$), the lifetime of the molecule is comparatively long ($\sim 10^{-8}$–10^{-9} s). Radiative transitions from this state to the various vibrational states of the ground electronic state give rise to the characteristic fluorescence spectrum of the molecule. In this, the wavelengths of emitted radiation, λ_f, are always longer than that of the exciting radiation, λ_i ($\lambda_f > \lambda_i$).

If the excitation energy does not match up with an allowed excited state, the molecule will re-emit the energy instantly and revert back to its ground state. This is the scheme for the Rayleigh scattering process where the excitation energy is said to have corresponded to a virtual state, m, and there is no change in the wavelength, i.e. $\lambda_i = \lambda_{Ray}$. However, there is some probability that the molecule may revert back to an excited vibrational state $v'' = 1, 2, 3$ etc. In this case the emitted radiation, known as Raman scattered radiation, has less energy (photon energy) than the energy of the incident (exciting) radiation, i.e.

$$E_i - E_r = \Delta E = hc/\Delta\lambda$$

The energy difference ΔE corresponds to the energy of the excited vibration state and is called the Raman shift. The Raman shift pertaining to a vibrational mode of a molecule is unique to that molecule and is given in wavenumber units as

$$\Delta\left(\frac{1}{\lambda}\right) \equiv \frac{1}{\lambda_i} - \frac{1}{\lambda_r} = \Delta v'\ (\mathrm{cm^{-1}}) \tag{5.19}$$

If allowed by selection rules, molecules may undergo direct vibrational transitions through the absorption of infrared radiation. The Raman scattering is therefore analogous to infrared absorption as far as the molecular transitions are concerned and they complement each other in molecular spectroscopic analysis. However, it is to be noted that all infrared-active molecules are not necessarily Raman active. These activities depend on certain changes in the properties of molecules during transitions. Raman spectroscopy may be viewed as a means to obtain infrared spectra using UV–visible light.

An interesting effect occurs if the excitation energy is very close to an allowed excited energy level as in Figure 5.9. The fluorescence and Raman scattering efficiencies could become abnormally high, giving rise to what are called 'resonance effects'.

5.3.1.1 *Fundamentals of optical absorption*

Modern absorption spectrometers provide a graph of absorbance, A, of a sample against wavelength. For convenience, the absorbance is defined as

$$A = \log_{10}(1/T) = \log_{10}(P_0/P)$$

The transmittance, T is defined by the Beer–Lambert law as

$$T = P_t/P_0 = \exp(-\mu l)$$

where P_t and P_0 are transmitted and incident power respectively, l is the path length through the sample and μ is the attenuation coefficient of a uniform medium. In a transparent medium the attenuation (loss) is primarily due to scattering by the molecules and particulates of the medium. If the wavelength-selective absorbing molecules at a concentration, c, are uniformly distributed in the medium and the attenuation of the beam by the medium is either negligible or compensated, the transmittance at a wavelength λ is

$$T = \exp(-\alpha cl) \tag{5.20}$$

where

$\alpha \equiv$ **absorption cross-section** (cm^2) when c is in number of molecules cm^{-3} and l is in cm

$\equiv$ **absorptivity** ($l\ g^{-1}\ cm^{-1}$) when c is in g l^{-1} and l is in cm

$\equiv$ **molar absorptivity** ($l\ mol^{-1}\ cm^{-1}$) when c is in mol l^{-1} and l is in cm

The absorbance is then given as

$$A = 0.434\alpha cl \tag{5.21}$$

The linear relationship between concentration and absorbance is only strictly valid for dilute solutions. At high concentrations of absorbing molecules (usually $>0.01\,M$), the average distance between them becomes small enough to affect each other's charge distribution and therefore their absorption properties. The other causes of deviation from linearity (Beer's law) are the changes in the refractive index of a solution with concentration, lack of high monochromaticity of the optical radiation (in the case of non-laser sources) and the presence of stray light.

5.3.1.2 *UV–visible absorption*

In an organic molecule, the double bonds contain both σ and π orbitals, whereas the single bonds pertain to σ orbitals only. The transitions involving electrons in the single bonds (σ–σ^*) correspond to vacuum ultraviolet wavelengths. Practical absorption spectroscopy, therefore, mostly involves unsaturated absorbing centres (double bonds) called 'chromophores' having absorption ($\pi \rightarrow \pi^*$) over the 200 nm to 700 nm wavelength band. In addition to σ and π electronic states, the unshared electrons in a molecule constitute the non-bonding state, n, which also participates in the absorption process, i.e. n $\rightarrow \pi^*$. Some inorganic species such as nitrate (313 nm), carbonate (217 nm) and nitrite (360 nm and 280 nm) also exhibit n $\rightarrow \pi^*$ transitions.

Compared with the molar absorptivities associated with $\pi \rightarrow \pi^*$ transitions (10^3–$10^4\,l\,cm^{-1}\,mol^{-1}$), the values for n $\rightarrow \pi^*$ are some 2 orders of magnitude lower. For many inorganic complexes, e.g. iron thiocyanate, and some organic complexes, e.g. quinone–hydroquinone, the molar absorptivities could be very high ($>10^4\,l\,cm^{-1}\,mol^{-1}$). These complexes absorb through the mechanism of transfer of the electron orbital in the electron donor part of the complex to an excited state in the electron acceptor part.

The wavelength at the peak of an absorption band of a certain type of chromophore is strongly dependent on the polarity of the solvent and the type and number of substituents.

Peaks associated with n → π* transitions are usually shifted to shorter wavelengths (a blue shift) with increasing polarity of the solvent. In the case of π → π* transitions, the reverse is normally encountered.

5.3.1.3 *Infrared absorption*

Absorption of infrared radiation takes place through allowed vibrational (and rotational) transitions (see Section 5.2.2). The infrared absorption spectrum encompasses a very wide wavelength band extending from about 0.78 µm to about 1000 µm. For practical reasons the spectrum is divided into near (0.78–2.5 µm), middle (2.5–50 µm) and far (50–1000 µm) infrared bands.

For most molecules, the energy difference between the successive levels of the fundamental stretching vibrational modes corresponds to strongest absorption. This energy difference lies within the infrared band between 2.5 µm and 15 µm. Therefore, most analytical applications of infrared spectroscopy have so far been confined to this band. Absorption lines beyond this band are due to bending modes or combinations and overtones of stretching modes, and the line strengths pertaining to these modes are usually relatively low. Despite this, because of their accessibility with esoteric small, compact and inexpensive diode lasers and good transmission through fibre optics, there is a good prospect for devising miniaturised absorption spectrometers for environmental pollution monitoring.

For molecules with more than two atoms, vibrational modes are most often influenced by each other. This is known as vibrational coupling and this changes the energy normally associated with the vibration corresponding to a particular mode and thus shifts the wavelength of the absorption peak. As a result the absorption peak corresponding to a functional group (e.g. C—O, C—C, C—H) cannot be specified exactly. However, these shifts of the absorption peak provide unique features for the positive identification of compounds containing large molecules.

5.3.2 Fundamentals of Raman scattering

Raman scattering, like Rayleigh scattering by molecules, takes place almost instantaneously as there are no real energy states involved in the transition process. However, in the Raman process, the molecules are temporarily left at an excited state and hence the scattered radiation has lower frequency, i.e. longer wavelength (Stokes line), compared with the frequency of the incident radiation. In some circumstances molecules already excited to vibrational states may absorb photon energy in the UV–visible wavelengths and make a transition to a vibrational level in the ground electronic state which is lower than its previous state. In this case the energy released will be higher than that being absorbed. The Raman lines due to such events are, therefore, at shorter wavelengths than the wavelength of the incident optical radiation and the lines are known as anti-Stokes lines. However, the probability of anti-Stokes scattering events is much less than that of the Stokes events. Normally the various Raman lines correspond to various vibrational modes observed in infrared absorption spectroscopy. In the Raman process both the incident and the scattered radiation are in the UV–visible wavelength band; therefore, it allows analysis of infrared spectra of molecules in the convenient UV–visible band and thus avoids the irritating problem of water vapour absorption in direct infrared spectroscopy.

The theory of Raman scattering is exceedingly complex. From detailed analysis it transpires that not all infrared-active modes are necessarily Raman active and the intensity of the Raman-scattered signal for any particular mode is inversely proportional to the fourth

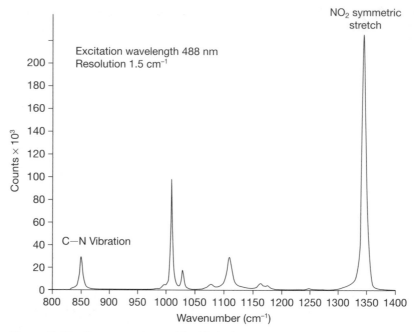

Figure 5.10 Raman spectrum of liquid nitrobenzene showing some major vibrational modes

power of the wavelength of the exciting radiation ($\propto \lambda^{-4}$). However, for molecules having strong isolated absorption bands, the Raman scattering could be anomalously strong for excitation at wavelengths close to these bands (resonance excitation). Inherently, Raman scattering is a weak interaction process, providing only 1 photon per 10^7–10^8 incident photons, whereas this could rise to 1 photon per 10^4–10^5 incident photons at resonance excitation (2–3 orders of magnitude enhancement).

Raman scattering is parametrised by a term called the cross-section, σ_R. This is defined as the total scattered power per unit of incident power density per molecule, i.e. $\sigma_R = P_R/(P_i/A)$. The dimension of σ_R is therefore area, A (m^2). For practical reasons, the cross-section is often quoted in terms of a differential cross-section ($d\sigma_R/d\Omega$) in units of m^2 sr^{-1}, where $d\Omega$ is the detection solid angle in steradians. However, for most practical purposes, the Raman scattering could be considered to be isotropic so that the total Raman scattering cross-section $\sigma_R \approx 4\pi \, d\sigma_R/d\Omega$.

The Raman spectrum of a molecular species is a unique signature, consisting of isolated lines having extremely narrow linewidths. An example of the Raman spectrum of liquid nitrobenzene is included in Figure 5.10. The intensity of Raman lines is proportional to the concentration of the species. Therefore, Raman scattering can be used for the non-invasive and remote identification and measurement of concentration of a target species.

Most molecular species are, to some extent, Raman active and emit their characteristic Raman lines. In a sample, the Raman lines from various vibrational modes belonging to various bonds of different molecules are expected to be very crowded and often overlap with one other. Therefore the laser source and high-resolution spectrometer have become indispensable in applied Raman spectroscopy.

Even at room temperature, there are a few molecules in the excited vibrational states according to the Boltzmann distribution law. These also participate in the Raman scattering process by making induced transitions from the excited state to a lower (or ground) state. In this case, the scattered radiation will have components which correspond to shorter wavelengths (anti-Stokes lines) with respect to the Rayleigh-scattered line (excitation wavelength).

The Raman shifts corresponding to the anti-Stokes lines are represented by a negative sign, i.e. $\Delta v'_i - \Delta v'_r = \pm\Delta v'$, where the v' are the energies of the lines in wavenumber (cm^{-1}) units. The anti-Stokes Raman cross-sections are usually 1–2 orders of magnitude smaller than those for Stokes lines. Therefore, these are not normally utilised for analytical and diagnostic purposes. However, the ratio of the strengths of the Stokes lines and the corresponding anti-Stokes lines is proportional to the absolute temperature through the Boltzmann distribution law and provides a means for remote and non-invasive temperature measurement.

5.3.3   Fundamentals of fluorescence

The optical energy absorbed by a molecule is partly dissipated through vibrational relaxation and partly re-radiated through competitive elastic (Rayleigh) and inelastic (Raman) scattering and also through fluorescence. In addition to non-radiative vibrational relaxation, the absorbed energy may also be dissipated through internal and external conversions and intersystem crossing, governing the fluorescence efficiency of molecules (see Figure 5.9).

Internal conversion results from the overlap of vibrational levels belonging to two closely located electronic states. In this situation the molecule may make a transition to a lower electronic state without any emission. If the lower electronic state happens to be the ground state, the deactivation takes place exclusively through vibrational relaxation and no fluorescence occurs as is the case in most aliphatic compounds. A conversion (transfer) to a very high-energy vibrational level of a lower electronic state may cause a rupture of the molecular bond, giving rise to the phenomenon called pre-dissociation.

In external conversion, the energy deactivation takes place through interactions between molecules of the fluorescing species and between these and other surrounding molecules (e.g. solvents). At high concentrations of fluorescent species, the electronic configurations of molecules will interact with each other, and, at higher temperatures, the rate of collisions increases. The overall effect of these is to lower the fluorescence efficiency (quenching of the excited states). A non-linear relation between the concentration of fluorescent species and emission intensity is the consequence of such external conversion.

For the majority of molecules, the electron spins are paired up even if one of the pair is excited to another electronic state (Pauli's principle). The energy levels are normally not split and are called singlet states. It is, however, possible that the electron may change its spin orientation during a transition to an excited electronic state. The total energy of the state is then split into three levels depending on their three possible spin moment combinations. The energy levels are called triplet states. The average lifetime of a molecule in a triplet state may range from 10^{-4} s to several seconds. The radiative deactivation from the excited triplet states to the electronic ground state gives rise to phosphorescence. Photoluminescence is the general name for fluorescence and phosphorescence.

5.3.3.1   *Quantum yield*

The efficiency with which a molecule in any environment and at any structural configuration exhibit fluorescence is parametrised by a quantity called quantum yield, η_f, defined as

$$\eta_f = \frac{\text{number of fluorescing molecules}}{\text{number of excited molecules}}$$

$$= \frac{\text{number of emitted photons}}{\text{number of absorbed photons}}$$

The value of η_f lies between 0 and 1, for example $\eta_f = 1$ for fluorescein and $\eta_f = 0$ for nitrobenzene.

Since fluorescence is always associated with other deactivation processes with their efficiencies given by the rate constant parameter, k, the fluorescence quantum yield can be expressed as

$$\eta_f = k_f/k_t \quad \text{where} \quad k_t = k_f + k_{ic} + k_{ec} + k_{ix} + k_d + k_p$$

k_f, k_{ic}, k_{ec}, k_{ix}, k_d and k_p are the rate constants for fluorescence, internal conversion, external conversion, intersystem crossing, photodissociation (pre-dissociation) and phosphorescence respectively. The magnitudes of k_f, k_d and k_p are dependent on the molecular structure, whereas the others are dependent on the chemical and physical environment of the fluorescent molecules. Therefore, those variables which give maximum value for k_f and minimum value for k_t provide the strongest fluorescence from a species.

5.3.3.2 *Structural effect*

Since absorption is a prerequisite for fluorescence, naturally chromophores in molecules or compounds having π, π^* energy states are likely to be most fluorescent. Compounds containing aromatic functional groups fall into this category. Because of the complex and high degree of interactions between various vibro-electronic modes, the absorption spectra and consequently the fluorescence spectra of these compounds are broad and usually mirror images of each other. The difference in wavelength between the peaks of the absorption and emission bands are sometimes called the Stokes shift (not to be confused with Stokes lines in Raman scattering).

Unsubstituted aromatic hydrocarbon rings usually fluoresce in solution, the quantum yield increasing with the number of rings and their degree of condensation. Structural rigidity, i.e. fused ring structure, has been found to favour fluorescence. For example, quinoline and fluorene having fused and rigid ring structures, exhibit strong fluorescence whereas pyridine and biphenyls, having no substitution or loosely bound substitution, are practically non-fluorescent.

Substitution on the benzene ring causes shifts in the wavelength of fluorescence (and absorption) maxima. It is therefore possible to tailor compounds to fluoresce over any desired wavelength band in the UV–visible bands. However, the substitution also affects the quantum yield. The substitution of heavy atoms decreases the quantum yield; for example, the effect is quite striking for halogen substitution and is called the heavy atom effect. For example, benzene molecules attached to fluorine atoms having atomic number 9 (fluorobenzene) exhibit good fluorescence. Substitution of bromines having atomic number 35 reduces the fluorescence yield by as much as half. Substitution of still heavier atoms, e.g. iodine (atomic number 53), makes the molecule (iodobenzene) practically non-fluorescent.

5.3.3.3 *Environmental effect*

The fluorescence of a molecule is affected by the temperature, pH and the type of the solvent and the concentration of the species. At high temperatures, the mobility in solid samples and collision rates in liquid and gaseous samples increase. This increases external conversion and consequently decreases the fluorescence quantum yield. Solvents containing heavy atoms promote intersystem crossing and thereby reduce fluorescence. The effect of pH of the solvent on the fluorescence of a compound is rather complex. The effect arises from the dissociation or protonation of acidic and basic functional groups and therefore alters the non-radiative rate constants and can also cause a shift in the wavelength of fluorescence. For example, the protonation of electron-withdrawing groups (e.g. carboxyl) results in shifts to longer wavelengths, while that of electron-donating groups (e.g. amino groups) gives rise to a shift to shorter wavelengths. It is therefore necessary to have close control of pH in analytical fluorescence spectroscopy.

Since fluorescence is a consequence of absorption, the dependence of fluorescence intensity (I_f), i.e. the power per unit area, on the concentration of the fluorescence species is related through Beer's law (equation (5.21)):

$$I_f = \eta_f I_a = \eta_f I_0 (1 - T) = \eta_f I_0 (1 - e^{-\alpha cl})$$

The exponential term can be expanded as a Maclaurin series, so that

$$I_f = \eta_f I_0 \left[\alpha cl - \frac{(\alpha cl)^2}{2!} + \dots \right]$$

For $\alpha cl < 0.05$, the higher-order terms can be neglected with only 2.5% error, so that

$$I_f = \eta_f I_0\, \alpha cl$$

Thus the fluorescence intensity is approximately linear with concentration for low concentrations of a particular absorbing species, distributed along a small path length l.

The quenching of fluorescence by other molecular species is an important consideration in analytical spectroscopy. Fluorescence quenchers such as molecular oxygen, acryl amide, nitrobenzene and iodine ions provide information about the location of fluorescent groups in a compound. The effect of quenchers on the steady-state fluorescence of a homogeneously fluorescing sample is described by Stern–Volmer's empirical relation:

$$I_f(0)/I_f(Q) = 1 + K_{sv}Q = \tau_0 \tau^{-1} \tag{5.22}$$

where $I_f(0)$ and $I_f(Q)$ are the intensities without and with quenchers respectively, Q is the concentration of the quenchers and K_{sv} is the dynamic Stern–Volmer constant. Similarly, τ_0 and τ are the fluorescence lifetimes without and with quenchers respectively.

5.3.4 Practical considerations

A versatile system for optical spectroscopy consists of six major components – a stable and tunable source of radiation, a transparent sample container, a wavelength selector, photodetector, electronics (or computer) for the acquisition, processing and storing of data, and a device for obtaining hard copy.

A layout of a spectroscopic system is shown in Figure 5.11. For absorption spectroscopy the photodetector is usually located behind the sample cell to receive the transmitted beam. For solid samples, absorption spectroscopy is enabled by recording the diffused reflection spectrum of the sample at a convenient angle. Emission spectra are most commonly recorded at a 90° angle with respect to the direction of the incident radiation.

In modern absorption spectrometers, a part of the radiation beam is diverted to pass through a reference cell and onto the same detector. Alternate detection of the signal and reference allows elimination of effects due to source intensity fluctuation, non-linear

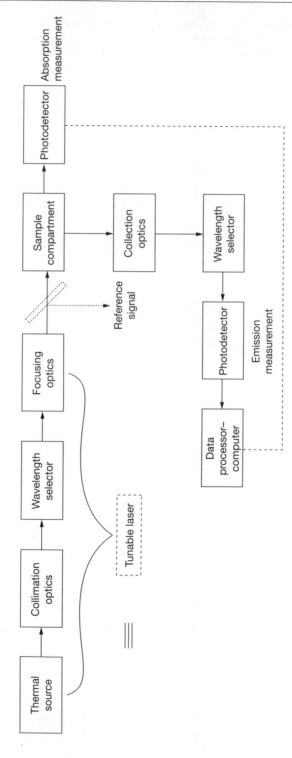

Figure 5.11 Layout of basic components for molecular spectroscopy

wavelength response of optical components and the photodetector and this also permits a direct measurement of the concentration of a species. It should be noted that, in absorption spectroscopy, a broadband source may be transmitted through the sample cell, and the ratio of the spectra of that source with and without the target species then provides the desired absorption spectrum of the sample.

For emission spectroscopy, a small fraction of the incident beam intensity may be diverted to a reference photodetector to enable compensation for source intensity fluctuation and variation in the wavelength. However, for quantitative measurements either an internal or an external reference standard is necessary.

It is to be noted that optical components used in a spectroscopic system need to have high transmission in the operating wavelength range of the source. For infrared wavelengths up to about 30 μm, NaCl, KBr and TiBr are suitable materials for optical components. For near UV and visible wavelengths fused silica, quartz or corex glass are suitable, whereas, for UV wavelengths below 180 nm, LiF optical components are needed.

5.3.4.1 *Radiation source*

Laser source Lasers are undoubtedly the ideal source for use in analytical molecular spectroscopy. Large spectral power densities available from lasers can induce strong signals in a target and therefore improve the signal-to-noise ratio or reduce data averaging time for the detection of such signals. This property also allows implementation of non-linear spectroscopic techniques such as saturation spectroscopy and multiphoton spectroscopy. Because of its small beam divergence, the laser source can be imaged most efficiently. Lasers can also provide output with extremely narrow linewidth (~ 0.1 cm^{-1}) and short pulse width ($\sim 10^{-12}$ s). These properties allow realisation of spectral resolution several orders of magnitude higher than that achievable by any wavelength selectors (monochromator–filter) and permit the measurement of ultrafast phenomena (e.g. relaxation processes of excited states). Lasers operating at any wavelength between ultraviolet and infrared wavelengths, some continuously tunable over a small range and others tunable line by line (e.g. CO_2 laser), are commercially available. An intense thermal source and a very high-resolution spectrometer can provide output with spectral purity comparable with that available from some lasers (Figure 5.11). Sophisticated collimating optics can be used to reduce the beam divergence down to levels equal to those of lasers. However, the available output power from such sources is far less than that from a very modest laser source. Therefore, lasers allow remote sensing of the environment by spectral detection.

Although lasers are the most effective source in spectroscopy, these are usually costly and often require complicated operating systems. Therefore, their uses are limited to laboratory instruments for specific applications. For example, in the case of dye or semiconductor lasers the range of tunability is very small and is limited by the optical properties of the specific dye or the electro-optical properties of the specific semiconductor material. Different dyes or semiconductor diodes are needed to extend this range.

For infrared absorption spectroscopy the CO_2 laser, tunable over 80 different lines in the range 9–11 μm, is most commonly used. For fluorescence and Raman spectroscopy the most commonly used lasers are the excimer laser, operating at a few discrete lines between 0.192 μm and 0.35 μm, and the frequency-quadrupled neodymium–YAG laser emitting at 0.25 μm.

Thermal source Commercial spectrometers, perhaps with the exception of Raman spectrometers, incorporate a conventional thermal light source. For infrared spectroscopy, inert solids, usually in the form of a bar or a wire, are heated electrically. These emit radiation at visible and infrared wavelengths. The intensity and the wavelength range of the radiation depend on temperature according to Planck's black-body radiation law. Commonly used sources are Nernst Glower (rare earth oxide), Globar (silicon carbide), incandescent wire (Nichrome), tungsten filament and high-pressure mercury arc. Among these only the last one is capable of giving output at wavelengths longer than 30 µm.

Sources for UV–visible spectroscopy are lamps containing argon, xenon, hydrogen or deuterium gases at high pressures. The power output of these lamps is governed by the operating current (temperature) and external cooling is often necessary for high output power. Sometimes the gases are mixed in the lamp to increase their operating lifetime.

5.3.4.2 *Wavelength selector*

Interference filters are commonly used to select a narrow band of wavelengths (lines) from a broadband source. These are basically transparent dielectric films sandwiched between two semitransparent metal films and held between two thin quartz plates. The thickness of the dielectric layer determines which particular line will give rise to constructive interference and hence high transmission through the filter. Since constructive interference takes place in different orders (equation (5.8)), these filters will transmit wavelengths at higher orders with respect to the designated transmission line. Other filters need to be used to suppress transmission at these unwanted bands. Commercial interference filters usually have a very narrow transmission band (1–5 nm) and are usable over the UV–visible and IR wavelength bands up to 14 µm. Transmission of common interference filters is usually 70–80%.

For diverse spectroscopic applications it is necessary to vary the wavelength position of the transmission band continuously over a wide range. The monochromator is a device designed for this purpose – called wavelength scanning. A schematic of the configuration of a monochromator is shown in Figure 5.12, and its principle of operation is described below.

The vital part of the monochromator is a ruled (or holographic) reflection grating which is the wavelength-dispersive element and is placed in the so-called Czerney–Turner configuration. The light to be spectrally analysed is collected by a lens and focused onto the entrance slit. The diverging beam is collimated (made parallel) by a concave mirror. The parallel beam is then reflected at an angle θ_i and spectrally dispersed by the grating. The spectrally separated beam is intercepted by a second concave mirror which focuses the spectrum onto the plane of the exit slit. The grating is mounted on a kinematic mechanism so that it can be rotated by an electrical signal. The rotation of the grating scans the spectrum across the exit slit, which permits only the spectral band corresponding to the angle θ_d (scanning angle) and its width. The widths of the entrance and exit slits are normally kept the same (w).

A monochromator, like a lens, is characterised by a parameter, called $F^\#$ (f number), defined as the ratio of the focal length of the mirror to its aperture (diameter), i.e. $F^\# = f/D$. In terms of this parameter the throughput (transmission efficiency) of a monochromator is expressed as

$$\mathrm{TP} \simeq Kf^2/\mathrm{RP}\,[1 + 4\,(F^\#)^2]$$

where the parameter K accounts for the spectral radiant power at the lens and the transmittance of the total optical system of the monochromator and RP is the resolving power of the system. From this, it is clear that there is an inherent trade-off between the resolution and the throughput of a monochromator.

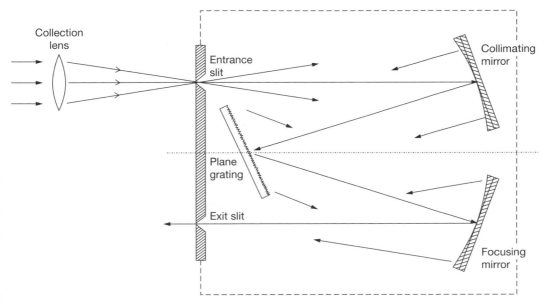

Figure 5.12   Schematic of a monochromator in the Czerney–Turner configuration

For the most efficient use of a monochromator, $F^{\#}$ of the signal collection lens needs to match up with that of the monochromator so that the collected light just fills the mirrors (and the grating). For any mismatch of $F^{\#}$, there will be either transmission loss due to overspreading or resolution loss due to underfilling of the grating.

For sensitive and high-resolution spectroscopy such as that for laser-Raman spectroscopy, monochromators with large focal lengths and gratings with large dispersions are usually necessary. In commercial monochromators, the dispersion is quoted for a particular grating as $nm\,mm^{-1}$ of the exit slit width. To minimise size, monochromators can be internally or externally coupled together to make double (two gratings) or triple (three gratings) monochromators.

The whole spectrum of a source can be recorded either photographically or using multi-element photoelectric devices called charge-coupled devices (CCDs) or diode-array detectors without the need for scanning. In this case, the exit slit is replaced by these multi-element detectors, and the monochromator is called a spectrometer.

5.3.1.3   Photodetector

For spectroscopic applications, the most important parameters of photodetectors (transducers) are the spectral response (usable wavelength range) and the detectivity (D^*). The other parameters to be considered are the response speed (for time-resolved measurements), the effective size of the detector element, the package for the element and the operating temperature and the method of cooling (to improve signal-to-noise ratio).

In infrared spectrometers, mercury cadmium telluride (HgCdTe) detectors are widely used for their high detectivity ($>10^{10}$ at $-30\,°C$) and good response over the $8-12\,\mu m$ region. The range of wavelength response can be extended up to $2.5\,\mu m$ by varying the composition of the alloy. These detectors, however, suffer from long response time (several microseconds). In the near infrared wavelengths indium gallium arsenide (InGaAs, 0.8–1.7 μm) or germanium avalanche (0.8–1.7 μm) photodiodes provide subnanosecond response and could be operated at room temperature.

In UV–visible spectrometers, photomultipliers are most widely used as photodetectors. The choice of the photomultiplier type is dictated by the wavelength response characteristics of the photocathode material.

Most spectroscopic instruments use side-on-type phototubes for wider spectral response and faster rise time (~2 ns) compared with those typically provided by the head-on-type photomultipliers.

The state-of-the-art photomultipliers and more robust solid state devices such as vacuum avalanche photodiodes do have adequate resolution to be able to discriminate between the arrival of a single photon or more than one photon at the same time without any noise. This has provided the most sensitive optical detection technology based on single-photon counting.

Another advancement in optical detection technology is multi-element detectors. In these devices the detector elements, silicon for the 0.2 μm–1.1 μm spectral range and HgCdTe or InSb for the infrared wavelength range, are fabricated as integral parts of the chips. These could be a linear array of typically 1024 individual pixels (detector elements) having a width of about 25 μm and length of about 25 mm (diode array detector) or a two-dimensional mosaic of detector elements (typically 27 μm square) known as CCDs. The elements are charged on receiving photons and the charge is sensed sequentially through internal electronic scanning (like the scan on a television screen). The advantage of multi-spectral capability is that it allows simultaneous acquisition of an extensive spectroscopic data set when coupled with a standard or imaging spectrograph.

Environmental monitoring instruments based on photometers generally use photomultiplier tube (PMT) detectors having large active area, high sensitivity, gain in excess of 10^7 and extremely low noise. The need for cooling and stabilised power supply for sensitive detection has, in the past, rendered optical environmental monitoring systems unsuitable for field applications. However, recent advances in solid state hybrid detector–amplifier and PMT technologies have enabled much smaller and rugged PMTs which can be operated at room temperature and with only a 12 V dc source (e.g. OPTO-8 from Hamamatsu Corp.).

5.3.4.4 *Data acquisition–signal processing*

The output of a photodetector is an electrical signal proportional to the incident intensity of the light. The designated signal is always accompanied by noise generated by various sources. The passive noise due to the photodetector's dark current, Johnson noise due to thermal fluctuation of the charge carriers in conducting elements, amplifier noise and the noise due to background illumination give rise to what is called the dc noise and its overall level can largely be biased off by electronic circuitry or minimised by cooling the detector system. The active noise arising from interaction processes, e.g. Rayleigh scattering or fluorescence as noise in the detection of Raman signal, can largely be eliminated by spectral and temporal filtering. However, the noise associated with the random and discrete nature of the radiation field arriving at the detector, known as 'shot noise', ultimately limits the sensitivity of photodetection. In terms of signal photon flux, S, the associated noise, given by Poisson's statistics, is $S^{1/2}$. This noise is generated by the incident photon flux and, in the case of a very low-signal and almost noise-free detector, primarily by the background photon flux.

The output voltage of a photomultiplier, although amplified internally (gain $\sim10^6$–10^7), could still be quite low for very weak signals. For very sensitive spectroscopic measurements these weak signals are amplified, processed by signal-to-noise improvement electronics and the data displayed in a convenient form in relation to other parameters such as wavelength and time using a computer. A variety of data-processing techniques are utilised in modern spectroscopic instruments which are summarised below.

Phase-sensitive detection This processing technique allows retrieval of a weak ac signal or a modulated dc signal from an overwhelming noise background or the measurement of a relatively clean signal having strengths spread over several orders of magnitude. In this technique the signal is detected synchronously with the reference pulse derived from the same source. Commercially, these devices are called 'lock-in amplifiers'.

The input signal with noise is first amplified and if necessary filtered (to get rid of 50 or 100 Hz from the mains pick-up) before being applied to the phase-sensitive detector. The amplifier and pre-amplifier are needed to bring the signal (with noise) to a level where the phase-sensitive detector performs optimally and there is an impedance match between the signal source and the phase-sensitive detector. The reference signal triggers a circuit which generates a square wave synchronously with the signal pulse at the same phase. The phase shifter compensates for any phase shift between the signal and the reference caused during detection.The reference signal is used to effect a full-wave rectification of the signal at the phase-sensitive detector output. This is then applied to a low-pass filter which smooths or averages the waveform and gives a dc level proportional to the mean signal amplitude.

Gated integration These are commercially known as boxcar averagers. A typical instrument measures radiant intensity by sampling a pre-selected time slice of the input signal and accumulating the equivalent charge on a capacitor repetitively until a pre-selected level is reached. Waveform reconstruction occurs as the narrow sampling gate is scanned via a delay unit. The same procedure is adopted for a reference signal taken from the modulated excitation source and is used for the normalisation of the signal intensity. An example of the signal-to-noise improvement obtained by data averaging in the Raman spectrum of atmospheric nitrogen, recorded using a boxcar integrator, is shown in Figure 5.13.

Photon counting The spectrophotometers based on single-photon counting detection techniques allow detection of extremely low-level signals from low-noise photodetectors. The technique is based on the concept that the radiation field is quantised and a single photon on a photocathode will give rise to a voltage pulse (after internal multiplication) having a well-defined amplitude at the output of the photodetector. What is then needed is electronic circuitry called a discriminator to allow only the signals corresponding to a single photon to be counted as 1 and 0, i.e. no signal for lower amplitudes. Such digital signals are then converted into analogue signals electronically. It is obvious that a high intensity will correspond to a higher count rate and by spectral scan of photon count rates the whole spectrum is obtained.

Since large errors caused by detector voltage drift, source intensity fluctuations and other system instabilities are avoided, the photon counting technique provides a signal-to-noise ratio only limited by the statistical nature of the photons and photoelectrons in the detector.

Fourier transform technique This is concerned with the transformation of an optical signal in the time domain (temporal) into the frequency domain (spectral) by a

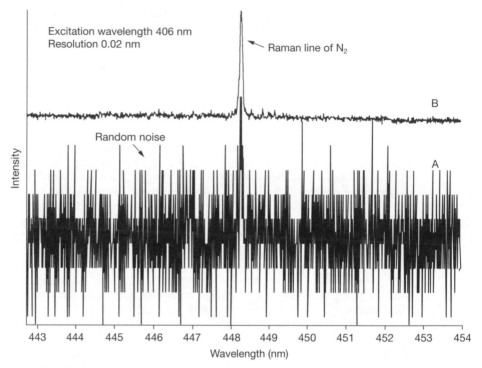

Figure 5.13 Example of signal-to-noise ratio improvement on signal averaging using a gated integrator for the Raman line of nitrogen

numerical procedure called a Fourier transform (FT). The technique is often referred to as time domain spectroscopy. FT spectrometers are primarily based on scanning Michelson interferometers (Figure 5.14(a)) and the principle of its operation is as follows.

The radiation whose spectrum is to be analysed is divided equally by a partially reflecting mirror into beams that travel distances of D_1 and D_2, perpendicular to each other, before they are reflected back and both brought to a focus at the photodetector. For a monochromatic source emitting radiation at λ, the phase difference between the two beams at the detector ($2\pi/\lambda \times$ path difference) will be zero when $D_1 = D_2$. The path difference, $2(D_2 - D_1)$, is called the retardation, δ.

If one of the mirrors (F) is fixed and the other (M) is moved linearly to a distance $\lambda/4$, i.e. to a path difference of $\lambda/2$ (two-way travel), the two beams will interfere destructively. In the same way, the beams will interfere constructively when M is moved to a distance of $\lambda/2$ (path difference of λ). Therefore, the linear motion of the mirror (M) at a constant speed, V_m, will produce a train of cosine waves equivalent to an interference fringe (Figure 5.14(b)). If it takes a time t (s) to move a distance $\lambda/2$ (cm), then $V_m t = \lambda/2$; therefore the frequency, f, of the electrical signal at the detector is related to the optical frequency ν ($= c/\lambda$) of the radiation by

$$f = 1/t = 2V_m/\lambda = 2V_m \nu/c$$

FT spectroscopy consists of recording the detector signal intensity $I(\delta)$ as a function of retardation, δ, and transforming this relationship to one that gives a power spectrum, $P(\nu)$,

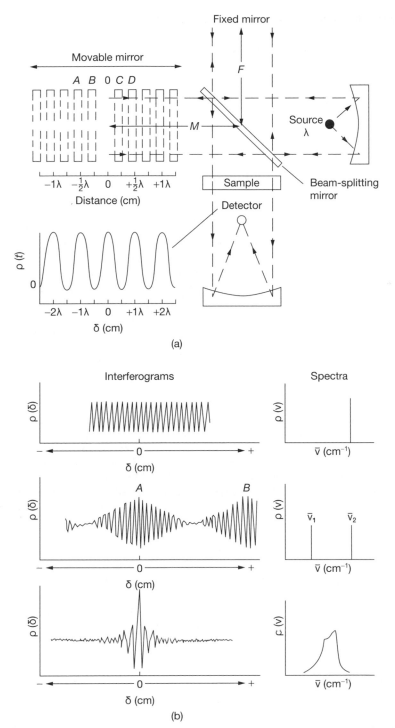

Figure 5.14 Schematic of a Michelson interferometer and an illustration of FT (Reproduced with permission from Skoog D A and Leary J L © 1992 *Principles of instrumental analysis*, Saunders College Publishers, Sidcup)

as a function of the optical frequency based on the above relation. This is carried out quite rapidly using sophisticated software in modern computers.

For monochromatic radiation, the FT of a sine (or cosine) wave at the detector gives a single line for the power spectrum corresponding to the wavelength, λ ($= c/v$), of the radiation source. For two discrete wavelengths (λ_1 and λ_2) in the source, the detector signal will be a superposition of two cosine waves, and, for a band of wavelengths, the electrical signal will be a complex superposition, but the transformation will retrieve the shape of the spectrum.

The resolution of an FT spectrometer is defined by the difference (in cm^{-1}) between two adjacent lines ($\bar{v}_1$ and $\bar{v}_2$) that can just be resolved by the instrument, i.e.

$$\Delta \bar{v} = 1/\lambda_1 - 1/\lambda_2 = \bar{v}_1 - \bar{v}_2$$

To resolve two lines, it is necessary to scan the time domain spectrum for a time long enough so that a cycle is completed. It can be shown that the resolution increases (resolution of smaller differences) in proportion to the reciprocal of the distance the mirror travels.

Modern FT spectrometers are capable of operation (with the exchange of suitable optics and detectors) over the far infrared through the visible wavelengths and their resolution can vary from $8\,cm^{-1}$ to less than $0.01\,cm^{-1}$. A few minutes are required to obtain a complete spectrum (limited by the optics) at the highest resolution (averaging over many scans). For quantitative analysis, both in absorption and in emission (e.g. Raman spectroscopy), the spectra of the samples are compared with those of a suitable standard.

5.3.4.5 *Fibre optics sensor*

Optical fibres are fine strands of glass, fused silica or plastic that transmit optical radiation. Each strand is coated with a material of lower refractive index than that of the transmitting medium. This allows light to travel along the strand through repeated total internal reflections and thus avoids loss due to partial transmission through the surface. The diameter of optical fibres ranges from $0.05\,\mu m$ to as large as $0.6\,cm$. For most practical applications, the strands are made into a bundle within a flexible cladding just like a multicore electrical cable. In a cable the ends of the fibres are fused and polished. Optical fibres are extensively used for optical communication (by British Telecom) and in the medical field to illuminate normally inaccessible parts and their use as light guides in spectroscopic and environmental diagnostics is rapidly increasing.

The sensors based on fibre optics (optrodes) are essentially suitable optical fibres having an immobilised enzyme or a reagent at the tip. When an analyte is probed by bringing the tip in contact, biochemical reactions change the fluorescence or reflectance of the immobilised reagent and give a measure of the concentration of the specific species in the analyte. Fibre optics is a very convenient device to transmit optical signals and allow manufacture of small and rugged spectrophotometers.

5.3.5 Instrumentation

5.3.5.1 *Absorption spectroscopy*

Modern UV–visible–near infrared absorption spectrometers (e.g. Perkin-Elmer Lambda-9) are versatile devices and cover a broad spectral range ($0.185\,\mu m$–$3.2\,\mu m$). For this, two different sources – a tungsten–halogen lamp for the visible and near infrared ranges and a deuterium lamp for the UV range – are required. Two monochromators with interchangeable holographic gratings and slits permit a highly spectrally pure radiation beam to be diverted to a beam splitter. One of the beams is passed through the sample cell and the other through a reference cell. Both the transmitted beams are focused onto the same detector. The detection of the reference and sample signals is carried out alternately by the use of a beam chopper. In such a spectrometer, the data acquisition, manipulation and the setting of experimental parameters are controlled by an internal microprocessor. For example, during a single scan, the source, extra-broadband filter (on a filter wheel), the gratings of the monochromators and the detector type (PM tube for UV–visible range and PbS for near infrared range) are automatically changed for the appropriate wavelength range.

These spectrometers are capable of providing a minimum bandpass of 0.05–5 nm in the UV–visible range and 0.2–20 nm in the near infrared range. The judicious design of the monochromator system keeps the stray light level below 0.0001% at UV–visible wavelengths and below 0.002% in the near infrared range.

The instrumental noise is often less than 0.001 A at 0 A (A is absorbance) measured with a 4 nm bandpass, at 500 nm and 1 s response (peak to peak over 3 min) time.

The normalised spectral absorbance data of a sample are digitally stored in the microprocessor's memory and can be processed later and printed as an absorption spectrum or transferred to another computer for further processing and storing.

5.3.5.2 *Emission spectroscopy*

A modern commercial emission spectrometer is equipped for the detection of luminescence and Raman scattering and for recording absorption spectra, excitation spectra and synchronous excitation or emission spectra. The operational principle of such a spectrometer is described below.

The spectrometer is divided into two separate sections based on two separate monochromators, called the excitation monochromator and emission monochromator, both configured in the modified Czerney–Turner scheme. The excitation monochromator selects a spectrally narrow beam from a broadband xenon source and diverts the beam onto the sample or the sample cell. A small fraction of the beam is directed by a partial reflectors to a photodiode for use as a reference signal for implementing compensation for source intensity fluctuation.

The emitted signal is collected by a mirror at an angle of about 90° with respect to the direction of the incident beam. This is then delivered to a photomultiplier having a broad spectral response, through the emission monochromator stage. In emission spectroscopy, the excitation is carried out at wavelengths shorter than those detected as emission signals.

The grating of the excitation monochromator is usually blazed (optimised) at about 300–350 nm and the emission monochromator is blazed at a suitable visible wavelength (typically 500 nm).

The signal detection in these instruments is usually carried out using photon-counting techniques. The digital data and the data for corresponding wavelength settings of either the excitation monochromator (for a selected detection wavelength) or the emission monochromator (for a selected excitation wavelength) are stored in the on-board microprocessor memory. The acquisition and initial processing of data and the control of experimental parameters such as slit widths, independent and synchronous scanning of the monochromators, source compensation and processing and recording (as hard copy) of data are carried out with an additional external computer supported by powerful software.

The recording of the emission spectrum of a sample with synchronous scanning of the excitation monochromator with no offset, i.e. $\lambda_{ex} = \lambda_{em}$, provides a reflection spectrum of an opaque sample. By introducing a fully reflecting mirror at the rear of the sample position and moving the sample cell so that the beam passes through the sample cell onto the sample collection mirror, an absorption spectrum can be obtained on synchronous scanning.

The spectrometer incorporates all-reflecting optics to minimise loss and obtain the broadest possible wavelength response. The spectral resolution is controlled by the width of the exit slit of the emission monochromator and is calculated from the quoted dispersion (nm mm^{-1} of slit width) for a particular grating. Although the gratings are blazed at specific wavelengths, these can be effectively operated over a broad band, e.g. 200–900 nm for the excitation and 200–950 nm for the emission. The data averaging time (integration time) (1 ms–160 s), step size (sampling interval), number of scans to be averaged, etc. are pre-selectable from the computer.

5.3.6 Applications

In the past, spectroscopy was confined to laboratory analysis of fairly pure samples, primarily for fundamental studies. The state-of-the-art spectrophotometers are now universally used for the analysis, monitoring and quality determination of substances and products in academic and industrial laboratories and have become indispensable and versatile tools. Recent advances in electro-optics and laser techniques have extended the use of spectroscopy into the realm of non-invasive real-time monitoring of the real-world environment and thereby enable on-line quality and process control in industries. Spectroscopic techniques, utilising wavelengths between mid-infrared and UV wavelengths, in both absorption and emission modes, are now routinely used in a variety of ingenious ways for the remote sensing of atmospheric and water-borne pollution and the meso-scale surveillance of the land mass. These aspects are described in detail in Chapter 8 on remote sensing.

5.3.6.1 *Absorption spectroscopy*

The application of infrared absorption spectroscopy may be categorised in terms of the earlier division of the spectral range into three regions. The near infrared band (0.78–2.5 μm) corresponds mostly to the first overtones of stretching vibrations and therefore their absorption strengths are much lower than those of the corresponding fundamental modes appearing in the mid-infrared band between 2.5 and 15 μm. Despite this, near infrared spectroscopy has been effectively applied for the quantitative analysis of water content (overtone O—H stretch at about 1.4 μm) in many organic substances such as acids, alcohol, phenols and oils. It has also been used for the detection of organic contaminants having C—H, N—H or O—H bonds in air, chimney stack emission and automotive exhaust.

The mid-infrared region is, by far, the most extensively used band for the identification of many organic compounds and measurements of their concentration. This is done by comparing the spectrum of the unknown sample firstly with the known bands of various functional groups such as C=O, C=C, C—H and O—H. The influence of vibrations associated with other atoms connected to these functional groups does not significantly alter the absorption bands of these functional groups. However, small differences in the structure and morphology of the substituents result in significant changes in the distribution of other absorption peaks in the spectrum, as shown in Figure 5.15. As a consequence, a close match between the spectrum of a known compound with that under test in this wavelength region (~8–15 μm), known as the 'fingerprint' region, along with the match in the group frequency region allow positive identification of the compound. This matching procedure

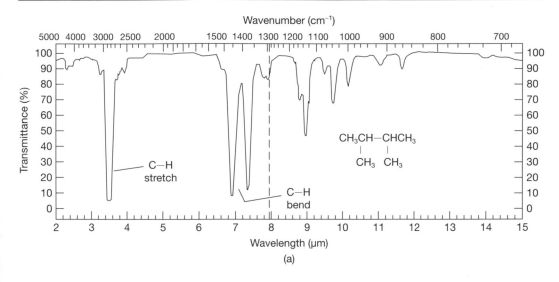

(a)

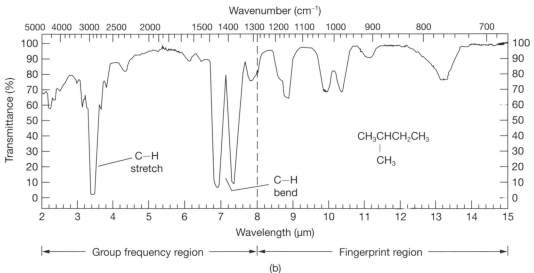

(b)

Figure 5.15 Absorption spectra of 2,3-dimethyl butane and 2-methyl butane, illustrating the matching (group frequencies) and mismatching (fingerprint region) of lines (From Roberts R M, Gilbert J C, Rodewald L B and Wingrove A S 1985 *Modern experimental organic chemistry*, Saunders College Publishing, Philadelphia, PA, reprinted in Skoog D A and Leary J L 1992 *Principles of instrumental analysis*, Saunders College Publishers, Sidcup with permission)

can now be carried out with commercially available computer software packages containing spectra of thousands of model compounds.

Absorption spectroscopy in the far infrared is only useful for the identification of metal complexes. Both stretching and bending vibrations of bonds between metal atoms and inorganic or organic ligands normally correspond to wavelengths greater than 20 μm. This region is not important for environmental diagnostic purposes.

125

5.3.6.2 *Emission spectroscopy*

Photoluminescence methods are inherently much more sensitive than absorption methods. In the latter, only the difference in optical intensities is measured, whereas in the former the emitted signal intensity is obtained by increasing the incident intensity (up to the saturation limit), by choosing an appropriate wavelength for excitation and by judicious choice of other experimental parameters. In general the sensitivities of fluorometric methods are 2–3 orders of magnitude higher than those of absorption methods. However, because of the high sensitivities, compounds containing many chromophores generally give composite spectra and fluorescence from cell walls and other optical components limits the applicability of this technique to relatively highly fluorescent species in a compound for identification and measurement.

Despite the lack of specificity, fluorescence spectroscopy is now routinely used to measure trace contaminants in the atmosphere and in waters. Inorganic species are also detected by measuring the fluorescence of chelates (these form with transition metals) or by measuring their quenching (reduction) action on the fluorescence of other species in a sample.

Fluorescence methods are finding important applications in detecting species in a sample at the end of a liquid chromatographic column and also in detecting species by fibre optic sensors. Techniques based on time-resolved and time-correlated measurements of fluorescence excited by extremely short pulses ($\sim10^{-9}$–10^{-12} s) are expected to find valuable applications in the diagnosis of species in real-world complex samples.

Although Raman scattering is very weak process, its unique wavelength specificity, extremely narrow linewidth and instantaneous response permit implementation of effective spectral and temporal (time gating) filtering and thereby enable detection with good signal-to-noise ratio. The technique, therefore, can be applied for unambiguous identification of a species in a mixture. The requirement of a small quantity of samples allows this technique to be applied for the analysis of biological samples and also samples of environmental interest. Since the Raman scattering process is excited by UV–visible wavelengths, the method is applicable to samples in aqueous solutions or gaseous samples in the presence of water vapour.

5.4 Atomic spectroscopy

5.4.1 Basic concepts

Atomic spectroscopy is the analysis of the wavelength dependence of absorption and emission of electromagnetic radiation by the atoms and ions of elements. In the UV–visible wavelengths, electrons in the outermost atomic orbitals are involved, and are the subject of this section. Transitions involving electrons in the inner shells give rise to absorption or emission in the vacuum UV (<180 nm) and those involving the innermost shell belong to the field of X-ray spectroscopy.

For many-electron atoms, the description of energy levels and possible transitions between them is not as simple as that for the hydrogen atom as described earlier. The charge cloud of the inner electrons shields the outer electrons from the nuclear force of attraction, the repulsive forces between electrons come into effect and the angular momenta and the spin momenta couple with each other in many possible ways. All these make the analysis of the atomic spectroscopy of many-electron atoms very complex.

However, absorption and emission transitions in many-electron atoms can, usually, be treated in terms of a change of the energy configuration of one outer electron only. On this basis, the transitions in alkali metals having only one valence electron have some similarity (not in the energy values) to those in the hydrogen atom, derived from the names of the series mentioned above. Detailed analysis including contributions from the electron spin reveals that the higher-energy p, d and f orbitals are split into two states. Therefore, the atoms containing one electron in the outermost orbital (two allowed by Pauli's principle) exhibit much the same general spectral features revealed in high-resolution spectral analysis.

For atoms with two or more electrons in the outer shell, further splitting of the energy levels arises owing to the coupling of the spin states. For example, in the ground state of helium, the two outer electrons are spin paired with no resultant spin momentum ($m_s = 0$). In the excited states, the electron spins are no longer paired and therefore not restricted by the Pauli principle and could both be $+\frac{1}{2}$ or $-\frac{1}{2}$. Therefore, the coupled momentum could be 0, +1 or −1, thereby splitting the energy level into a triplet; however, the selection rule does not allow transitions directly to triplet states.

A large number of lines are emitted or absorbed by many elements. Often, high-resolution spectroscopy using sophisticated equipment is needed to identify elements Despite this drawback, atomic spectroscopy has become a powerful technique for the identification of trace elements in samples. This is because each element gives its own set of spectral lines and high-resolution spectroscopy allows clear identification of elements. In qualitative analysis, it is not necessary to examine all the lines in the spectrum but only the 'persistent lines', i.e. the lines which are last to disappear when the concentration of the element under the test is progressively reduced. For quantitative determination of the concentration of an element in a sample, the usual procedure is to compare the unknown spectrum with the spectrum of a selected standard. In modern spectrometers this is done by electronic recording of the intensities and computer-aided comparison with stored data.

5.4.1.1 *Linewidth*

The atomic absorption or emission lines are never extremely (infinitely) sharp. The symmetrical spread in the wavelength (or the wavenumbers) about a peak value is known as the 'linewidth' and it is governed, inherently, by the lifetime of the excited state, i.e. by the law of statistics of transition process, dictated by the uncertainty principle. In practice further broadening (spread) of the linewidth is caused by the motion of the atoms (Doppler broadening) and collision between atoms (pressure broadening). For convenience of measurement, the linewidth is the width, in either wavelength or wavenumber units, measured at half the maximum signal (Figure 5.16). The relationship between the linewidth, $\Delta\lambda$, in wavelength units and Δv in wavenumber units is $\Delta\lambda = \lambda_0^2 \Delta v$ where λ_0 is the wavelength at the peak value. Natural linewidths at visible wavelengths are generally about 10^{-5} nm.

The Doppler broadening is exhibited by those atoms that move either towards or away from the radiation detector. Motions perpendicular to the line of sight of the detector do not contribute to Doppler broadening. In a hot environment, such as in a flame, the atoms will have a distribution of velocities (Maxwell–Boltzmann distribution) and the natural linewidth will be broadened by as much as two orders of magnitude. It is to be noted that the number density of atoms (or molecules) N_j in the excited state j at thermal equilibrium at a temperature T is given by the Maxwell–Boltzmann distribution law as

$$N_j = (g_j/g_0)N_0 \exp(-E_j/kT) \tag{5.23}$$

127

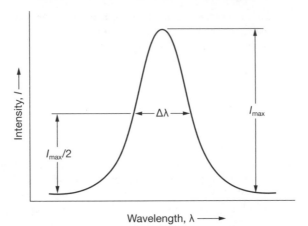

Figure 5.16 Scheme for defining the linewidth ($\Delta\lambda$) of a narrow absorption or emission spectrum

where N_0 is the number density of atoms in the ground (unexcited) state and g_j and g_0 are the statistical weights of the jth and ground states respectively ($g = 2J + 1$, J being the magnetic quantum number) and k is the Boltzmann constant.

5.4.2 Practical considerations

There are three aspects of atomic spectroscopy, namely AAS, atomic emission spectroscopy (AES) and atomic fluorescence spectroscopy (AFS). Of these AAS is by far the most widely used technique for environmental analysis and is described in detail later in this section. Although these spectroscopic methods work on different principles and require different types of source for excitation, the detection and analysis of spectra require high-resolution spectrometers, sensitive photodetectors and efficient atomisation (vaporisation) devices in all cases. In solid or liquid states the interaction of surrounding media with the target atoms is so strong that a few measurements are informative. Atomisation or vaporisation is therefore an essential part of atomic spectroscopy.

5.4.2.1 *Atomisation*

The atomisation for AAS is most conveniently carried out by naked flames and this is still the most popular method, although, for some applications, resistive heating in a furnace, known as electrothermal atomisation, is being effectively and conveniently used for this purpose.

For AES and AFS, inductively coupled plasma (ICP), direct current plasma (DCP) or electrical arc–spark methods are commonly used for atomisation. The atomisations based on flame, electric arc, ICP and DCP are of the continuous type. In these, a solution of the sample is continuously converted into a mist of fine droplets by a jet of compressed air; the process is known as nebulisation. The flow of gas carries the sample into a heated region (e.g. flame) where it is atomised. In the electrothermal atomisation process, and sometimes in the electrical arc atomisation method, samples are analysed in discrete mode. In this, a measured volume of the sample solution is introduced into the heating device and the temperature is raised suddenly to accomplish atomisation of the whole sample within a brief period. The spectra are therefore time dependent.

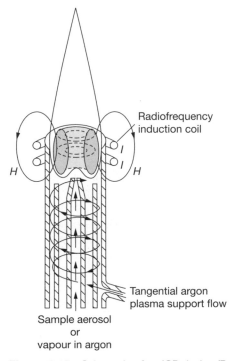

Radiofrequency induction coil

H

H

Tangential argon plasma support flow

Sample aerosol
or
vapour in argon

Figure 5.17 Schematic of an ICP device (Reproduced with permission from Skoog D A and Leary J J 1992 *Principles of instrumental analysis*, Harcourt Brace College Publishers, Sidcup)

The atomisations based on plasma and electric arc use very high-energy devices providing temperatures in the range between 4000 °C and 6000 °C. The advantages of using these devices are that at such high temperatures complete atomisation takes place and good emission spectra are obtained for most elements with a minimum of inherent interference. Since temperature of atomisation varies widely with the type of element, these devices enable excitation of many elements in a small quantity of the sample and the detection of their spectra simultaneously, using a multichannel detector system. In addition, the plasma devices can be used to identify and measure non-metals such as chlorine, bromine and sulphur over a concentration range much wider than is possible in other devices.

5.4.2.2 *Inductively coupled plasma device*

An ICP device is commonly known as a torch. In this, a high temperature is generated by the closed-circle resistive motion of the ions and electrons in an argon plasma as shown in Figure 5.17. The nebulised sample is carried along a quartz tube to the head of the torch by argon gas. Ionisation is initiated by a spark from a Tesla coil and the closed-circle motion of the electrons and ions is sustained by applying a strong oscillating magnetic field from a water-cooled induction coil. The induction coil is powered by a radio-frequency generator typically operating at about 2 kW and 27 MHz.

The samples which can be dissolved in a suitable solvent are introduced into the ICP devices by the nebulisation process. Solid samples are initially either vaporised by an

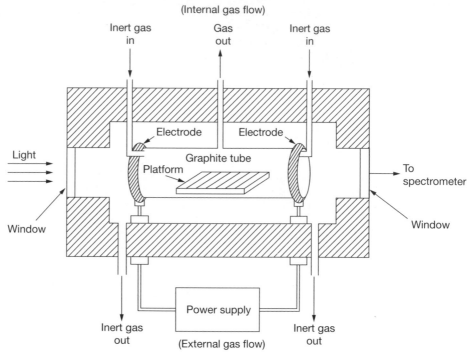

Figure 5.18 Schematic of a typical electrothermal atomic furnace

electric spark or made into fine powder, to be carried to the active region of the torch by argon gas. Electrothermal vaporisation is sometimes used before a sample is introduced into the ICP for AES.

5.4.2.3 *Electrothermal device*

An electrothermal atomiser is basically a graphite furnace designed for the discrete introduction of samples, usually as liquids, and for exposing these to high temperatures (>3000 °C) within a short time (<2 s). A schematic of the design of a typical commercial electrothermal atomic furnace is shown in Figure 5.18.

The atomisation takes place in an open-ended cylindrical graphite tube. The tube, usually of 5–6 cm internal diameter, is the resistive element and is heated by a high electrical current. The sample is introduced through a small (2 mm) orifice using a micropipette onto a replaceable graphite platform (L'vov platform). Inert gas, usually argon, is passed along the two open ends of the tube to be transported out through the orifice used also for sample injection. This ensures that external air is excluded in the sampling region and vapours generated from the sample matrix during the initial stages of the heating cycles are carried away. The outside stream also excludes air and protects the tube from incineration due to excessive heat.

The power supply controlling the furnace can be programmed so as to, in steps, dry the sample after injection, ash it at an intermediate temperature (~500 °C) and atomize it at a higher temperature. Absorption or, in some instances, emission is measured during the atomisation period.

There are many advantages of electrothermal atomisers over the other devices. These include increased sensitivity (100–1000-fold), decreased sample size ($10-15\,mm^3$), applicability to solid samples, operational cost-effectiveness and safety. However, there are a few disadvantages of electrothermal atomisers. For example, they suffer more from interference than the flame atomisers, sample handling is problematic, complicated programming is required and they are expensive to install.

5.4.2.4   *Flame atomisation*

In this, a solution of the sample is sprayed into a flame by a nebuliser. The flame converts the mist of tiny liquid droplets into a mixture of atoms, ions and a variety of other molecular species belonging to both the sample and the analyte and many other products of reaction between the fuel, oxidant and the sample. The temperature profile in the flame and the flow rate of the fuel–oxidant mixture are two important factors in atomisation for AAS.

Natural gas in air gives a temperature around 1800 °C whereas acetylene in oxygen gives a temperature in excess of 3000 °C. These are generally regulated experimentally for ideal atomisation conditions. The maximum temperature is located somewhere above the primary combustion zone. Therefore it is important to focus the same part of the flame onto the entrance slit of a monochromator for both calibrations and measurements.

5.4.2.5   *Flame atomic absorption spectroscopy*

The reasons for the immense popularity of AAS based on flame atomisation and hollow cathode tube source are its simplicity, versatility and remarkable lack of interference. A schematic of a typical AAS spectrometer is shown in Figure 5.19.

In this configuration, a probe beam from a hollow cathode tube is collimated and is alternately transmitted through the gaps between the blades and reflected from the mirrored blades of a chopper. The beams, one passing through the atomised samples in the flame, called the signal beam, and the reflected beam, called the reference beam, are alternately detected by two detectors. The signals are averaged at the chopping frequency, called the modulation frequency, in a phase-sensitive detector. This type of electronic detection system is also known as a lock-in amplifier, as it locks the signals and the references at the

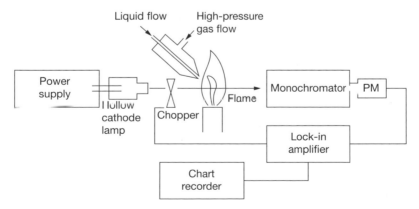

Figure 5.19   Schematic of an atomic absorption spectrometer incorporating a hollow cathode tube

chopping frequency and therefore rejects optical signals at any other modulation frequencies including dc levels. The number of light pulses averaged in a sampling period depends on the electronic time constant of the lock-in amplifier and the chopping rate. The ratio of the signal to reference intensities is thus displayed as a dc voltage level, averaged over the sampling (integration) time. The double-beam configuration compensates for source intensity drift and warm-up and source noise but does not compensate for noise in the atomisation process which puts the limit to the detection sensitivity.

5.4.2.6   The hollow cathode lamp

In a hollow cathode tube a cylindrical cathode and a tungsten anode are separated by a cylindrical glass shield encapsulated in a glass tube containing 1–5 mbar of inert gas (Ar or Ne). The cathode is made of the metal whose spectrum is to be examined. Therefore, different lamps are needed for the analysis of different elements.

At about 300–500 V applied between the electrodes, ionisation of the inert gas takes place. The ions acquire sufficient kinetic energy to strike off metal atoms from the cathode surface (sputtering). Some of these atoms are in excited states and emit their characteristic radiation. Since the absorption and emission lines of an element have the same wavelength, the use of an appropriate source gives rise to the so-called 'lock-on-key' effect and is responsible for the very high selectivity of AAS. However, for effective utilisation of AAS, the interference caused by the emission of radiation by the flame and by other excited molecular or ionised species needs to be eliminated.

5.4.2.7   Reduction of interference

The radiation from the flame at the transmission wavelength window corresponding to the monochromator setting interferes with the absorption of the source radiation. This interference is eliminated by phase-sensitive detection, as only the modulated light from the source at the chopping frequency is detected by the system.

The absorption by molecular species and scattering from particulates contribute as non-specific absorption noise in AAS. For a broadband source, the specific atomic line absorption is usually much weaker than that for an atomic line source (hollow cathode lamp), although the amount of non-specific absorption is the same. Therefore, simultaneous use of two types of source and signal subtraction has been found to reduce such interference noise.

5.4.2.8   Detection limit

The detection sensitivity in AAS depends on the type of element and also on the process of atomisation. For example, mercury can be detected at a level of $500 \, \mathrm{ng \, ml^{-1}}$ with flame atomisation, whereas the limit is lowered to about $100 \, \mathrm{ng \, ml^{-1}}$ with electrothermal atomisation. In general, for many elements the sensitivity of detection lies in the range 1–$20 \, \mathrm{ng \, ml^{-1}}$ (0.001–0.02 ppm) for flame atomisation and 2×10^{-3} to $1 \times 10^{-2} \, \mathrm{ng \, ml^{-1}}$ (2×10^{-6} to 1×10^{-5} ppm) for electrothermal atomisation. The limit of detection by AFS and ICP-based AES varies widely for different elements. Absorption spectroscopy of elements based on electrothermal atomisation provides the best sensitivities for the detection of aluminium ($0.005 \, \mathrm{ng \, ml^{-1}}$), lead ($0.002 \, \mathrm{ng \, ml^{-1}}$) and cadmium ($0.0001 \, \mathrm{ng \, ml^{-1}}$). Emission spectroscopy based on flame atomisation gives the highest sensitivities for the detection of arsenic ($0.0005 \, \mathrm{ng \, ml^{-1}}$) and mercury ($0.0004 \, \mathrm{ng \, ml^{-1}}$) (Fuller, 1977).

5.4.3   Applications of atomic absorption spectroscopy

The atomic spectroscopies are sensitive methods for the quantitative determination of over 60 metals or metalloid elements. The accuracy and sensitivity of atomic spectroscopy are primarily governed by the methods of sample preparation, sample presentation and also the accuracy in the production of calibration curves. The emission or absorption lines of non-metallic elements generally are located in the vacuum UV wavelength region below 200 nm and therefore cannot be accessed by conventional spectroscopy. Atomic spectroscopic techniques have been applied to the analysis of metals in such materials as lubricating oil, sea water, biological samples, graphite and many other organic and inorganic materials. AAS based on both flame and electrothermal atomisations plays an important role in international geological survey. Atomic spectroscopies are extensively used for metal content determination in drinking water and sea waters, agricultural soil, plants and vegetation, drugs, etc. However, the techniques require complicated and expensive sample preparation and presentation procedures and consequently are invasive and do not offer real-time monitoring of environmental contaminants for on-line processes or quality controls in industries.

5.5   Nuclear magnetic resonance spectroscopy

5.5.1   Basic concepts

The magnetic properties of atomic nuclei are utilized in nuclear magnetic resonance (NMR) spectroscopy. The nucleus has a positive charge, and, like the orbital electrons of an atom, this also executes a spinning motion about an axis (see Figure 5.7). As such, this can be thought of as a sphere with distributed charge on its surface and rotating (spinning) about an axis and therefore equivalent to a current in a loop. According to Maxwell's theory, this current will create a magnetic field directed along the axis of the spinning motion. Such a magnetic field is quantified by a quantity called magnetic dipole moment, μ, which is related to the magnitude of the current in the loop and the size of the loop, i.e. the nucleus, and consequently the nuclear charge and the angular velocity (ω) of the spinning motion. The dipole moment is therefore proportional to the angular momentum, $P_\omega (= m\omega$, m being the mass), of the spinning nucleus: $\mu = \gamma P_\omega$. The constant of proportionality, γ, is called the gyromagnetic ratio.

An external magnetic field will influence the spinning motion of the nucleus through its magnetic dipole moment. If the direction of the magnetic field makes an angle θ with the direction of the dipole moment, the spinning nucleus will be made to precess (gyrate) about the direction of the applied field, i.e. the tip of the dipole moment vector may be thought of making an orbit about the field direction making a solid angle θ (sr). The angular frequency (ω_0) of this precession is proportional to the strength of the applied magnetic field, B_0, and is given by $\omega_0 = \gamma B_0$. This angular frequency, when converted to linear frequency, ν_0, using the identity $\nu_0 = \omega_0/2\pi$, gives the Larmor frequency, $\nu_0 = \gamma B_0/2\pi$.

To explain the mechanism of absorption of EM radiation by the nucleus, it is necessary to invoke a quantum mechanical description of the spin and magnetic quantum numbers. The spin angular momentum of the nucleus is contributed by its constituent protons. Since the probabilities of clockwise and anticlockwise spin motions are equal, the spins in a nucleus with even numbers of protons are paired up and there is no net spin moment. However, for nuclei with odd numbers of protons, one will remain unpaired and the nucleus as a whole will have a net spin moment. In quantum concept this is described by

133

magnetic quantum number $m = +\frac{1}{2}$ for clockwise spin and $m = -\frac{1}{2}$ for anticlockwise spin. For such nuclei the potential energy is given as $E = -mh\gamma B_0/2\pi$ so that $E_{+\frac{1}{2}} = -h\gamma B_0/4\pi$ and $E_{-\frac{1}{2}} = +h\gamma B_0/4\pi$.

The difference in energy between the two nuclear energy states created by the external magnetic field is given by $\Delta E = h\gamma B_0/2\pi$. A transition between these two states (spin flip) will be brought about if the frequency, v_0 of the incident electromagnetic radiation coincides with this energy difference ΔE, in frequency units ($\Delta E/h = \Delta v$). This then leads to the absorption of the EM radiation. For proton NMR (hydrogen nucleus), the frequency range, depending on the strength of the magnetic field (1.4–14 T), corresponds to the radiofrequency waveband 60–600 MHz.

The absorption of radiofrequency energy by nuclei takes place because of a small difference in the populations of the two spin states. If the rate of absorption is higher than the rate at which the absorbed energy is dissipated by the nuclei, saturation results. It is therefore necessary to ensure that the rate of relaxation is higher than the transition.

In a practical sample, the target nuclei are surrounded by many other types of nuclei. Because of vibrational and rotational motions, these nuclei will constitute a continuum of magnetic components which interact with the magnetic components of the target species (spin–lattice relaxation) and provide a relaxation pathway. When two nuclei possess identical precession rates but different quantum states (m), the absorbed energy by the target species will be rapidly transferred to the other one (spin–spin relaxation). The high relaxation rates, however, give rise to line broadening (shorter lifetime) which prevents high-resolution measurements of spectral lines.

5.5.2 Practical considerations

NMR spectrometers may be classified into two basic types depending on the mode of operations, i.e. the CW type and the pulsed type. The CW types are less sensitive and are suitable for the analysis of samples having high abundance of the target nuclei and large values of gyromagnetic constant. Pulsed NMR generally operates at a much higher magnetic field strength (>3 T) than that used in CW types. Field strengths as high as 14 T are achieved in modern commercial pulsed NMR spectrometers. Such high magnetic fields are generated by utilizing superconducting material technology. The high magnetic field strengths in pulsed systems allow very high sensitivity and are commonly used to analyse samples with the designated nuclei of very low abundance in the test samples, such as ^{13}C in fossils.

The introduction of digital electronics and computers has now enabled the application of FT techniques in NMR spectroscopy. This has improved the sensitivity and resolution of NMR spectroscopy and facilitated the acquisition, processing and interpretation of data. These techniques are described below.

5.5.2.1 *Continuous-wave nuclear magnetic resonance spectroscopy*

A layout of the basic components of a typical CW (broadband) NMR spectrometer is given in Figure 5.20. For its operation, a CW radiofrequency signal from a stable oscillator (generator) is applied to a balanced radiofrequency bridge circuit. One of the arms of the bridge circuit consists of a coil acting as an inductance and into which the test tube containing the sample to be analysed is placed.

The radiofrequency field, generated in the coil, is partly absorbed by the target nuclei in the sample at the resonance condition. The dissipation of radiofrequency power from

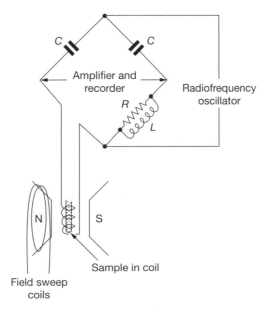

Figure 5.20 Schematic of a typical CW NMR spectrometer

the coil appears as an apparent increase of the resistance in the coil. A signal due to the imbalance created in the bridge appears across the detector section of the bridge, which is amplified and detected. A spectrum is obtained either by sweeping the magnetic field (in the case of an electromagnet) for a fixed frequency radiowave input or by scanning the frequency of the radiowave at a fixed magnetic field.

5.5.2.2 Pulsed nuclear magnetic resonance spectroscopy

In a pulsed NMR spectrometer, the oscillator (usually a continuous crystal oscillator) is followed by a pulser switch and a transmitter to provide intense and repetitive pulses of selected pulse width and pulse separation. A coil connected to the radiofrequency transmitter placed between the poles of a magnet is used to deliver the radiofrequency power to the sample in a test tube placed inside it. The same coil is used to detect signals emitted by the excited nuclei during their relaxation process.

Input radiofrequency signals in a pulsed NMR spectroscopy are typically 1–10 µs wide and are separated by 1 s or more. The absorption of the signal at the Larmor frequency of the nuclei can be compared with a standard for quantitative measurement, and the pulse decay time can be utilised to analyse the environment of the nuclei through their temporal relaxation history.

Great improvements have been made by incorporating Fourier techniques in the use of pulsed NMR spectroscopy. The time domain radiofrequency signal, called the free induction decay (FID) signal, emitted by the excited nuclei during their relaxation is detected by the same coil that provides the excitation and analysed by a phase-sensitive detector. The FID signals from many successive pulses are averaged, digitised and stored in a computer. The time domain signals are then converted to frequency domain signals using a digital FT algorithm.

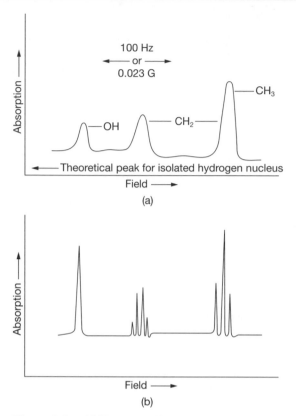

Figure 5.21 NMR spectra of ethanol at (a) low resolution and (b) high resolution showing the spin–spin splittings (Reproduced with permission from Skoog D A and Leary J L © 1992 *Principles of instrumental analysis*, Saunders College Publishers, Sidcup)

5.5.2.3 *Performance of nuclear magnetic resonance spectrometers*

The performance of an NMR spectrometer is described by its sensitivity and resolution. To achieve both high resolution and high sensitivity, high magnetic fields are needed. In addition to the strength, the magnets need to provide highly homogeneous and reproducible fields. Ideally the homogeneity required is few parts per billion within the sample area, but this is not currently met in practice.

The most commonly detected nucleus in NMR spectroscopy is ^{1}H, although ^{13}C and ^{19}F are also used for some samples. Practical samples have hydrogen nuclei in a variety of configurations which alter their Larmor frequency. Usually much interference noise is imposed on the very low (a few nanovolts) signals received in the coils. Signal averaging is, therefore, necessary to improve the signal-to-noise ratio. The configurational effect of the proton (^{1}H) NMR signal is demonstrated in Figure 5.21. At low resolution, the hydrogen atoms in the hydroxyl, methylene and methyl give well-separated absorption peaks (chemical shift) having areas in proportion to the number of hydrogen nuclei in the compound (e.g. 1 in OH, 2 in CH_2 and 3 in CH_3). The same spectrum, obtained with a high-resolution spectrometer, reveals fine structure superimposed on the absorption bands.

5.5.3   Applications

NMR spectroscopy is a very powerful tool not only for structural analysis of organic and inorganic species but also for sensitive quantitative determination of many species relevant to environmental monitoring and diagnosis including metal–organic and biochemical molecules. For quantitative measurements, an appropriate internal or external standard is needed. NMR spectroscopy is very useful for the analysis of multi-component systems. For example, analysis of benzene, heptane, ethylene glycol and water in various mixtures is routinely performed by commercial NMR spectrometers. Quantitative analysis of functional groups such as hydroxyl groups in alcohols and phenols is most conveniently carried out using this technique. This is also extensively used for elemental analysis, that is the determination of total amount of magnetic nuclei, e.g. hydrogen, ^{19}F and phosphorus, in organic mixtures.

Although very versatile and sensitive, the application of NMR spectroscopy is not widespread. Because of the bulkiness, high cost and operational complexity of such equipment, the use of NMR spectrometers is confined to a few large industrial organisations, government laboratories and some educational institutions.

5.6   X-ray spectroscopy

5.6.1   Basic concepts

X-rays are EM radiation of extremely short wavelengths (i.e. high energies). The band of wavelength covered by X-rays is very large, although most analytical spectroscopies are confined to the X-ray wavelength band between 0.1 Å (angstrom) and 25 Å. The photon energies (hc/λ) in this wavelength band correspond to transitions involving the innermost atomic orbitals, i.e. K and L shells. X-rays can knock out electrons of innermost shells, are scattered by the atoms and can induce fluorescence in elements. X-rays are also diffracted by a regular array of atoms in the same way as UV–visible radiation is diffracted from gratings.

5.6.1.1   *Absorption*

When X-rays pass through a material (e.g. metal foil), the intensity of the beam diminishes with thickness according to Beer's law: $\ln(I_i/I_t) = \mu_m \rho x$, where x is the sample thickness, ρ is the density of the sample, μ_m is the mass absorption coefficient and I_i and I_t are the intensities of the incident and transmitted X-ray beams respectively. Beer's law holds for absorption (attenuation) of the beam through scattering process in the bulk of the sample.

The wavelength dependence of the mass absorption coefficient is exclusively governed by the type of element (mass number, z) and is not influenced by its physical or chemical states. Actual absorption of the X-rays takes place when the photon energies are sufficiently high to knock out an electron of the K or L atomic orbitals. The energy is then distributed between the kinetic energy (KE) of the ejected electron and the potential energy of the ion thus created. The probability of absorption is highest when the energy of the X-ray photons coincides with the ionisation energy of the electrons (minimum KE). Examples of X-ray absorption spectra of two elements, i.e. silver (Ag) and lead (Pb), are shown in Figure 5.22. Sharp discontinuities (absorption edges) are unique characteristics of X-ray absorption spectra. The peaks correspond to the ejection of K and L electrons. The sharp discontinuities are understood from the fact that, immediately above the peak absorption wavelengths,

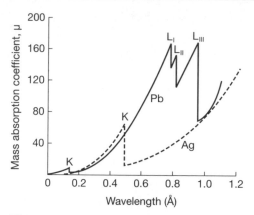

Figure 5.22 Partial X-ray absorption spectra for lead and silver

the photons have the lowest probability of interaction with the atoms. However, as the wavelength increases (decreases in photon energy), the probability of absorption (primarily due to scattering) increases until the wavelength corresponds to the energy of another shell (L, M, etc.).

5.6.1.2 *Fluorescence*

The removal of an electron from a K or L shell leaves the atom in an excited ionic state. The ions return to their ground state through a series of electronic transitions from higher orbitals (M → L, L → K, etc.). These transitions give rise to emission in the X-ray wavelength band with characteristic spectral features of the target. The wavelengths of fluorescence lines are always at longer wavelengths than those of exciting radiation. This is because absorption requires complete removal of an electron (with some kinetic energy) whereas the emission is due to transitions between shells. For efficient fluorescence, the wavelength of excitation needs to be optimised so that it lies just beyond the absorption edge of the target element and the wavelength can be effectively discriminated against the X-ray fluorescence.

5.6.1.3 *Diffraction*

X-rays are scattered by atoms when the wavelength does not correspond to the absorption edge of the target element. In semitransparent (at X-ray wavelengths) crystals such as lithium fluoride (LiF), sodium chloride (NaCl) and ammonium dihydrogen phosphate (ADP), the atoms exist in an orderly manner along successive layers of lattice. The radiation scattered by the atoms of the first layer at an angle θ (same as the incident angle), will interfere constructively or destructively with the radiation scattered from atoms at other layers (equal spacing, d) depending on the phase relationship (Brag diffraction), exactly in the same way as interference of UV–visible radiation from multilayer coatings. Thus, for constructive interference the condition is $n\lambda = 2d \sin\theta$, where n is the order of diffraction. It is, therefore, evident that the diffraction from a crystal can be used to select wavelengths in X-ray spectroscopy by varying the angle of diffraction.

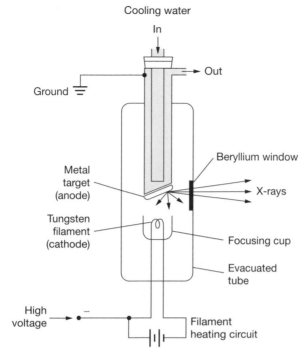

Figure 5.23 Schematic of a typical X-ray tube

5.6.2 Practical considerations

The components in X-ray spectrometers are functionally similar to those in UV–visible or infrared spectroscopy. The components therefore needed are a suitable X-ray source, a wavelength selector, a sample holder, a radiation detector, data acquisition electronics and a readout device. These are described below.

5.6.2.1 *Source*

X-rays can be generated either by the bombardment of a metal target with high-energy electrons or by irradiating a target with a primary source to obtain X-rays as fluorescence. Some radioactive materials also emit X-rays during the decay process.

However, for analytical spectroscopy, the most common source of X-rays is the X-ray tube, as shown schematically in Figure 5.23. This operates on the basis of electron bombardment of a suitable metal target (tungsten, copper, molybdenum, etc.) usually embedded in the surface of a heavy cylinder of copper. The target usually needs to be cooled by passing water. Modern tubes operate at considerably lower powers and do not require cooling. The heater circuit is used to control the intensity of the emitted X-rays, and the accelerating high voltage between the heater element and the anode (target) determines the photon energies (wavelengths) of the emitted X-rays.

The spectral characteristics of emitted X-rays depend on both the accelerating voltage and the atomic number of the target material. The shortest wavelength is generated when the bombarding electrons are instantaneously decelerated (KE = 0) in a single collision.

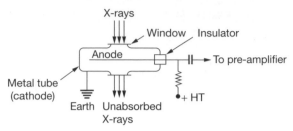

Figure 5.24 Schematic of a typical gas-filled X-ray detector

The minimum wavelength of the X-ray is related to the accelerating voltage by $\lambda_0 \simeq 1.14 \times 10^4/V$, where λ_0 is in angstrom units and the accelerating potential V is in volts. The bombarding electrons lose their energies through many collisions with the atoms of the target and therefore continuous spectra, independent of the target material, result. The intensity and wavelength position of emission lines superimposed on the continuous spectra depend on the accelerating potential and the atomic number of the target. For example, the characteristic K emission (fluorescence) lines of tungsten ($z = 74$) appear at wavelengths of 0.18 and 0.21 Å at 70 kV potential. For a molybdenum target ($z = 42$), the sharp K emission lines are obtained for voltages above 20 kV.

5.6.2.2 *Detector*

Three types of detector encountered in X-ray spectrometers are gas-filled detectors, scintillation counters and semiconductor detectors. The gas-filled detectors work on the principle that X-rays produce a large number of ions and electrons in an inert gas which will cause a current flow in an applied field. A typical gas-filled detector is shown schematically in Figure 5.24. The variation of the current with applied voltage is not linear. A gas-filled detector is operated as an ionisation chamber or as a proportional counter or as a Geiger counter depending on the magnitude of the accelerating voltage.

In modern scintillation detectors, activated (with ~0.2% thallium iodide) sodium iodide crystals are used as scintillators when X-rays are incident on these. Each X-ray photon gives rise to several hundred scintillation photons in the visible wavelengths (~400 nm) in the bulk of the crystal. These photons are then detected by a photomultiplier.

Semiconductor detectors are, by far, the most widely used detectors in X-ray spectroscopy. These are fashioned from a wafer of crystalline silicon or germanium in a way that it consists of a p-type layer that faces the X-rays, an intrinsic zone and an n-type layer. Absorption of X-ray photons in the intrinsic band gives rise to highly energetic photoelectrons, which in turn lose their kinetic energy to raise thousands of electrons to the conduction band. In an applied field, a current pulse is generated for each X-ray photon and the magnitude of this pulse is directly proportional to the energy of the absorbed X-ray photon. These detectors are therefore very suitable for photon counting detection and energy-dispersive measurement by pulse height analysers.

5.6.2.3 *Wavelength selector*

In X-ray spectroscopy the selection of a narrow band of wavelengths (line) can be accomplished either discretely or continuously. To obtain a narrow spectral line, needed in diffraction or fluorescence measurements, thin metal foils having appropriate absorption edges in relation to the emission lines of the target are used. The choice of narrow lines selectable by using such filters is limited because of the small number of suitable target–filter combinations that are available.

The wavelength-dispersive element (monochromator) in this is a crystal of suitable material mounted on a rotatable platform at the centre of an evacuated sphere. The choice of crystal is primarily based on the wavelength range of operation (depending on the lattice spacing, d). The arm containing the detector is also mounted on a goniometer-type arrangement with mechanical linkage such that, during a wavelength scan ($n\lambda = 2d\sin\theta$), the detector arm rotates at twice the speed of rotation of the crystal along the vertical axis.

The incoming beam could be fluorescence induced by the primary X-rays or a broadband beam passing through an absorbing target. The incoming beam and the monochromated beams are both collimated by small metal tubes.

5.6.2.4 *Data acquisition*

In X-ray spectroscopy this is most conveniently carried out using photon counting electronics (see Section 5.3.4.4). For weak sources of radiation, as encountered in most analytical X-ray fluorescence and absorption measurements with a low-voltage X-ray tube, photon counting provides better signal-to-noise ratio and more accurate intensity data than those obtained from average current measurements. The photon counting technique also allows energy-dispersive measurements in X-ray spectroscopy utilising pulse height analysers. A single-channel pulse height analyser has a voltage range of perhaps 10 V with a window of say 0.1 V. The window can be electronically adjusted to scan the entire voltage range, thus providing an energy dispersion spectrum, since the pulse height is directly proportional to the energy of the X-rays.

5.6.3   Applications

The application of X-ray absorption spectroscopy in environmental diagnosis is rather limited. This is primarily because the absorption peaks are usually quite broad and, in samples containing many elements, the spectra will be diffused owing to both overlap and matrix effects (bulk scattering). However, X-ray absorption methods can be effectively applied for the determination of heavy trace elements in a sample containing only light elements, for example lead in petrol and halogens in hydrocarbons.

The X-ray diffraction technique is extensively used to study the arrangement and spacing of atoms in crystalline materials. The most remarkable applications of this technique have been the studies of structures of complex natural products such as steroids, antibiotics and vitamins. For diagnostic purposes, the diffraction method is only applicable to the identification and measurement of crystalline compounds such as NaCl and KBr in a solid mixture.

5.6.3.1 *Fluorescence spectroscopy*

X-ray fluorescence spectroscopy has proved to be an effective method in commercial, industrial and environmental analytical applications. It allows rapid (in minutes) quantitative determination of all but a few lighter elements in complex samples. For quantitative determination, the X-ray spectrometers need to be calibrated against standards and proper correction procedures for matrix effects need to be implemented.

This technique is now routinely used for on-line quality control in the manufacture of metals and alloys. It is also quite extensively used in the determination of lead and bromine in aviation gasoline, calcium, barium and zinc in lubricating oil and pigments in paint samples.

This technique is finding increasing application in the diagnosis and monitoring of atmospheric pollutants. For example, paper discs impregnated with orthotolidine, silver nitrate and sodium hydroxide will absorb chlorine, sulphides and sulphur dioxide respectively. These discs then can be subjected to X-ray fluorescence analysis to monitor the species in the atmosphere.

X-ray fluorescence provides a relatively simple, non-destructive means of environmental analysis. The analysis can be carried out either using a minute quantity of the sample or on a massive object. The accuracy and precision of this method are comparable with those achieved by other chemical or physical analytical methods. However, the sensitivities of this technique are limited to the detection of species at concentrations of 100 ppm, whereas optical spectroscopic methods allow detection sensitivities in the sub-ppm level.

5.6.3.2 *Particle-induced X-ray emission*

The energies of X-ray photons emitted from a material are characteristic of the atoms from which they originate. A plot of number of emitted X-ray photons (intensity) against the corresponding energies (or wavelengths) of the X-rays constitutes the particle-induced X-ray emission (PIXE) spectrum, giving the sharp and well-defined lines representing the elements involved. In practice, the target under investigation is bombarded by a beam of charged energetic particles (ions) to induce characteristic X-rays instead of an electron beam. This is to reduce the high background noise generated as bremsstrahlung emission when an electron beam is used (resulting from the smaller mass of electrons compared with that of ions).

This technique allows automatic, non-destructive and quantitative analysis of environmental pollution. With the use of multichannel detection devices with appropriate computer algorithms and standards, parts per million sensitivity for heavy elements in light matrices can easily be achieved. However, the main disadvantage of this technique is the high capital cost of a particle accelerator facility.

Major environmental pollutants having atomic numbers higher than that of silicon can be measured in a wide range of media using the PIXE technique. However, suitable sample collection, preparation, presentation and, in some instances, pre-concentration schemes are needed.

The PIXE technique is used most widely for the analysis of air pollution. In this a known volume of air is drawn through a filter in order to collect particulate aerosols. The filter selected must contain only low atomic number elements, resistant to the ion beam and compatible with the vacuum environment of the target chamber. The most commonly used filter materials are Mylar and Teflon. The commercially available aerosol samplers have many special features and designs to suit particular requirements. The most common types have two filters: a coarse one to prevent large particles entering the apparatus and a fine one to collect the particulate for analysis. Normally the particles of sizes between $0.2\,\mu m$ and $10\,\mu m$ are collected as these are respirable.

Since water cannot be analysed under vacuum, it must be pre-treated or pre-concentrated in a suitable substrate. Non-soluble matters can be extracted from water through filtering and drying. The water sample may be chemically treated to produce complexes of the elements to be studied. Measurement of species in aqueous solution is indeed quite difficult, but it provides ppb sensitivity. The external beam PIXE technique can be used to overcome the difficulty. In this the ion beam is directed to the target chamber through a thin Teflon window. The chamber is filled with helium at about atmospheric pressure so that the liquid can be placed in the chamber instead of a vacuum chamber.

Soil, plant leaves and biological samples (e.g. hair, nails, bones) can be prepared by freeze-drying, grinding and compressing into small pellets.

5.7   Reference

Fuller C W 1977 *Electrothermal atomization for atomic absorption*, The Chemical Society, London, pp 65–83

5.8   Bibliography

Section 5.1

Fowles G R 1989 *Introduction to modern optics*, 2nd edn, Dover Publications, New York
Hawke J 1995 *Lasers; theory and practice*, Prentice Hall, Englewood Cliffs, NJ
Hecht E 1998 *Optics*, 3rd edn, Addison-Wesley, Reading, MA

Section 5.2

Coles B R and Caplin A D 1976 *The electronic structure of solids*, Edward Arnold, London
Zewail A (ed) 1992 *The chemical bond: structure and dynamics*, Academic Press, New York

Section 5.3

Brown J M 1998 *Molecular spectroscopy*, Oxford University Press, Oxford
Colthup N C, Daly L H and Wiberley S E 1990 *Introduction to infra-red and Raman spectroscopy*, 3rd edn, Academic Press, New York
Ferraro J R and Marquette K N 1994 *Introductory Raman spectroscopy*, Academic Press, New York
Guilbault G G 1990 *Practical fluorescence*, 2nd edn, Marcel Dekker, New York

Section 5.4

Lajunen L H J 1992 *Spectrochemical analysis by atomic absorption spectroscopy*, Royal Society of Chemistry, London
Softley T P 1994 *Atomic spectra*, Oxford University Press, Oxford

Section 5.5

Gunther H 1995 *NMR spectroscopy; basic principles, concepts, and applications in chemistry*, Wiley, New York
Sanders J K M and Hunter B K 1993 *Modern NMR spectroscopy*, 2nd edn, Oxford University Press, Oxford

Section 5.6

Agarwal H, Tamir B K and Shawlow A L 1989 *Introduction to X-ray spectroscopy*, 2nd
    edn, Springer, Berlin
Johansson S A, Malmqvist K G and Campbell J L 1995 *Particle induced X-ray emission
    spectroscopy (Pixe)*, Wiley, New York
Lachance G R, Claisse F and Chessin H 1995 *Quantitative X-ray fluorescence analysis;
    theory and applications*, Wiley, New York

Chromatography

6.1 Introduction

Chromatography has been widely used for the separation, identification and quantification of chemical components in mixtures, some of which are too complex to be resolved by other classical chemical methods. The first experiments used finely divided calcium carbonate in glass columns to separate the green chlorophyll components present in plant extracts. Because the separated components appeared in coloured bands the term chromatography was applied.

Chromatography is based on the distribution of material between a stationary phase and a mobile phase. The materials are carried through the stationary phase by the mobile phase and separation is achieved by the different migration rates due to different distribution ratios between the two phases for the different molecules.

6.1.1 Classification of chromatographic methods

A number of different classification systems are commonly in use depending on the methods of presenting the two phases and their composition. The stationary phase can be supported in a narrow tube and the mobile phase forced through the column by pressure differences or gravity, column chromatography, or the stationary phase is supported on a flat plate or between the pores of a paper sheet, planar chromatography. In the latter case the mobile phase moves by capillary action or gravity or in some unusual systems by electrodynamic pumping.

An alternative chromatographic methods classification is based on the nature of the mobile phase. There are essentially two systems using gas and liquid mobile phases with a recent addition of supercritical fluids. The last of these materials are gases held above their critical temperature and subjected to very high pressures. The stationary phase for these systems can be a solid or an immobilised liquid held on an inert solid. Ideally the solid support would take no part in the separation process, serving only as a carrier for the stationary liquid. Sometimes the nature of the solid support does have an effect, particularly if the immobilised liquid is held by an absorption process involving indirect chemical bonding to the solid surface.

6.1.2 Elution chromatography

In these systems two or more components dissolved in a suitable solvent, sometimes one component of a mixed mobile phase, are added to the top of the column and washed through the column by addition of successive portions of the mobile phase solvent. The

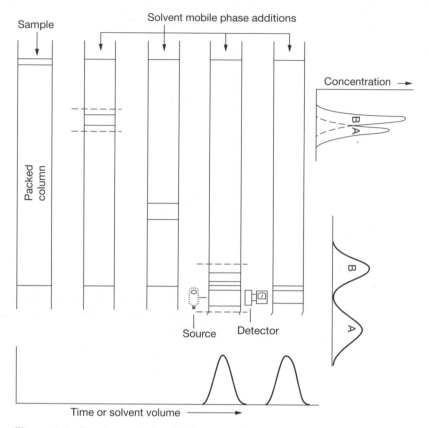

Figure 6.1 Development of a simple open column chromatogram showing the position of the two components, A and B, in a mixture as a function of solvent volume passed. The expanded sections shows the relative A and B concentrations at these positions. The lower trace is the detector signal as a function of time or volume of mobile phase passed

components are successively distributed between the two phases as they pass through the column. The material initially added to the top of the column distributes itself between the two phases and addition of further mobile phase forces the solution containing the dissolved material down the column where further partition occurs between the mobile phase and a fresh portion of the mobile phase. Meanwhile the original part of the column is the site for a fresh equilibrium to be established between the material held on the stationary phase and a fresh sample of the mobile phase. Further additions of the solvent carry the solute material down the column in a continuous series of transfers between the stationary and mobile phases. The solute is effectively picked up by the solvent from the sample region closest to the original addition point and is deposited on the sample region furthest away from the original addition. Because solute movement only occurs in the mobile phase, the rate of migration is proportional to the flow rate of the stationary phase and the time the solute spends in the mobile phase. Solutes strongly retained on the stationary phase spend a limited amount of time in the mobile phase and hence migrate very slowly through the column. Weakly held solutes spend considerable time in the mobile phase and hence

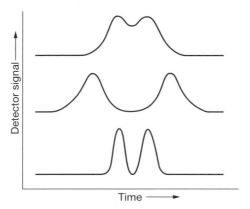

Figure 6.2 Problem of separation in a chromatogram. Adjacent peaks can be separated either by increasing the difference in retention time or by reducing the peak width

migrate at a rate which is comparable with the rate of mobile phase movement. Ideally this difference in migration rates causes the components to split into bands as the mobile phase passes down the column. Passage of a sufficient quantity of a suitable mobile phase through the column causes the bands to pass out of the end of the column, where they can be detected by a suitable device, sensitive to the solute at that concentration, placed at the end of the column. The detector signal, as a function of time (or volume of added solvent), shown in Figure 6.1, constitutes a chromatogram. Collection of the mobile phase in discrete portions whose volume is controlled by the detector output can be used as a preparative method for the preliminary screening of complex natural samples. Alternatively, when the materials are slow moving on the column the stationary phase is removed and the distinct bands separated. This can only be achieved when there is an easy method of distinguishing the presence of the components, e.g. by colour or radioactivity or fluorescence. Further discussion on this topic will be given elsewhere under sample preparation procedures.

The position of the peak in the chromatogram, called the retention time (or volume), can be used to identify the species, but there is no absolute method of predicting the characteristic peak position for any chemical species and hence some form of additional information or calibration is required. This may be provided either by running a known standard sample and comparing the retention times or by the use of a detector system with identification capabilities. Combination systems using chromatography and specific detector systems are discussed in another section. The peak area is proportional to the quantity of the chemical present and hence chromatography can be used quantitatively, but, again, not in an absolute manner, thus requiring careful calibration procedures. As the material moves through the column the distribution of the material between the mobile and stationary phases changes and separation between the components increases. Unfortunately the peaks broaden and hence the separation may not be perfect. Resolution can be maintained either by increasing the separation between broad peaks or reducing the peak width, as indicated in Figure 6.2. Provided that the breadth of the peaks can be controlled, then, by careful selection of experimental parameters, particularly the length of the column, a successful chromatographic analysis may be achieved. Before any analytical method can be accepted, extensive trials to determine the optimum operating conditions may be necessary, although chemical reasoning regarding the solutes and the two phases may reduce the extent of the trials.

Several chemical and physical variables influence the separation of the chromatographic bands, their breadths and the rate at which they move through the column. The effectiveness of a column for separating two solutes depends on the relative rates at which they are eluted; a property which depends on the partition ratios of the solutes between the stationary and mobile phases. For any species, A, an equilibrium is established between the two phases, which can be written as

$$A_{\text{mobile phase}} \rightleftharpoons A_{\text{stationary phase}}$$

and an equilibrium constant, K_A, called the partition coefficient, is given by

$$K_A = \frac{\text{concentration of A in stationary phase}}{\text{concentration of A in mobile phase}} = \frac{[A_s]}{[A_m]}$$

This is an ideal thermodynamic situation, which may not be achieved in practice, under normal operating conditions, since the mobile phase flow rate may be too high to provide sufficient time for the equilibrium to be fully established.

Examination of a chromatogram starting, at $t = 0$, with the injection of the sample onto the column, shows a first peak due to species not retained by the column. Their migration rates will be close to the average motion of the mobile phase and should appear at the column outlet shortly after passage of one column volume of mobile phase through the column. Subsequent peaks are due to solutes which are retained by the column, and the retention time, t_R, is the time taken for the solute to reach the detector. Retention time is a function of the distribution ratio and the relative co-volumes of the mobile, V_m, and stationary, V_s, phases. This is usually expressed as a capacity factor for the solute, which for the example solute A given above is

$$k'_A = K_A \frac{V_s}{V_m}$$

Suitable separations are achieved when the capacity factor lies in the range from 1 to 30. When two species A and B are moving through the column then the sensitivity factor α is given by

$$\alpha = \frac{K_B}{K_A}$$

Conventionally B is the more strongly retained (bound) species and hence α is always >1.

Chromatographic theories must account for the following phenomena:

1. different migration rates
2. Gaussian peak shape
3. peak broadening

The peak shape can be explained through a random walk mechanism for the migration of the molecules through the column. Each molecule undergoes a large number of transfers between stationary and mobile phases during the movement through the column. Transfer between phases requires energy and the molecules acquire this energy from the surroundings. Energy distribution follows Gaussian curves and thus residence time in each phase is very variable for any molecule. As a result there will be a Gaussian spread in residence times and hence the band profile will also be Gaussian. The breadth of the band increases as it moves through the column because more time is available for diffusion and spreading to occur by a random walk migration accompanying the Gaussian distribution. Zone breadth is directly related to column residence time and inversely related to mobile phase flow velocity.

6.1.3   Efficiency of a chromatographic column

Column efficiency is described by the following parameters derived from comparison with distillation theory:

1. number of theoretical plates N
2. plate height H (height of equivalent theoretical plate, HETP)

Both quantities are related by the expression

$$N = L/H$$

where L is the column length in the same units as H, usually centimetres.

Column efficiency increases as N increases and H decreases. This is not an exact model because, unlike distillation theory, true equilibrium conditions are never established during the column operation. N, the number of theoretical plates, is related to the square of the retention time t_R^2 and hence to the length of time the mobile phase is in contact with the stationary phase; a function of the mobile phase flow rate. Efficiency, determined by H, is a function of flow rate, passing through a minimum at $\approx 4\,\text{mm}$ for a gas–liquid system whereas the minimum value is $\approx 0.15\,\text{mm}$ for the more efficient liquid–liquid system. It is operationally difficult to use long columns for the latter systems.

6.1.4   Optimisation of column performance

The aim of the optimisation process is to completely separate all components in the mixture in the minimum time. Generally it is important to reduce zone broadening (peak width) to a minimum and/or to alter relative migration rates. Zone broadening is affected by the diameter of the column and the diameter of the particles used to fill the column. Very small particles tend to be favoured, particularly in gas–liquid systems, because of the higher surface area and the increased speed of establishing the distribution conditions. Similarly the stationary liquid phase should be as thin as possible in order to reduce solute transfer time. Improved efficiency is thus achieved by:

1. reduced packing particle size
2. reduced column diameter
3. reduced column temperature
4. reduced stationary phase film thickness
5. increased capacity factor (solvent composition)

Resolution increases with increase in capacity factor but retention time passes through a minimum at a capacity factor of ≈ 2. Other factors for improving selectivity and hence the capacity factor are:

(a)   stationary phase composition
(b)   special chemical factors (mainly involving limited chemical bonding)

The general problem for real samples is that the conditions optimised for the earlier-eluting materials produce long elution times, with the associated band identification difficulties, for the later-eluting materials. A partial solution to these problems is to use variations in eluting conditions during the experiment, i.e. either changing operating temperatures or varying the mobile phase composition, gradient elution, Figure 6.3, or both. Sometimes this idea is carried a stage further by using two successive chromatographic separations, with the unseparated components from the first stage being collected and subjected to a second process with radically different column operating conditions which will produce a separation.

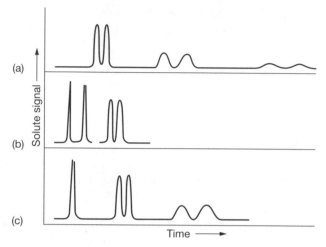

Figure 6.3 General elution problem in chromatography. Simple isocratic systems produce either poor resolution of the early eluting peaks or very long analysis times

6.1.5 Calibration

Neither qualitative nor quantitative chromatographic determination of a species is absolute and hence calibrations with known standards are required. For qualitative work then only the retention times for the postulated material are required and only a single injection of each standard is required. The retention time of the standard and the unknown should agree. For best practice the standard should be added to the sample matrix and a repeat measurement taken. No new peaks should appear and there should be a suitable increase in the peak area of the unknown. Quantitative measurements can only be achieved by careful calibration with a series of standard solutions of known concentration of the species in question. For each component in the mixture a series of standards are run and the peak area as a function of concentration is determined. Sensitivity for each species may be different but graphs of peak area against concentration should be a straight line passing through the origin. Sample volume injected is a major source of error and hence the use of a mechanically activated sample loop to inject a reproducible quantity of the sample at a specified rate is desirable for very accurate work. The peak areas for the sample to be measured must be within the range of the standards used in the calibration. However, this calibration method does not take into account the effects of the sample matrix on the separation and sensitivity of the determination. High-precision measurements require the use of an internal standard substance with both the standards and the samples. The analysis parameter is then the ratio of the analyte peak area to the standard peak area. The internal standard peak should be well separated from the other peaks in the chromatogram but as close to the analyte peak as possible. Under these circumstances high precision is easily obtained. After this brief review of the basic principles of chromatography, attention will now focus on the specific methods of chromatography, identified by the nature of the mobile and stationary phases.

6.1.6 Applications of chromatography in environmental analysis

Chromatography has applications in preparative and analytical chemistry. The separation and purification of specific reagents using large-scale liquid–liquid chromatography is an established technique for biochemical diagnostic packages but it is not the major use of chromatography. Two major interest are in the identification and quantification of the components in a sample presented for analysis. Chromatography is used to recognise the presence or absence of components in mixtures containing limited numbers of known possible species. Because chromatography provides only a single piece of information, the retention time t_R, about each species present in the mixture, only limited identification is possible in completely unknown mixtures. Comparison of t_R with those of known standards allows some identification. Ideally the standard should be added to the sample matrix and measurements repeated. No new peaks should appear and there should be a suitable increase in one of the original peaks. Additional information from a selective detector can help with this problem. Absence of a band can be definitive by comparison with known standards. The sensitivity of chromatographic methods makes them ideal for measuring the levels of toxic materials such as pesticides and combustion products down to less than the critical levels.

Quantitative information can be obtained from the chromatogram since the height and area of the observed peak are proportional to the concentration of the solute responsible for the peak. Absolute values are not possible because of the variation in detector sensitivity with solute. In order to make peak measurements reproducible, careful control of column temperature, eluate flow rate and rate of sample injection must be achieved, while column overload must be avoided. Peak height is affected by peak broadening but peak area is largely unaffected by peak broadening, if the background level for the broad peaks is carefully chosen. Very narrow early peaks can cause an area measurement problem, but, with modern electronic integration systems, these problems can be overcome. Gradient operating conditions can also reduce this problem.

6.2   Gas phase chromatography

In gas phase chromatography the material to be analysed is applied to a column, containing a non-volatile stationary phase coated on an inert support, either via a fixed loop, as described earlier, or from a gas-tight syringe. The analytes are converted into the vapour phase in the hot environment and swept through the column by a stream of inert carrier gas, i.e. helium or argon. Distribution occurs between the inert carrier gas mobile phase and the stationary phase. No interaction occurs between the analyte and the mobile gas phase. The elution rate is thus dependent on the volatility of the analyte, the interaction between analyte and stationary phase, but is independent of the mobile phase.

In **gas–solid** chromatography separation occurs because of differences in the absorption equilibria between the gas phase and the solid surface. It is usually applied to the more volatile species: CO, N_2, O_2, CH_4. Adsorbants are usually porous polymers and some of the more polar molecules may be permanently held on the solid or will be evolved over a period of time. Surface adsorption on solid supports, particularly acidic silica systems, may be a problem for analysis of polar molecules using liquid stationary phases.

Gas–liquid chromatography (GLC) involves the use of a non-volatile liquid immobilised on the surface of a solid support by either adsorption or chemical bonding. Elution rates from the column are influenced by the following parameters:

(a)   volatility and polarity of the solute
(b)   polarity of the stationary phase
(c)   temperature of the column
(d)   physical properties of the solid, i.e. particle size and film thickness
(e)   nature and flow rate of the carrier gas

Analytes with a low vapour pressure, at the column temperature used, will be retained at the point of addition or will elute very slowly. In the worst situation they will slowly bleed from the column over an extended period of time. Volatile compounds will pass more rapidly through the column. Polarities of the analyte and stationary phase affect the partition ratio between the gaseous and stationary phases. Like polarities move more slowly than unlike. Temperature controls the analyte vapour pressure and hence the rate of elution with small variations sometimes having a dramatic effect.

Relative volumes of the gaseous and stationary phases also affect the rate of elution. Reduced stationary phase volume increases the speed of elution and also the resolution. Stationary phase film thickness is a function of stationary phase volume and the surface area of the inert support, which depends on the supports particle size. Thin films, i.e. low stationary phase volume, control the extent of the equilibrium and the speed at which it is achieved, thus affecting the resolution. Column length and diameter also affect the elution rate.

The carrier gas transports the sample through the column. If the flow rate is increased, retention time is reduced, but so is resolution as indicated by the reduction in HETP as function of mobile phase flow rate. Generally at low flow rates, high molecular weight carrier gases give better separations (N_2 or argon). At high gas velocities, low molecular weight gases (H_2 or He) are better. Sometimes the detector used determines the optimum carrier gas to be used.

6.2.1 Column operation and specification

Packed columns use a 2–3 m length of glass or metal tube 2–4 mm in diameter, wound in a coil (15 cm diameter) and filled with a solid support coated with the stationary phase. Supports can be silicate with surface areas of $>1 \, m^2 \, g^{-1}$. These suffer from physical adsorption due to the presence of acidic groups on the silicon surface. Acidic sites can be reduced by chemical reaction with chlorinated silanes. The stationary liquid phase must have a low volatility, i.e. boil at 100 °C above the highest temperature used. Arrangements of typical stationary phases besides the simple packed column are:

wall-coated open tubular (WCOT)
porous layer open tubular (PLOT)
support-coated open tubular (SCOT)

Table 6.1 *Typical stationary phases for GLC*

Apiezon grease
Squalene
Silicone oil (methylphenylsilicone)
Siloxanes (methyl and dimethyl siloxanes)
Substituted methyl siloxanes with attached side groups
Carbowaxes (hydroxypolyethers)
High molecular weight hydroxycarboxylic acids (casterwax)
Substituted Teflon-based systems

The polarity of the systems increases from top to bottom of the table and subtle variations in polarity can be achieved by modifying the side groups on the siloxanes and silicones

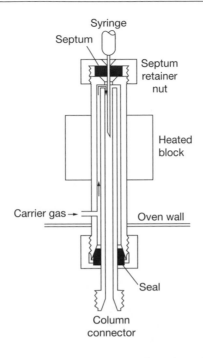

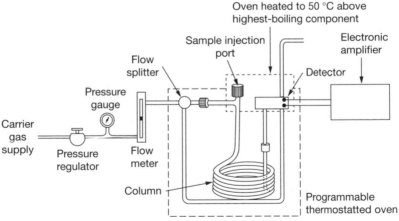

Figure 6.4 Schematic gas chromatograph using a packed column with an expanded section through the heated inlet port

Thermal stability, chemical inertness and solvent properties are selected such that the capacity and selectivity factors are within the suitable range. In order to maintain reasonable separations, polarities of the stationary phase and the solute should be similar. Under these circumstances retention time correlates with boiling point. Solutes with identical boiling points, but different polarities, require a stationary phase that selectively retains one species in order to achieve a separation. Typical stationary phases are indicated in Table 6.1. Figure 6.4

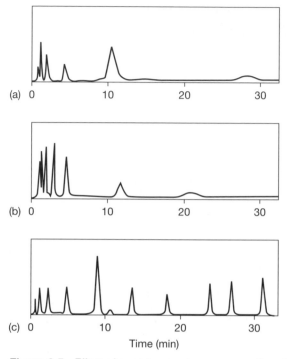

Figure 6.5   Effect of oven temperature on separations by gas–liquid chromatograms: (a) isothermal at 50 °C, (b) isothermal at 150 °C and (c) programmed gradient from 30 to 180 °C

shows a schematic diagram of a typical gas chromatography system fitted with a packed column and a syringe inlet port.

Capillary columns have a thin layer of the stationary phase coated either on the wall or on a support which has been coated onto the wall. Typical columns have 30 m of a fine fused silica capillary, 300 μm inner diameter, wound into a coil of 20 cm diameter. Column temperature is controlled in a thermostatted oven, which can follow an agreed temperature programme, in order to optimise separation over the whole range of retention time. Figure 6.5 shows the effect of varying oven temperature on the separations achieved.

6.3   Detectors for gas–liquid chromatography

Ideally the detector should respond rapidly to the minimum concentration of solute as it leaves the column. Ideally typical response times should be ~1 s and concentration in the 1 ppm range. Other requirements are:

1. linear response
2. stability
3. uniform response to a variety of chemicals or to a specific class

A number of the detectors available for GLC will be discussed with the attention given to operating principles rather than manufacturers' devices.

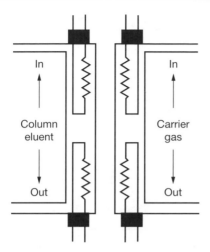

Figure 6.6 Thermal conductivity (katharometer) detector for gas chromatography

6.3.1   Thermal conductivity detectors (katharometers)

These respond to changes in thermal conductivity of the gas stream due to the presence of analyte molecules. An electrically heated filament achieves a steady temperature dependent on the thermal conductivity of the surrounding gas. H_2 and He possess very high thermal conductivities so that 10^{-8} g of organic material present in the column output will cause a change in gas conductivity, which alters the filament temperature and resistance. If the filament is one arm of a resistance bridge and a similar filament, inserted in the gas stream prior to the injection port, forms the other arm, as shown in Figure 6.6, then carrier gas effects can be reduced and sensitivity increased. Katharometers can also detect H_2O, N_2 and CO_2, which other detectors either cannot detect or have very low sensitivity for.

6.3.2   Flame ionisation detectors

Organic compounds, when injected into H_2 air flames, are pyrolysed to produce ionic intermediates and electrons which can conduct electricity through the plasma. Application of an electric field causes the charged species to be attracted to and captured by a collector electrode. Amplification of the ion current yields a measurable signal. Sensitivity is 10^{-12} g cm^{-3} for most organics but negligible for H_2O, CO_2, SO_2, NO and NO_2.

Phosphorus and nitrogen compounds can be detected with enhanced sensitivity and specificity using a flame ionisation detector (FID) in which the flame is surrounded by an alkali metal salt (KCl or CsBr) ring. Interaction between the ions in the plasma and the alkali metal halide bead produces a change in the distribution of the ions, which can be detected by using a multiple electrode system in which the electrodes are connected in different combinations for the two elements.

6.3.3   Electron capture detectors

Column eluent is passed over a radioactive source, usually ^{63}Ni, or other source of electron emission, which ionises the carrier gas according to the equation

$$N_2 + \beta^-(^{63}Ni) \rightleftharpoons N_2^+ + 2_e^-$$

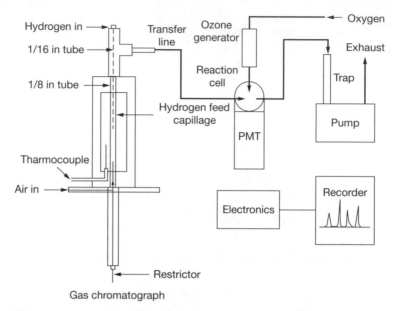

Gas chromatograph

Figure 6.7   Sulphur flame-less photometric detector. The SO$_2$ originally produced is converted to SO by reaction with hydrogen and then reacts with ozone with the emission of a light photon detected in the photomultiplier tube (PMT)

and produces a standing electrical current between two electrodes held at different potentials. The outer case is usually held at ground potential for safety reasons. Current decreases in the presence of compounds which capture electrons and interfere with the normal ionisation processes. Electron capture detectors (ECDs) are particularly sensitive for compounds with electronegative functional groups, such as halogens, nitro groups and carbonyls ($>$C$=$O), but response is non-linear. ECDs are poor detectors for amines, alcohols or hydrocarbons.

6.3.4   Flame photometric and thermionic detectors

Compounds containing S and P when combusted in a hydrogen-rich flame emit characteristic radiation of 394 and 526 nm respectively. Photomultipliers can detect this output, producing a highly specific and sensitive S and P detector. Alternatively, in the chemiluminescence detector the sulphur oxides produced by the thermolysis are passed into a reaction cell where ozone, generated immediately prior to use, reacts to produce SO$_3$, which emits characteristic radiation. A typical system is shown in Figure 6.7. Similar methods are used for some nitrogen-containing compounds, which on pyrolysis produce nitrous oxide which can react with ozone, emitting light.

6.3.5   Mass spectrometers

Column eluate can be identified specifically by use of a mass spectrometric detector. Usually the spectrometer is set to scan each peak as it is produced from the column over a pre-set mass range. This enables unknown compounds in mixtures to be identified after separation. Coupled data systems allow the mass spectra to be readily compared with standard spectra. Mass spectrometer (MS) discrimination between ions is based on their mass-to-charge ratio. Movement of the ions in electric and/or magnetic fields allows them to be characterised by time of flight, flight path or similar property. More details will be given in other sections. Other mass-selective detectors, e.g. the ion trap detector, can also be used and these will be briefly discussed in the next section.

6.3.6   Ion trap detectors

These analyse the eluate by mass-selective processes. Although not possessing the sensitivity or resolution of a conventional MS, the ion trap detector has sufficient performance to discriminate most of the materials eluting from a standard gas chromatograph with the added advantage of a higher operating pressure. The analyte from the gas chromatograph is passed via a transfer line fitted with an open-split interface and a flow restrictor into the analysing chamber of the ion trap. The analyte molecules are ionised by impact with an electron beam generated from a filament and focusing lens system. The ions are then trapped in a region of the analyser defined by the electric fields generated from three carefully shaped and positioned metal electrodes subjected to rf oscillating voltages. The three-dimensional hyperbolic field generated traps the ions in aperiodic oscillations. Careful selection of these rf voltages can lead to all ions of interest being trapped in the cavity. Starting at low storage voltages, as the voltage is increased to higher levels, the oscillations of the ions increases in amplitude and those with greater mass-to-charge ratio are ejected from the cavity, as a tight collimated beam, on the axis of the cavity towards the electron multiplier and the ion detector. Careful control of the rf voltages applied to the electrodes enables a number of repeat scans can be obtained for each eluting peak from the gas chromatograph, thus improving the spectral signal-to-noise ratio. The high ionisation potential of helium has two beneficial effects on this detector system. Careful control of the electron beam energy produces very few helium ions and hence few additional ion reactions. Also the collision of the ions with the neutral helium atoms reduces their kinetic energy, dampening their oscillation amplitude during storage and focusing the ions into the central region of the ion trap. Careful control of the rf storage voltages means that additional ion reaction products are ejected from the cavity before the analysis commences. The electron multiplier typically used is an open-ended system with a continuous dynode chain, which is more flexible in response to the varying demands of the different ions than the conventional fixed plate photomultipliers. Typical mass ranges detected are 20–700 Da with a resolution of 0.5 Da.

6.3.7   Application of gas–liquid chromatography to environmental analysis

GLC can be used as a method of separating complex mixtures and in analysis as both a qualitative and a quantitative method. In qualitative analysis GLC is used as a criterion for purity of organic compounds, e.g. drugs and pesticides. GLC is good for confirming the presence, or absence, of a compound in a mixture. A typical chromatogram produced for a mixture of pesticides is shown in Figure 6.8. Notice that not all pesticides can be analysed by gas chromatography. Some of the more thermally sensitive, or less volatile, compounds must be analysed by other methods such as high-performance liquid chromatography (HPLC).

Identification of a peak can be completed if a suitable standard can be found. The chromatogram of the sample and added standard must contain no new peaks and enhancement of an existing peak must be observed.

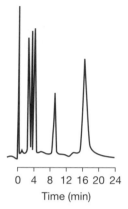

0 4 8 12 16 20 24
Time (min)

Figure 6.8 Separation of some volatile thermally stable pesticides by gas chromatography

6.3.7.1 *Quantitative analysis*

Since the system sensitivity varies depending on the compound analysed, quantitative analysis requires accurate system calibration for each compound in the matrix used for the measurement. Simple calibration can be in terms of peak height but peak area is preferable. The area of the peak is determined for a series of known concentration standards of the compound tested and a calibration curve determined. Ideally this should be in the matrix used for the determination. The highest precision uses an additional internal standard added to both the calibration and the determination. This additional standard should be sufficiently close to the determinand in elution time to be subject to almost identical conditions but must not interfere with the peak integration for the determination. Standard preparation is particularly important for high-volatility samples.

6.4 Liquid phase chromatography

Liquids can be used as the mobile phase in column and planar chromatography systems. Column systems can be divided into four classifications depending on the nature of the stationary phase:

1. partition liquid–liquid chromatography
2. adsorption liquid–solid chromatography
3. ion exchange
4. size exclusion or gel permeation

Requirements for liquid systems are similar to those for gas systems, except that higher pressures, $>1000 \, \text{lbf in}^{-2}$ or $7000 \, \text{kPa}$, are required to force the liquid through the column packing at typical flow rates of $0.5 \, \text{cm}^3 \, \text{min}^{-1}$. Packing material particle size and eluate flow rate have important effects on separation. For $1 \, \mu\text{m}$ packing particles, pulse free pressures of $6000 \, \text{lbf in}^{-2}$, i.e. $41 \, \text{MPa}$, are required.

Solvents must contain no dissolved gases, to avoid gas bubble formation. Gas bubble volume changes with decreasing pressure as the liquid passes through the column, altering the effective flow rate and contact between mobile and stationary phases. These processes distort the elution pattern. Either a constant composition, isocratic, solvent mixture or a programme-controlled variable composition mixture (gradient) solvent is used for the elution. The latter is used when changes in polarity are required to improve the separation and provide adequate resolution for separation of the rapidly eluting compounds and reducing the retention time of the later-eluting species. Samples are usually admitted to the column by a closed loop injection system, since this both avoids the difficulties with a syringe injection at the solvent pressures used and also eliminates variable volume sample aliquots.

Since the columns must withstand high pressures they are made of stainless steel tubing of inner diameter 4–10 mm and 30 cm long filled with 5–10 μm diameter particles, usually silica based, coated with a thin organic film.

6.4.1   Partition chromatography

The stationary liquid is retained on the column either by physical adsorption or by formation of covalent bonds with the support. Bonded phases tend to involve reactions between silica particles and organochlorosilanes as shown in the scheme

$$\left.\begin{array}{r}\text{silica}\\[20pt]\text{surface}\end{array}\right\}\quad -Si-OH + Cl-\underset{\underset{CH_3}{|}}{\overset{\overset{CH_3}{|}}{Si}}-R \quad \rightarrow \quad -Si-O-\underset{\underset{CH_3}{|}}{\overset{\overset{CH_3}{|}}{Si}}-R$$

By varying the nature of R from simple hydrocarbons to amines, esters, nitrates etc. a range of bonded phase polarities is available. Bonded phase systems tend to be more stable than simple physically adsorbed phases, which require re-coating on a regular basis. Gradient elution is possible with bonded phase columns, but it slowly elutes the stationary phase from surface adsorbed columns.

So far discussion has concentrated on polar bonded stationary phases with non-polar solvents such as hexane, historically identified as the normal phase. In reversed phase, the stationary phase is non-polar and the mobile phase is polar. This change in polarity has a dramatic effect on the order in which materials elute from the column. In normal phase systems the least polar material is eluted first and increasing mobile phase polarity decreases elution time. In reversed phase systems, the most polar compound elutes first and increasing the mobile phase polarity increases the elution time.

Successful partition chromatography requires a proper balance between intermolecular forces among the three participants: solute, mobile phase and stationary phase. Optimum separation occurs when the polarities of analyte and stationary phase are matched but differ from that of the mobile phase. Poor separation occurs when mobile phase and analyte are matched but differ from the stationary phase. Under these circumstances the stationary phase cannot compete with the mobile phase in the equilibrium distribution and hence retention time t_R is too short for effective separations.

6.4.1.1   *General polarity series*

Molecular polarities generally follow the series

   alkanes < alkenes < aromatics < halides < sulphides < ethers < nitrocompounds
   < esters = aldehydes = ketones < alcohols = alkynes < sulphoxides < amides
   < carboxylic acids < water

6.4.2   Absorption chromatography

Liquid–solid chromatography involves the surface of a finely divided polar solid, usually alumina or silica, which shows the greatest flexibility and capacity, as the stationary phase. The analyte competes with the mobile phase for sites on the surface of the packing. The only variables are rate of flow and polarity of the mobile phase. Adsorption chromatography is important for the determination of high molecular weight, water-insoluble, organics.

6.4.3   Ion exchange chromatography

The stationary phase is an ion exchange resin formed from high molecular weight co-polymers of styrene and vinylbenzene with functional groups attached to the aromatic ring. Functional groups are either acidic, typified by $-SO_3H$, sulphonic acid, or the weaker carboxylic acid, $-CO_2H$, or basic groups based on the quaternary ammonium group $-N(CH_3)_3^+OH^-$. These polymer beads, typically $5\,\mu m$ in size, are insoluble in aqueous media. Ions are introduced at the top of the column either through a pneumatic valve or by electromigration from solution, which offers some sample clean-up, and absorbed there by exchange from solution onto the column. Analytes are then eluted, in order of decreasing affinity for the resin, by passage of a mobile phase consisting of an excess of like-charged ions.

Acidic groups are capable of exchanging H^+ for a metal cation M^{x+} as indicated below:

$$xRSO_3H + M^{x+} \rightarrow (RSO_3^-)_x M^{x+} + xH^+$$

Basic groups can exchange OH^- for other anions in the sample according to the equation

$$xRN(CH_3)_3^+OH^- + A^{x-} \rightarrow [RN(CH_3)_3^+]_x A^{x-} + xOH^-$$

Consider a typical cation, e.g. calcium. When the solution is added to the column an equilibrium is established

$$Ca^{2+} + 2RSO_3H \rightleftharpoons Ca^{2+}(RSO_3^-)_2 + 2H^+$$

for which an equilibrium constant can be written

$$K = \frac{[Ca^{2+}(res)][H^+(aq)]^2}{[Ca^{2+}(aq)][H^+(res)]^2}$$

where the bracketed value, $[X]$, denotes concentration of X in moles per litre. Under normal circumstances $[H^+] \gg [Ca^{2+}]$ in both resin and aqueous phases and hence the constant K' can be written as

$$K' = \frac{[Ca^{2+}(res)]}{[Ca^{2+}(aq)]}$$

K' is the affinity of the resin for the ion. K' depends on the radius and charge of the ion, its bonding energy with the resin and also on the size and bond strength of the hydrated ion. Polyvalent ions were more strongly absorbed than monovalent ions. Similar equilibria can be written for anion behaviour:

$$xRN(CH_3)_3^+OH^- + A^{x-} \rightleftharpoons [RN(CH_3)_3^+]_x A^{x-} + xOH^-$$

Elution is by addition of a mobile phase which competes with the analyte ions for resin sites and causes the initial reaction to be reversed, i.e. for anion A^{x-} addition of base yields

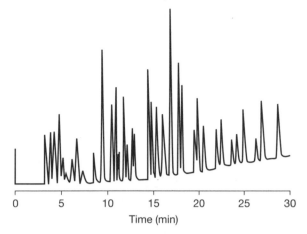

Figure 6.9 Separation of 36 anions by ion exchange chromatography using gradient elution starting with a borax buffer and moving to a carbonate–bicarbonate buffer, enabling the first 11 to be resolved and the later peaks to appear in <2 h. Note the sloping baseline with gradient eluant. Reprinted courtesy of Dionex Inc., USA

$$[RN(CH_3)_3^+]_x A^{x-} + xOH^- \rightarrow xRN(CH_3)_3^+OH^- + A^{x-}$$

Cations are recovered by the use of a dilute acid mobile phase.

Three problem areas of ion chromatography are associated with cations, carbonate and carboxylic acids. Cations are not held on anion exchange resins and co-elute in the void volume and hence they interfere with the rapidly eluting anions, namely fluoride. Carbonate, being a weakly acidic anion, which is invariably present in samples due to atmospheric absorption, elutes at pH < 8.5 between fluoride and chloride and excess of carbonate can mask chloride levels. Some systems use carbonate as the eluant and hence avoid this difficulty but cannot detect CO_3^{2-}. Short-chain (C1–C5) carboxylic acids elute between the fluoride and chloride ions, interfering with determinations of both these species. Choice of an isocratic process using a weak eluant in low millimolar concentrations can resolve the short retention time peaks but at a cost of long retention times for divalent species such as phosphate and sulphate. In the gradient approach the eluant strength is increased during the separation. If steps are taken to balance the solution conductance throughout the separation, then baseline drift with change in eluant composition can be eliminated and the analysis time reduced (Figure 6.9). Typically, when testing a water sample for anions, a gradient eluant starting with a borate-based buffer and moving to a carbonate-based eluant can resolve the early-eluting organic acids from fluorides and chlorides. Hydroxides do not interfere since they are neutralised by the acid functions of the buffer.

An alternative solution to the co-elution problem is the use of coupled ion chromatographic columns. The early eluants in the void volume from the first column are transferred automatically to a second column which produces a parallel chromatogram with a separation of the co-eluants from one single injection. Ion exclusion columns can be used for the weakly acidic anions $pK_a > 2$ and anion exchange columns for the strongly acidic anions $pK_a < 2$. The eluate for the ion exclusion column is usually a sulphonic acid and for the anion column the sulphonic acid sodium salt.

161

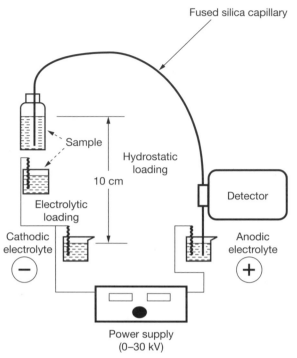

Figure 6.10   Block schematic of an ICE apparatus showing the two arrangements for the introduction of the sample onto the column

6.4.4   Ion capillary electrophoresis

The problems of co-elution in ion chromatography can be overcome by using a different discriminatory property of the analytes, namely their mobility, as ions, under an applied electric field. If the ions are contained in an electrolyte solution in a capillary tube and subjected to an applied electrical potential, sample ions will migrate along the capillary according to their electrical mobility and are detected usually by photometric methods. A typical system is shown schematically in Figure 6.10. The power supply, capable of providing up to 30 kV, applies an electrical potential to the capillary, which forces the ions to migrate to their corresponding electrodes. The higher the field, i.e. voltage, the faster the ions migrate. The carrier electrolyte provides the ions to support the separating current through the capillary system and thus is effectively the eluant of the capillary electrophoresis (CE) system. The carrier electrolyte must be selected on the basis of significantly different physical parameters from the analytes to allow the detection of the analytes by difference. The capillary is normally fused silica with an inner diameter of 50–75 μm and an outer diameter of 375 μm. Although the outside wall of the capillary is polymer coated to provide strength, flexibility and opacity, in simple systems there is no inner surface coating, stationary phase or packing. Since the fused silica is transparent in the UV spectral region, an uncoated part of the capillary can be placed directly in the detector, thus avoiding some of the resolution degradation associated with a separate measuring cell. Conductivity detection is under development but the conductance of the carrier electrolyte must be carefully matched to give reliable results.

A normal injection valve cannot be used for sample introduction into the capillary tube because of electrical insulation problems. Two alternatives are currently used as indicated in Figure 6.10. In direct hydrostatic or gravity injection the end of the capillary tube is immersed in the sample and the sample tube raised above the running electrolyte level by a predetermined height, typically 10 cm, for a predetermined time. The height step is normally fixed and the time can be varied. The volume of the sample to be injected is usually very small, nanolitres rather than the µl of conventional chromatography. Electromigration is also used to inject the sample into the capillary. The capillary end and an electrode are placed into the sample solution and a predetermined voltage applied for a predetermined time, forcing the ions to migrate into the capillary. Since this mode depends on the mobility of the ions and is biased in favour of the more mobile ions, while rejecting neutral species, it effectively provides some sample pre-treatment and clean-up.

Selectivities of CE are different from those of ion chromatography. Cations which interfere with ion chromatography anion determinations are moved in the opposite direction and hence do not interfere and vice versa for anions. Interferences can be experienced at low voltages with ions which form significant numbers of ion pairs. This effect appears mainly as a modification of retention times and is a function of concentration. This may be a significant problem when searching for low concentrations of analyte in the presence of high concentrations of ion pair forming species. Ions migrate according to their mobility, which is related to their equivalent conductivity. Those with a relatively high charge-to-mass ratio will migrate more rapidly. Notice that solvation of the migrating species can reduce the ion mobility. Strongly acidic inorganic ions will migrate rapidly while weakly acidic anions will migrate more slowly.

CE exhibits high efficiency in terms of specificity and also speed. A comparison between the relative performance of CE and ion chromatography is shown in Figure 6.11, for the separation of a particularly difficult mixture of 15 mono- and polyvalent anions. Some of the components, which co-elute over a period of 30 min in ion exchange chromatography (IEC) are clearly separated in <3 min by CE. A number of alkyl sulphonate homologues are easily separated by CE; see Figure 6.12. The components differ only in the length of the hydrocarbon chain and could not be separated by any of the other liquid techniques. The late-eluting peaks show some asymmetry which is due not to column effects but to differences in the mobility of the carrier electrolyte ions compared with the analyte ions. If the carrier is more mobile than the analyte the peak exhibits tailing but if the carrier is less mobile than the analyte then the leading edge of the peak is drawn out.

Typical detection limits of CE for eight anions, based on a 30 s hydrostatic injection, are compared with the ion chromatography detection limits (3 times background noise signal) in Table 6.2. Typical accuracy is 2.5% at the 1 ppm range. Note that CE was carried out at 20 kV following 30 s hydrostatic injection time using the absorption at 254 nm for detection. A partial list of anions for which standard CE methods are available is given in Table 6.3 to indicate the range of anions the technique can accommodate.

Electrically neutral species cannot be separated by normal ICE process, but a variation of ICE using a surface active agent electrolyte can be used to separate some neutral species. The supporting electrolyte forms an electrically charged micelle around the neutral species and the micelle moves under the influence of the applied electric field. The neutral species solubility in the micelle is an exchange equilibrium process between the micelle and the supporting liquid, and hence there is also an element of chromatography in the process. The neutral species outside the micelle do not migrate and hence behave as if they were trapped on a stationary phase and exchanging with the mobile phase of the micelle. Careful selection of the micelle-forming species is required to prevent interference with the

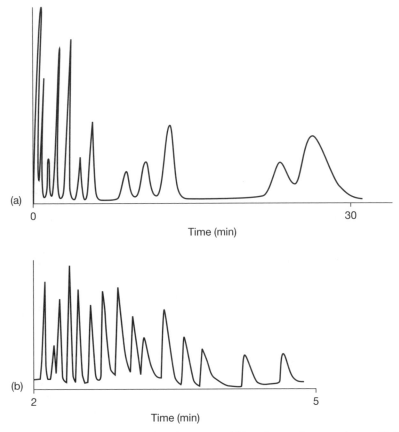

Figure 6.11 Comparison of performance of (a) an ion exchange column with (b) an ICE column for the analysis of a mixture of 15 anions. Note the difference in time scales as well as the number of peaks resolved. Reprinted courtesy of Dionex Inc., USA

detection method which normally must be by some means other than simple conductivity methods. Sometimes electrochemical methods, such as voltammetry, can be used provided that the analyte is easily oxidised or reduced. Retention times for electrokinetic chromatography are a function of the electrical mobility of the micelles, which is a function of their effective size. The nature and average number of molecules in the micelle, the residual charge on the micelle and the distribution coefficient of the analyte between the solution and the micelle all affect the mobility of the analyte. This distribution coefficient depends on the nature of the hydrophobic hydrocarbon chain component in the surface active agent and its interaction with the analyte molecules. It is almost impossible to predict retention times for analytes and it is most important to maintain constant reproducible conditions in the capillary. The presence of ions in the solution, particularly highly charged and polarisable species, can alter the stability of the micelles and hence interfere with the separation process. For reliable results the use of internal standards added to samples is most important.

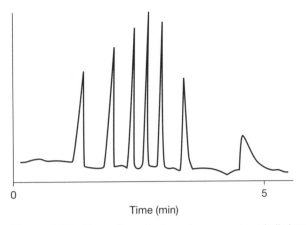

Time (min)

Figure 6.12 Separation of a homologous series of alkyl sulphonates by ICE. The difference between them is in the length of the hydrocarbon chain in the molecule. This mixture would not be resolvable by ion exchange methods or conventional HPLC

Table 6.2 *Detection limits for an anion mixture using CE and ion chromatography*

Anion	Detection limit, CE (ppm)	Detection limit, ion chromatography (ppb)
Bromide	0.38	2
Chloride	0.15	3
Sulphate	0.17	6
Nitrite	0.33	2
Nitrate	0.35	5
Fluoride	0.10	1
Phosphate	0.38	6
Carbonate	0.36	–

6.4.5 Solid phase extraction

Solid phase extraction provides an alternative method to simple solvent extraction for sample clean-up and concentration, sometimes eliminating solvent completely. Typical ingredients are the materials which form the basis of HPLC. Extraction columns are packed with a variety of bonded silica absorbent particles, which will selectively retain specific types of chemical compounds from a surrounding matrix. The material trapped on the column can then be eluted with a small quantity of a solvent. As an example 200 cm³ of an aqueous sample can be passed through a 3 cm long, octyl surface coated, silica column and the analytes released into 2 cm³ of methanol with >95% efficiency. Alternatively the silica particles can be embedded in a polytetrafluoroethylene micronet forming a membrane which can extract analytes from large dirty samples and release them into small

Table 6.3 *Ions determined by CE*

Inorganic

Arsenate	Azide	Borate	Bromide
Carbonate	Chlorate	Chloride	Chlorite
Chromate	Fluoride	Fluorophosphate	Iodide
Molybdate	Nitrate	Nitrite	Perchlorate
Phosphate	Phosphite	Selenate	Selenite
Sulphate	Tetrafluoroborate	Thiocyanate	Thiosulphate
Tungstate	Vanadate (ortho and meta)		

Organic

Acetate	α-Hydroxybutyrate	α-Ketoglutarate
Benzoate	Citrate	Formate
Glycolate	Maleate	Malonate
Mellitate	Oxalate	Phthalate
Propionate	Salicylate	Succinate
Sulphonates (butyl, ethyl, hexyl, methyl, pentyl, propanyl)		
Tartrate	Trimesate	

volumes of eluant, with the added advantage that the sample is physically filtered at the same time.

A recent development, which is particularly useful for chromatographic applications, is the solid phase micro-extraction (SPME) system. A short length, 2 cm, of a thin, 15–150 μm, silica fibre, coated with a suitable absorbent, typically polydimethylsiloxane, is mounted on a thin steel rod and contained within a normal chromatographic syringe needle; see Figure 6.13. The end of the fibre support rod is fastened to the plunger normally used to fill the syringe with sample. When the plunger is raised the fibre is withdrawn inside the hypodermic needle. When the needle is inserted through the septum of a sealed solution sample (Figure 6.14) the fibre can be lowered into the sample by depressing the plunger. After a suitable elapsed time, typically 2–15 min depending on the sample and the analyte, the fibre is withdrawn into the needle, which is then removed from the sample and inserted through the septum of a standard gas chromatography injector, where the analytes are rapidly thermally desorbed. Because of the small volume of the fibre and good thermal conductivity of the metals, the thermal lag of the system is small and chromatograph peak broadening is minimal. The fibre must be desorbed at temperatures above that of the highest-boiling analyte. If the analytes have a range of boiling temperatures, then it may be necessary to desorb the fibre into a cryofocusing device, held at 70 °C below the boiling temperature of the lowest-boiling analyte, prior to operation of the injector. Under these conditions analytes in the 100 ppt concentration range can be detected with an accuracy of better than 10%. The SPME system can also be used for head space analysis above a dirty sample, e.g. river waters or a sewage sludge sample, when it will trap the volatile species. Unfortunately if the fibre is immersed in dirty samples material may cling to the fibre and cause problems in the chromatographic injector. The normal solid phase extraction disc should be used on these samples prior to SPME.

Two experimental procedures, using SPME, which have been approved by the US EPA as standard methods for the analysis of water, are method 524.2 for the determination of

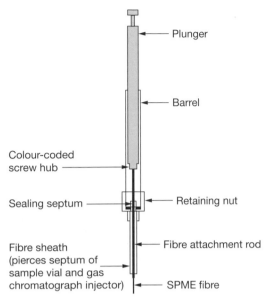

Figure 6.13 Cross-section through an SPME syringe (Reprinted courtesy of Varian, Inc.)

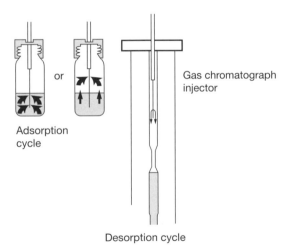

Figure 6.14 Adsorption and desorption cycle for SPME syringe system (Reprinted courtesy of Varian, Inc.)

volatile compounds and method 625 for the determination of semi-volatile compounds. The volatile compounds are mainly halogenated species and substituted single aromatic molecules. Detection limits are at the ppt level when the fibre is desorbed directly into the carrier gas stream of a gas chromatograph–MS (GC–MS) analysis system described in Chapter 7. The semi-volatile compounds are mainly polyaromatic hydrocarbons and some phthalate esters which are used as plasticisers in plastic materials. Again detection levels are in the ppb range when GC–MS is the analysis sytem. Some of the esters which are poorly absorbed on the polydimethylsiloxane fibres can be analysed using carbon fibres with a range of pore sizes of 5–50 nm. These carbon fibres are also very good at absorbing small molecules but they require higher than expected temperatures to desorb the adsorbates completely. The SPME fibres can be repeatedly used for a number of determinations provided that they are thoroughly cleaned between determinations. While the currently approved SPME fibre extraction methods are based on desorption into the GC–MS inlet, a number are currently being evaluated using desorption into the liquid eluant system used for HPLC–MS analysers. This technique uses specially developed injection ports which can accommodate the SPME fibre and produce the liquid-tight seal at column operating pressures. The technique will be particularly useful for heat-sensitive compounds such as explosives which are not easily measured by GC–MS.

Approval is also being sought for numerous other determinations which usually involve liquid–liquid extraction as part of the initial separation process. These include methods for pesticides, both organophosphate and organochlorine, surfactants and phenols. The SPME methods reduce the amount of environmentally unfriendly solvents required.

6.4.6 Size exclusion chromatography

This technique depends on differentiation of molecules on the basis of molecular size and has found application to high molecular weight mixtures. The column packing consists of small, 10 μm, silica or polymer particles containing a network of uniform dimension pores into which the solute and solvent molecules can diffuse. While within the pores, the analyte molecules are effectively trapped and removed from the flow of the mobile phase. The average residence time for the analyte molecules thus depends on their effective size. Molecules which are significantly larger than the average pore size are excluded and not retained; they pass through the column at the same rate as the mobile phase. Molecules smaller than the average pore size can penetrate into the pores and are trapped for the greatest time. Intermediate-size molecules exhibit an average penetration dependent on their size. Fractionation is thus directly related to molecular size and shape. Chemical or physical interaction between the stationary particles and the analyte is undesirable and impairs efficiency. When the packing is hydrophilic for use with aqueous phase samples the system is gel filtration. If the packing is hydrophobic for use with organic-based samples then it is referred to as gel permeation. Size exclusion chromatography has applications in polymer weight and molecular size distribution studies.

6.4.7 Super-critical fluid chromatography

At the critical temperature a substance cannot be condensed into the liquid state by the application of pressure. Super-critical fluids (SCFs) have different properties from either liquids or gases and are capable of dissolving large non-volatile molecules. SCF carbon dioxide can readily dissolve polycyclic hydrocarbons (aromatic and aliphatic), alkyl phthallates and other difficult-to-dissolve species. Materials such as carbon dioxide, ethane and nitrous oxide have critical temperatures around ambient and are easiest to use. Since the effective density of the SCF is very sensitive to pressure variations, it is important to measure and control the operating pressure very accurately. Retention times and capacity factors are very sensitive to density changes. Retention time for hexadecane in CO_2 SCF decreases from 25 to 5 min when the pressure rises from 70 to 90 atm. Gradient-type elution can be achieved by simple pressure regulation.

Typical columns are fused silica open tubular columns coated with a 0.4 μm, chemically bonded, polysilane film. Operating conditions for CO_2 SCF are 75 atm pressure and 31 °C. For best results traces (<1%) of a polar organic molecule, e.g. methanol, are added. Freons, chlorofluorohydrocarbons, are also used. SCF chromatography is more efficient than conventional solvent liquid chromatography by a factor of 3 at the optimum flow rate for HPLC. Since the columns operate around ambient temperatures separation of some thermally sensitive compounds is possible.

Detection of analyte is usually by FID since the SCF can be allowed to expand naturally into the hydrogen flame where the analyte anions cause conductivity changes. Separation methods using SCF extractions are also being extended to environmental analysis since again they offer non-polluting clean-up processes.

6.4.8   Planar chromatography

Included in this category are thin layer, paper and electrochromatographies. The stationary phase is a thin layer of material, which is either self-supporting or deposited on an inert glass or plastic plate. The stationary phase is trapped moisture or an active surface of the layer material. The mobile phase passes through the stationary phase by capillary action with or without the assistance of gravity or an electrical potential. Planar chromatography is essentially two dimensional since the third dimension, film thickness, is negligible compared with the other two. Thin layers are typically silica or kieselguhr. Paper can be used for basic systems, but quality control of materials is very difficult and hence these systems are used for rapid preliminary screening before adopting one of the more rigorous methods.

Planar chromatography can be a two-stage elution using solvent movement in two different directions perpendicular to each other. The plate is dried between the two solvents, but by judicious change of solvent difficult mixtures can be resolved. Separation of a mixed amino acid sample, using a two-stage development on a silica plate, is widely used in biochemistry. Some of the materials not separated by the first mobile phase are resolved when the plate is turned through 90° and processed with the second solvent. Matching of the solvent polarity with that of the solute is again critical. Common solvents can be arranged in a polarity series as follows:

> alkanes < alkenes < aromatics < halides < sulphides < ethers < nitrocompounds
> < esters = aldehydes = ketones < alcohols = alkynes < sulphoxides < amides
> < carboxylic acids < water

6.5   Detectors for liquid chromatography

Detectors in the systems described so far are for normal UV–visible absorption or UV fluorescence which imposes severe limitation. No single highly sensitive universal detector system is available for liquid chromatography. Because the greater part of the eluate is the liquid phase then simple flame systems are not easily available. In SCF chromatography, however, by allowing the carrier to expand through a pressure restrictor, the column output can be fed into a conventional FID. Detectors available are indicated in Table 6.4 with their typical limit of detection. A brief discussion of their operation is given later.

6.5.1   UV–visible absorbance

Many of the materials important to environmental chemistry exhibit absorption bands in the UV–visible region of the spectrum due to electronic excitation processes. Some organic

Table 6.4 *Performance of detectors for liquid chromatography*

Liquid chromatography detector	Limit of detection, normal operation	Limit of detection, special conditions
UV–visible absorbance	100 pg–1 ng	1 pg
Fluorescence	1–10 pg	10 fg
Electrochemical	10 pg–1 ng	100 fg
Refractive index	200 ng–1 µg	10 ng
Conductivity	500 pg–1 ng	500 fg
MS	100 pg–1 ng	1 pg
Fourier transform infrared	1 µg	100 ng
Light scattering	10 µg	500 ng
Optical activity		1 ng
Photo-ionisation		0.1–1 ng

These figures are based on a limit of detection of 3 × noise level and for 10 µl sample volume

Table 6.5 *Absorption edges for common solvents*

Solvent	Lower wavelength limit (nm)	Solvent	Lower wavelength limit (nm)
Water	180	Ethanol	220
CCl$_4$	260	Hexane	200
Cyclohexane	200	Acetone	330
Diethyl ether	210	Dioxane	320
2-Ethoxyethanol	320		

species are particularly responsive depending on the nature of some of the groups within the molecules. Most of the major absorbances occur in the 170–320 nm region. A number of inorganic materials also absorb in this region. The problem arises with the natural absorptions of the common solvent materials. The lower-wavelength limits for common solvents are given in Table 6.5. The UV–visible absorbance can only be used when the solvent does not mask the analyte absorbance since even blank corrections can lead to erroneous results. Optical absorbance can be measured with either spectrophotometers or simple photometers. Details of the instruments' operations have been given in Chapter 5. The output from the chromatographic column is introduced into a cell with silica windows and the longest optical path length possible in order to maximise sensitivity. Addition of a chemical reagent post-column can be used to develop a more intense absorbance and increase sensitivity. Since reagent addition is made after the eluate has left the column low-pressure pumps can be used. Post-column reagents are often used for inorganic species, which do not exhibit appreciable absorbance in the UV–visible region, particularly when highly absorbing charge transfer complexes are readily formed. Chromate and other species of chromium in drinking water can be determined by post-column reaction and spectrophotometry. Normally the system only measures at discrete wavelength using two or three detectors. If the complete absorbance spectrum is required for additional identification purposes then the flow though the detector cell has to be stopped for the time required by the spectrometer to scan the spectral region of interest. This can cause problems for the high-pressure pumps used in HPLC systems. An alternative spectral method is to make use of diode array detectors. A number of detectors are set to detect specific wavelengths, from a dispersive system, in the region of interest and hence can effectively continuously record a complete spectrum, albeit with poor resolution, without requiring an interruption in the column flow. This is not a serious disadvantage since most UV–visible spectra are normally poorly resolved. Data can be handled post-collection in order to optimise the sensitivity for all the analytes. Development of Fourier transform spectrometers has improved the speed of spectral analysis.

6.5.2 Fluorescence

Exposure of molecules to suitable wavelength radiation can cause excitation of electrons to higher energy levels. The excited molecules can dissipate the excess energy either by simple collision and vibrational transfer of energy to the solvent molecules or by emission of radiation whose energy corresponds to the difference in energy of the excited and ground state molecule, **fluorescence**. The processes are described in more detail in other chapters. Because the fluorescent radiation is propagated in all directions, the detector can be protected from the incident beam and stray scattering from the sample by observation normal to the incident beam. This is important because the quantum efficiency for fluorescence emission may be low. Many organic molecules, particularly those with rigid structural units, i.e. aromatic rings and cages, show good fluorescent behaviour, which can lower the detection limit below that of conventional absorption spectrometry, 1–3 orders of magnitude. Inorganic species, which are not efficient fluorescers, can react with reagents post-column to improve their sensitivity. Again, species which contain an aromatic ring, e.g. 8-hydroxyquinoline, are particularly useful since they readily form strongly fluorescent complexes with main group metal cations which normally exhibit particularly poor UV–visible absorption spectra. Similar reagents are available for difficult anions. Over 100 organic and biochemical molecules, including a number of drug metabolites, can be effectively determined by liquid chromatography with fluorescent detection. Since fluorescent intensity is proportional to concentration quantitative measurements are readily performed. On some instruments, fluorescence and UV–visible absorption are recorded together, since only an additional detector mounted normal to the absorption optics and examining a region of radiation output not normally used is required. Details of instruments and operation principles are given in Chapter 5.

6.5.3 Electrochemical

Three types of electrochemical detectors are possible in liquid chromatography. Measurement of the electrode potential is only of limited value since electrodes sensitive to each analyte are necessary. A number of methods are available, based on the hydrogen electrode, for detecting materials which alter the pH of the eluate, but the sensitivity and specificity of these methods are very poor. Conductivity and amperometry are more important as detectors. Both of these depend on the electrical current between two electrodes immersed in the solution. In one case the electrodes are maintained at fixed potentials or pulsed between carefully defined limits, conductimetry and amperometry; in the other, the potential is varied in a cyclical manner between positive and negative potentials, cyclic voltammetry. In voltammetry the cell current is measured as a function of applied potential using a reference electrode to control the applied voltage.

In ion chromatography systems, although UV–visible absorption or UV fluorescence can be used to detect some species, the method imposes severe limitations and electrical conductivity detectors are more widely used. Electrical conductivity is universal for charged species. While neutral species cannot interfere directly with the detection process, they may have indirect effects either by dilution of the analyte ions or by formation of donor complexes with the charged species, thus reducing their conductivity. Direct measurement of the small changes in eluate solution conductivity due to the presence of the sample ions can be used for the determination but since it is a low-sensitivity method it requires low-capacity columns and low-conductance eluates. Bridge circuits using two balanced conductance cells, with high surface area and small separation electrodes, one before and one after the column, can be used to offset the eluant conductivity and thus improve sensitivity.

Alternatively the eluant conductivity can be overcome by the use of chemical eluant suppressor columns between the separation column and the detector. The suppressor column is a second ion exchange column that effectively converts the ions of the eluting solvent into molecular species, with a limited ionisation and hence conductance, without affecting the conductivity of the analyte ions. Many cations are separated by elution with dilute hydrochloric acid solutions; if the solution leaving the separation column is passed through an anion exchange column in the hydroxide form, then the Cl^- from the acid eluant will be replaced by OH^- from the resin. This will react with the H_3O^+ in the eluant to give low-conductivity water as the product from the suppressor column:

$$Cl^-(aq) + resin-OH^-(s) \rightleftharpoons resin-Cl^- + OH^-(aq)$$

$$H_3O^+(aq) + OH^-(aq) \rightleftharpoons 2H_2O$$

Analyte cations will be unaffected by the suppressor column.

For anion separations the suppression column is packed with an acid form of a cation exchanger. If the eluant is sodium carbonate or bicarbonate then the reaction in the suppressor is

$$Na^+(aq) + resin-H^+(s) \rightarrow resin-Na^+(s) + H_3O^+(aq)$$

H_3O^+ combines with bicarbonate or carbonate to form the weak carbonic acid, which is essentially undissociated in aqueous solution:

$$HCO_3^-(aq) + H_3O^+(aq) \rightleftharpoons H_2CO_3(aq)$$

Suppressor columns require regeneration to the acid and basic forms on a regular basis. Switching between two columns of each type, one of which is being regenerated, while the other is in use, can reduce instrument down-time. More recently membrane suppressors have been developed. In these the column eluate passes inside a hollow membrane tube which transports H_3O^+ from the acid solution in contact with the outside surface of the membrane. The process is shown schematically in Figure 6.15. If the H_3O^+ concentration outside the membrane is above a critical level to provide an adequate H_3O^+ concentration inside for the neutralisation reaction, the suppressor will function. Only a constant supply of fresh acid to the outside is required for continuous operation. Further developments using H_3O^+ and OH^-, generated electrochemically, are now appearing and eliminate the need for a supply of standard suppression solutions.

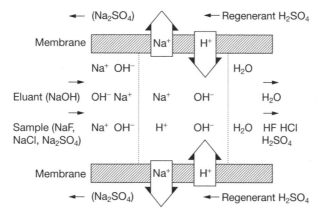

Figure 6.15 Principles of a membrane suppressor used to enhance sensitivity in ion exchange chromatography conductivity detection systems. Reprinted courtesy of Dionex Inc., USA

6.6   Use of liquid chromatography in environmental analysis

The various liquid chromatographic methods have extensive use in environmental analysis for materials which are thermally sensitive and cannot survive the thermal treatment received during GLC analysis or have insufficient volatility to provide and adequate vapour phase concentrations required to give clean separations. Particularly important examples of the former are the analysis of some pesticide residues, nitro compounds and phenols implicated in some pollution incidents and human illnesses. Those materials which may undergo reactions in the heated columns used for GLC are unaffected by the mild conditions used for HPLC provided that the solvents are carefully chosen. The vapour pressure problems are typified by amine salts, which produce poor GLC separations but can be easily handled by HPLC systems provided that the solvent system is chosen with care. The sensitivity and response of these systems are very dependent on other components in the sample and each has to be treated carefully with extensive use of internal standards to eliminate matrix effects.

6.7   Bibliography

Bruno T J 1991 *Chromatographic and electrophoretic methods*, Prentice Hall, Englewood Cliffs, NJ

Grobb R L (ed) 1995 *Modern practice of gas chromatography*, Wiley, New York

Meyer V R 1994 *Practical HPLC*, Wiley, Chichester

Robards K, Haddad P R and Jackson P E 1994 *Principles and practice of modern chromatographic methods*, Academic Press, London

7

Integrated (hyphenated) instrumental analysis

7.1 Introduction

The term 'integrated analysis techniques' is applied to the use of combinations of two or more instruments in which the output of one instrument forms the input for the second (i.e. in tandem) and subsequent instruments. Such arrangements are by convention indicated with the use of a dash between the constituent machines, e.g. gas chromatograph–mass spectrometer (GC–MS). Sequences of three or four instruments, e.g. the application of tandem and triple MSs, are not uncommon, but the majority of systems are based on combinations of two instruments. Combinations are not specifically restricted to chromatography but since these systems are the most common and best developed, with the greatest number of variants, they form a suitable starting point for the discussion.

7.2 Chromatography–mass spectrometry

An MS is the most selective detector which can be applied to the analysis of the eluate from a chromatographic system. Both GC and high-performance liquid chromatography (HPLC) are exceptional at performing their specific analytical tasks, as is the MS. Unfortunately they operate in incompatible environments. Both GC and HPLC chromatographic systems operate most effectively at pressures in excess of atmospheric, 1–2 atm for GC and >200 atm for HPLC, and the MS operates at pressures very much less than atmospheric, <0.1 Pa (10^{-3} mmHg). The solution to this problem is either to design an MS which will operate at, or near, atmospheric pressures or to provide an interface between the chromatographic system and the MS, which will convert the output of the former into a suitable form to provide an input for the latter. Before discussing available interfaces limitations imposed by the requirements of the two machines will be considered.

7.2.1 Requirements for combined systems

Choice of mobile phase and column materials is important. Normal chromatographic column and carrier selection is based on optimisation of the separation and detection of the components in the sample. When using the eluate from the column as input into an MS other requirements have to be considered. The mobile phase must be chemically inert, must not interfere with the mass spectral pattern of the analytes and should produce only a minor contribution to the total ion current monitor signal of the MS. Mobile phases should not contribute to the machine's natural background spectrum and should reduce to a minimum the possibility of corona discharge in the high-voltage region of the instrument.

The column stationary phase selected must have minimum volatility in order to prevent column bleed, either by coating the walls of the transfer system between the chromatographic column exit and the MS with material which can produce an additional fractionation of the eluate or by transferring into the MS to give a continuous background spectrum.

Eluant flow rate is governed by the vacuum pumping requirements and capabilities of the MS system. Capillary columns usually have the lowest eluate flow rate but also produce the smallest sample for analysis. A high percentage of the carrier phase must be removed at the interface between the chromatographic system and the MS, in order to enable both machines to function at their optimum. Since GC and HPLC have different requirements each will be discussed separately.

7.2.2   Interfacing a gas chromatograph

The GC operates at >1 atm pressure but the ion source and analyser of the mass spectrometer operate at $<10^{-3}$ mmHg pressure. Low pressures are necessary in the ion source to prevent burn-out of the hot wire filament source and interfering ion reactions in the source region and reduce the possibility of discharges in the high-potential regions used to eject the ions into the analyser. In the analyser low pressure reduces the background spectrum and prevents distortions in the ion fragmentation pattern due to ion molecule reactions. The mean free path between collisions must also be large to prevent distortion of the ion path during analysis. Only helium satisfies all of these criteria. It is an unreactive small inorganic atom with a simple mass spectrum and a high ionisation potential. In special applications other less satisfactory carrier gases such as CO_2, N_2, Ar and CH_4 may be used, particularly if chemical ionisation sources are used, obviating the need to introduce additional gases into the ionisation chamber of the MS. Ionisation of the carrier gas can be performed by low-energy electrons and analyte ions produced by reaction with the carrier ions. This is important for sensitive analytes, when the chemical ionisation process is less damaging than simple electron beam ionisation. There are two methods of overcoming this incompatability between the GC and the MS: either use an MS which can operate at near atmospheric conditions or design a suitable interface. The interface must operate at a temperature above that of the GC column in order to prevent condensation of the analytes on the interface device, which would interfere with the real time scale of the analysis. Interfaces operate on two principles: division of the GC eluate into two streams or separation of the analyte and mobile phase molecules. Splitting of the eluate into a major stream, which can proceed to a conventional GC detector, and a minor stream, which can be input into the MS, can be achieved by the use of a fine metering needle valve controlling the admission of carrier gas to the MS, a version of the split–splitless input port used on many GC systems to cope with wide variations in analyte concentrations. A number of different arrangements have been used to minimise the dead volume between the valve and the MS. Alternatively a differential pumping system either in the MS or on the transfer interface can easily divide the effluent flow allowing 1–10% of the gas phase to enter the analyser tube.

7.2.3   Molecular separators

Using interface devices based on molecular separators capable of fractionating the GC eluent has the additional advantage of increasing the ratio of sample to carrier gas in the

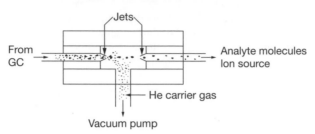

Figure 7.1 Schematic single-stage jet separator used to interface GC output to MS inlet

stream entering the MS. Molecular separators divide into three groups and the final choice depends on the application. In the jet separator the gas stream expands through a high-velocity, supersonic, jet into a vacuum and after a short travel the molecules pass into a second orifice and thence to the analyser. In travel across the region between the two orifices the molecules are subject to the vacuum, which causes the original stream to broaden, with the lighter carrier gas molecules forming the periphery and hence not meeting with the second orifice. The heavier analyte molecules remain closer to the axis of the stream and hence pass through the second orifice; Figure 7.1. Two-stage devices, with the second stage operating at the lower pressure, achieve higher separation factors. The two stages must be isolated except through the orifices. Separation is a function of the pressure differential, the distance between the two orifices and the dimensions of the pipe linking the two stages. With typical orifice dimensions of 10^{-6} m and separations of around 10^{-4} m, approximately 50% of the original analyte reaches the MS with separation factors of approximately 50. The separator is usually made from stainless steel but silylated glass appears to perform better with thermally labile analytes at the elevated temperatures used.

Ultrafine sintered glass tubes, with typical pore sizes of 10^{-6} m, enclosed in a vacuum tube allow the preferential effusion of the carrier gas out of the stream, while allowing the analyte to pass into the MS. Access to the effusion device is restricted by a glass capillary on the inlet and the ouput to the MS is similarly restricted. Since the pore sizes of the glass frit are fixed, the device cannot be optimised for all GC separations and hence variable conductance effusion devices have been developed. The simplest involves placing a tapered restrictor in the exit tube (Figure 7.2(a)) which, by control of the pressure in the sintered glass tube, alters the rate of effusion to the external pump. A more sophisticated variable conductance device allows effusion to occur through a small variable slit formed between a flat cover plate and two annular rings about 2 cm apart (Figure 7.2(b)). The opening is varied by the application of mechanical pressure on the top plate by a micrometer spindle. When the plate is completely closed all the GC output passes into the MS.

The third molecular separator interface is based on molecular diffusion through semi-permeable membranes. Thin silicon polymer membranes mounted on an inert support are widely used, but are unusual in that they reject the helium carrier gas, which is totally insoluble in the polymer at moderate temperatures, while allowing the organic molecules to pass through into the vacuum side. Exit gas from the GC passes into a chamber, one wall of which is the thin 10^{-6} m semi-permeable silicone membrane. Both single- and double-stage systems as shown in Figure 7.3 are in use. Besides the dimensions of the polymer film exposed to the gases the conductance of an organic material is a function of the specific solubility and diffusion constant of the vapour molecules and the operating

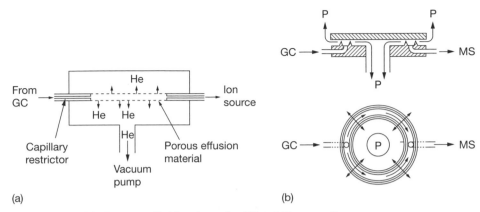

(a)

(b)

Figure 7.2 Effusion-controlled interfaces for GC to MS connections: (a) variable restricted exit porous tube; (b) variable diaphragm effusion device

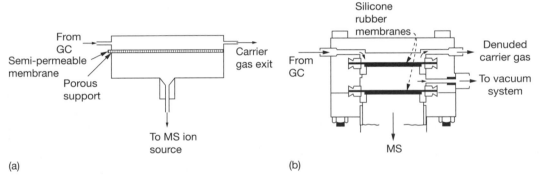

(a)

(b)

Figure 7.3 Interfaces between GC and MS based on diffusion through membranes: (a) basic principles; (b) two-stage process

temperature. These parameters ensure that the separation is different for each compound. If the diffusion is slow then the detected chromatographic peaks will exhibit tailing. Optimum diffusion occurs at higher temperatures but the solubility of the vapour in the membrane polymer decreases and the separation will decrease as the temperature increases. Ideally the membrane should operate at 50–70 °C below the boiling point of the analyte and the optimum results are obtained when the separator is operated in a temperature programme which matches that used for the GC operation. Inclusion of the separator in the GC oven improves the peak shape but optimum results are obtained when the separator is operated ~20 °C hotter than the GC column. Separator design is important if the cavity clean-up time between peaks is to be maintained below that at which peak broadening occurs (typically 3 s maximum). The cavity volume must be as small as possible with respect to the carrier gas flow and ensure full mixing with no dead spots which will delay the clean-up.

The spiral membrane separator (Figure 7.4) was designed to overcome some of these difficulties. Silicone membranes used range from simple dimethylsiloxane polymers to diphenylsiloxane polymers, each showing different separation characteristics and helium permeability as a function of temperature. Maximum safe operating temperature for silicone

177

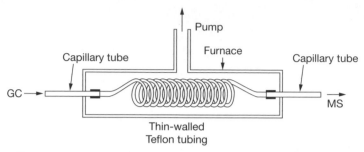

Figure 7.4   High-temperature diffusion interfaces using a Teflon tube spiral

membranes is 230 °C. Above this temperature a Teflon membrane is used and the system separates the carier gas by allowing it to diffuse from the sample. Below 200 °C helium permeability through the Teflon is negligible but above 250 °C sufficient passes through the membrane to produce a significant sample fractionation. Typically a 2 m length of 0.05 cm outer diameter thin-walled, 10^{-2} cm, Teflon tube, wound in the form of a coil, is housed in a heated glass or stainless steel jacket which is pumped by rotary pump to $<10^{-2}$ mmHg. Helium diffuses through the Teflon leaving the analytes in the gas stream admitted to the MS. Sample enrichments can be as high as 200 but only at low flow rates when considerable peak distortion and time lag occur. Because the Teflon separator only functions at high temperatures, with comparatively long contact times, they are unsuitable for samples containing heat-sensitive materials.

   If hydrogen is the carrier gas, a thin palladium–silver alloy membrane operated at 200–250 °C allows the hydrogen to pass through without transmitting any of the organic materials thus giving 100% sample yield with enrichment factors of >20 for flow rates of 2 cm^3 min^{-1} at atmospheric pressure. Unfortunately the system hydrogen and palladium at >250 °C is a good catalytic reduction system and some organic molecules may be reduced. This reduction is usually quantitative so that, by a combination of product identification by MS with retention time on the GC column, the identity of the original material is established. The appearance of ethylbenzene in the MS could be due to the reduction of styrene in the separator but these two compounds have different distinctive retention times on most common GC columns. Some materials present in the eluate, particularly sulphur compounds, can poison the surface of the membrane and reduce efficiency over a period of time. At temperatures above 300 °C when the hydrogen carier gas is almost completely removed then dehydrogenation can occur at the palladium surface.

7.2.4   Separator efficiency

Separator efficiency is defined by four parameters:

(a)   chromatographic carrier gas flow rate to the separator, V_{GC}
(b)   carrier gas flow rate to the MS, V_{MS}
(c)   quantity of sample entering the separator, m_{GC}
(d)   quantity of sample entering the MS, m_{MS}

Efficiency is defined simply as the ratio of the mass of compound entering the MS to mass of compound leaving the GC, i.e. m_{MS}/m_{GC}, and the separation factor by the ratio of the concentration of the compound entering the MS to the concentration leaving the GC, i.e.

$$\frac{m_{MS}/V_{MS}}{m_{GC}/V_{GC}} = \frac{m_{MS}}{m_{GC}} \frac{V_{GC}}{V_{MS}}$$

(gas volumes at 760 mmHg).

7.2.5 Separator selection

It would be an almost impossible task to select a single separator which would provide the optimum performance for all samples under all conditions. Since the performance of all separators depends on the flow conditions, sample thermal stability and MS conditions (vacuum level and pumping speeds) then selection of an appropriate separator must take into account these conditions.

When the flow rates of the GC and the MS are high then separation is not critical and most separators are suitable. Although single-stage systems are adequate, two-stage systems offer greater flexibility. The most difficult conditions are high GC flow rate and low MS flow rate. High efficiency and high separation factors are important. The only single-stage system capable of meeting these demands is the hydrogen–palladium separator. Two-stage, silicone membrane, separator systems are capable of yields in the 30–50% range with separations of about 400 at V_{GC}/V_{MS} of ~500.

Hybrid separators using different processes for each stage in a two-stage system are also used, the intention being to benefit from the changing conditions and separation requirements for each stage. First-stage membrane separators are often coupled to effusion or jet separator second stages. Good efficiencies are achieved and the timelag of the two-stage membrane separator is reduced. A combination of jet and effusion separators, while not producing the same efficiencies as the membrane–jet hybrid, does have the merit of being very physically and thermally robust.

7.3 Atmospheric pressure mass spectrometers

The ion mobility spectrometer (IMS) offers a compromise solution to the incompatibility between the operating pressures of GC and MS. The IMS operates essentially at similar pressures to GC equipment with a limited mass spectral capability. The IMS operation is based on the mobility of ions under controlled electrical conditions, which is a function of the charge and mass of the ions and the magnitude of the applied electrical field. A basic schematic of the instrument is shown in Figure 7.5. Sample eluate from the GC is introduced into the ionisation region of the IMS where the β radiation from a radioactive source, usually ^{63}Ni, produces ionisation in the analyte molecules and a limited number of ions from the carrier gas. These ions are directed by an applied electric field towards the shutter grid in front of the spectrometer drift region. A suitable pulse applied to the shutter grid allows a pulse of ions to pass into the drift region where they move, under the influence of the voltages applied to the grid electrodes, towards the ion collector detector against the flow of a drift gas which passes down the drift region in the opposite direction from the collector out through the shutter grid and into the waste GC gas stream. The electrical mobility of the ion is a function of its mass and charge, as well as the applied potential and the drift gas flow rate. Resolution of the system is very limited and the apparent mass is not a true measure of the analyte mass since the analyte ions may form adducts with the drift gas atoms, producing ion clusters, and also there may be some residual ions associated with any oxygen and water vapour in the sample or the GC gas stream which can also

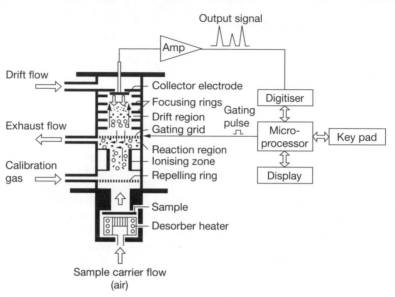

Figure 7.5 Schematic IMS and typical output

form cluster species with the analyte ions masking the mass effects of the analyte. These residual ions will produce additional peaks in the spectrum. The ion mobilities must then be calibrated using known standards. Conditions must be constrained within closely specified limits and if high-purity dry carrier gases are used interferences can be reduced and improved resolution obtained particularly with helium carrier gas. Materials added to the sample, before it passes into the ion source of the IMS, can influence the ion chemistry within the source by reducing the residual ion clusters and modifying the major ion peaks associated with the analyte. This may be important for analysis of complex systems since by careful selection of the additive the instrument background can be reduced, increasing the sensitivity, and resolution of compounds not separated by the GC can be achieved. Although the mass resolution of the IMS is poor, because it is connected to a GC it can give a good indicator of the mass of analytes as they are eluted from the column. A portable system for field surveys is available.

Other atmospheric pressure ionisation sources can be used to condition the GC output for mass spectrometry. Passage of the column eluate through a region of high potential around a point source can produce a corona-type discharge which ionises analyte molecules. These ions are focused by the electric field onto a pinhole in a metal screen. Un-ionised species are removed by an inert gas curtain blown across the pinhole screen. Ions passing into the analyser are processed by a first-stage quadrupole MS, which selects the parent ion, then travel through a further quadrupole reaction–separation section and finally into the third quadrupole MS where fragmentation patterns are used to identify the analyte. A schematic of an atmospheric pressure ionisation MS is shown in Figure 7.6. Sensitivity and resolution are lower than for a full double-quadrant MS but are more than adequate for most environmental work. Cost is usually less than for a full-deflection MS. Combination systems which use an IMS to select ions for a quadrupole analyser are also common. Often the sample is a solid, which can be laser desorbed in the instrument.

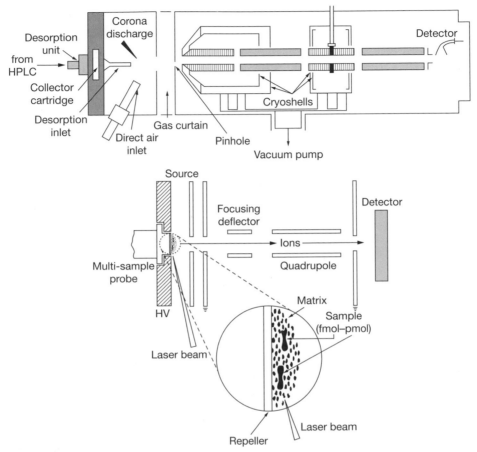

Figure 7.6 Atmospheric pressure ionisation spectrometer and laser desorption option used for solid samples

7.4   Interfacing high-performance liquid chromatography systems with mass spectrometers

Again the difficulty is the incompatibility of the two machine's working environment. HPLC uses liquids with bulk densities of $1\,g\,cm^{-3}$ and MS uses gas phase at densities of $\ll 1\,mg\,cm^{-3}$. Again the essence of using these systems is the design of the interface between the two machines.

Problems associated with interfacing HPLC systems to MS are greater than for GC systems. The mass of carrier material to be removed from the column eluate is 2 orders of magnitude greater for HPLC than for GC. The earliest methods involved trapping the analyte on a cooled surface, possibly at liquid nitrogen temperatures, with bulk solvent being being removed by high vacuum pumping. This was not a real-time system since the fractions would be continuously collected on the storage device, usually in the form of a rotating disc or a wire, which, after all the bulk solvent had been removed, would be desorbed from the surface by local flash heating. The analytes can be separated by HPLC,

181

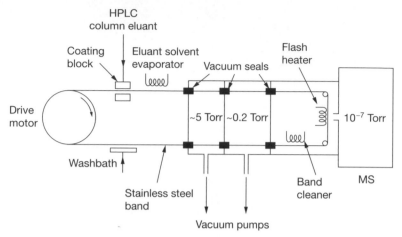

Figure 7.7 Interface between HPLC and MS using continuous deposition and three-stage vacuum locks

independently of the mass spectral analysis. A continuous, almost real-time, interface using a moving belt absorbent to collect the column eluate and passage through a series of vacuum locks is available. A series of infrared heaters remove the solvent in the intermediate locks and the flash evaporator produces analyte samples at the entrance to the ion source of the MS as shown in Figure 7.7.

Thermospray interfaces permit the continuous introduction of the liquid eluate into the MS inlet. The HPLC eluate passes through an electrically heated block, which vaporises the solvent and produces a supersonic jet. The jet can be separated by the differential pumping which occurs for the lighter elements, removing the eluant and leaving the heavier analyte molecules on the axis of the jet to be taken into the MS. A two-stage heater can be used to improve the separation. The jet separation can be improved by use of a skimmer device based on the difference between the momentum of the analyte and that of the solvent molecules; Figure 7.8. The number of skimmer stages enables the selectivity to be adjusted to match the different analytes produced during the chromatography by variation of the vacuum applied to each stage.

The jet contains some desolvated ions which can be focused directly into the MS ion source with excess vapour being pumped away through an additional high-vacuum system. Production of analyte ions can be increased by the appliance of an electrical discharge between an insulated electrode and the grounded body of the interface. The ions can also be focused into the MS inlet by a repelling electrode (Figure 7.9). The solvent, being un-ionised, passes into the extraction pump.

Alternative methods of producing analyte ions are the addition of a chemically reactive reagent or an electron beam produced from a heated filament. In order to reduce filament burn-out it is operated at reduced pressure. The ions move under the influence of an electric field into the inlet orifice of the MS.

7.5 Use of stable isotopes in environmental analysis

Stable isotopes can be used in a number of areas for environmental diagnostics. In principle a stable isotope can be used in the same applications as radioisotopes, which are discussed

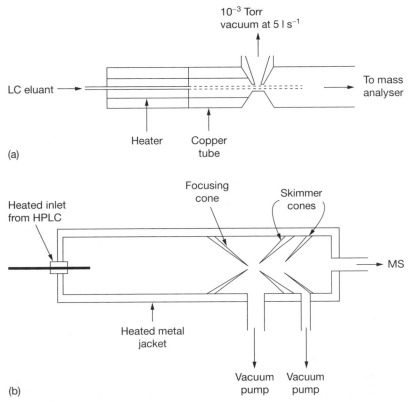

Figure 7.8   Thermospray interfaces for HPLC and MS using jet separator systems: (a) simple system; (b) skimmer system. Reprinted by permission of *Spectroscopy Europe*, Liquid chromatography/mass spectrometry in environmental analysis by Clench M R, Scullion S, Brown R and White J, *Spectroscopy Europe*, **6**; 16. Copyright *Spectroscopy Europe* 1994

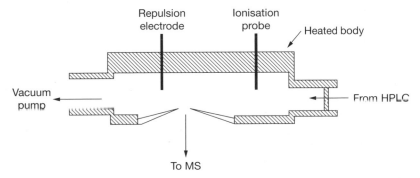

Figure 7.9   Ionisation separators using electrostatic focusing for ion selection. Reprinted by permission of *Spectroscopy Europe*, Liquid chromatography/mass spectrometry in environmental analysis by Clench M R, Scullion S, Brown R and White J, *Spectroscopy Europe*, **6**; 16. Copyright *Spectroscopy Europe* 1994

in Chapter 11. Stable isotopes have the advantage of no radiation hazard as well as their unlimited lifetime in the environment. There is one major drawback in that the best detectors for stable isotopes are several orders of magnitude less sensitive than radiometric methods where almost single atom events can be observed. Looking at typical detector sensitivities radiometric methods can detect the presence of 2×10^{10} atoms of tritium (heavy hydrogen), $\cong 10^{-14}$ g, whereas the best mass spectrometers, using only the most sensitive ion peak, detect typically a picogram. When considering nitrogen contaminations the only radioisotope of nitrogen, ^{13}N, has a half-life of 10 min so, although it will have a high sensitivity, it will not last long enough for a reasonable environmental experiment. Environmental studies often involve the two stable isotopes ^{14}N and ^{15}N or their ratio.

The principal methods of determining stable isotope contents and ratios are mass spectrometry and emission spectroscopy. The former has been discussed in the earlier sections and the latter uses the relative intensities of the isotopic wavelength-shifted emission spectral lines discussed in the sections on emission spectroscopy in Chapters 4 and 5. Other techniques based on microwave frequencies such as nuclear magnetic resonance, electron spin resonance and nuclear quadrupole resonance are too insensitive to be of value in environmental applications. Also, stable isotopes are also easily determined by neutron activation and radiometric methods as discussed in Chapter 11.

While there are a large number of isotopes which can be used in environmental studies the most widely used are those associated with air and water contamination, i.e. carbon, oxygen, nitrogen, sulphur and chlorine. A selection of the most widely used elements and their isotopes is given in Table 7.1. Only brief descriptions of some of the processes investigated with stable isotopes will be given to indicate the range of techniques available.

Simple measurements of water system capacity and throughput can be obtained from measuements of the dilution of tracers, both stable and radioisotope, added to the system. This is standard isotope dilution analysis. Similarly, by examining the distribution of added isotopes, dispersal of material within a system can be determined and any stagnant regions identified. This is particularly important when examining systems such as sewage treatment plants when insufficiently agitated regions may allow sewage components to pass through the system untreated. These simple processes can be followed by isotope ratio methods but are more easily followed by using simple tracer salts for which there is a highly sensitive analytical technique, e.g. sodium salts.

Stable isotopes have particular applications in the field of contamination origin and material ages. In studies using oxygen isotopes old waters contain a higher ^{18}O:^{16}O ratio than new waters, e.g. recent precipitation. The ratio enables the proportion of ground and rain waters mixed in any water source to be ascertained. Helium isotope ratios can also indicate the age of water samples.

The nitrogen isotope ratio ^{15}N:^{14}N can provide evidence of the origin of nitrogen species present in pollution incidents and has been very successful in identifying the source of nitrate polluting species, the most common pollutant in ground waters due often to widespread use of nitrate fertilisers. A similar indication of the origin of sulphur species in air, water and soil can be obtained from the sulphur isotope ratio present in samples containg sulphur compounds. Note that some anaerobic bacteria which use sulphate as a source of oxygen lead to a change in both the sulphur and the oxygen isotope ratios whereas cellulose-producing bacteria cause a 45% increase in the heavy carbon isotope content. Plants can also produce changes in isotopic ratios. Cellulose produced by photosynthesis in trees can show a 28% increase in the ^{18}O:^{16}O ratio. Lipids on the other hand are generally isotopically lighter. This information can be useful in establishing the metabolic pathways for pollutant uptake in plants and animals. Changes in the oxygen and carbon ratios can

Table 7.1 *Some stable isotopes used in environmental measurements*

Element	Isotope	Natural abundance (%)
Hydrogen	^1H	99.985
	^2H	0.0015
Helium	^3He	0.00013
	^4He	99.99987
Boron	^{10}B	19.78
	^{11}B	80.22
Carbon	^{12}C	98.89
	^{13}C	1.11
Nitrogen	^{14}N	99.63
	^{15}N	0.37
Oxygen	^{16}O	99.76
	^{18}O	0.20
Sulphur	^{32}S	95.0
	^{34}S	4.22
Chlorine	^{35}Cl	75.53
	^{37}Cl	24.47
Iron	^{54}Fe	5.82
	^{56}Fe	91.66
	^{57}Fe	2.19
Lead	^{206}Pb	23.6
	^{207}Pb	22.6
	^{208}Pb	52.3

indicate the difference between natural and anthropogenic sources of pollution. The carbon isotope ratio $^{13}C:^{12}C$ in carbon monoxide can be indicative of its origin. Four processes, i.e. vegetative oxidation of non-methane hydrocarbons, atmospheric methane oxidation, biomass burning and fossil fuel combustion, produce carbon monoxide with distinct carbon ratios so the CO distribution can be monitored and the source of pollution identified. These processes may be influenced by atmospheric conditions.

Beryllium isotope ratios can be an indicator of environmental change and can show locations where river erosion can affect marine sediments in coastal waters. Boron isotope ratios can be an unusual indicator in sewage pollution. The age of water can be determined by examination of helium and other inert gas isotope ratios while for waters in contact with uranium-containing minerals the ratio of lead isotopes can show both origin and age of the water. Such processes have demonstrated that some spring waters are several thousand years 'old', between falling as rain to appearing above ground again.

7.6 Fourier transform infrared–chromatographic connections

Interface devices are not required since both systems operate at comparable pressures. Few of the common carrier gases absorb in the infrared (IR) region of the spectrum but the

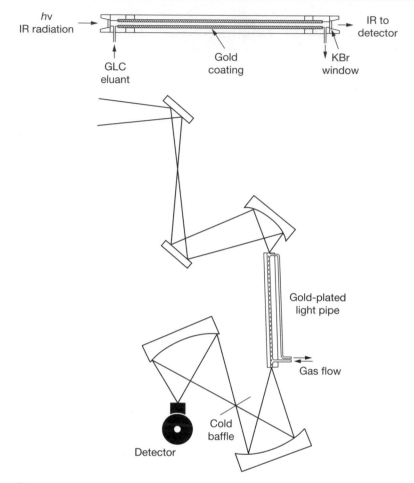

Figure 7.10 Principles for interfacing IR spectrometer with GC. The IR beam is focused onto a gold-plated long path length cell

sensitivity of the IR detectors does not match that of other GC detectors. IR spectroscopy offers a selectivity that is readily automated with Fourier transform (FT) machines. Data storage and handling require modest computing facilities to provide powerful analysis capability. To overcome the sensitivity limitations, small beam dimensions and long path length cells ensure that the small volumes issuing from the column can be measured. In a typical arrangement (Figure 7.10) the IR beam is focused down onto the front window of the cell (Figure 7.11) and the internal gold coating on the cell walls ensures that the beam is contained within the cell by multiple internal reflection and effectively increases the path length of the cell. An alternative is the use of a multiple-reflection long path length cell. The IR cell and the transfer line between the cell and the GC must be heated to avoid condensation of the eluate with the associated background problems.

Most solvents used in HPLC are strong IR absorbers and therefore the requirements are for large beam dimensions and small path length cells, in order to be able to measure the

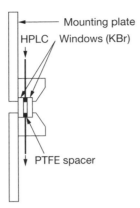

Figure 7.11 Cell for recording IR spectrum of HPLC eluants

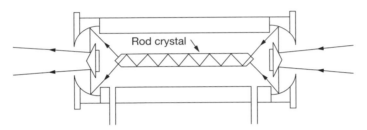

Figure 7.12 ATR cell used for low-volume HPLC eluant analysis

column eluate directly without any additional manipulation (Figure 7.11). An attenuated total reflection (ATR) device (Figure 7.12) can be used with micro volumes issuing from microcolumns or low concentrations of analytes. For standard organic HPLC solvents simple NaCl, KCl or KBr plates are suitable but with aqueous solvents thin PTFE or polythene windows can be used. Some insoluble inorganic materials, e.g. silver halide plates, can be used but they are not readily polished and suffer, in the case of silver halide plates, from opacity when exposed to UV–visible radiation. Zinc selenide and germanium can be easily grown as single crystals and are commonly used in the ATR cells. Because of the strong IR absorption by water it would only be used when no other solvent is available. Since eluate volumes from micro HPLC systems are small, solvent elimination techniques are possible: either the trapping of the material on a cold absorbing surface followed by extensive vacuum pumping to remove the solvent or rapid vaporisation of the solvent leaving the analyte on the surface. In the first case the complete chromatograph has to be collected spread over the surface and then analysed at a later date, i.e. it is not a real-time method. In the second case, if the eluate is deposited on a moving surface, this can be passed directly into the spectrometer for recording a low-resolution spectrum or the IR beam can be reflected off the surface into the spectrometer. The advent of FT machines has enabled real-time spectral recording. Typical dispersive machines require 3 min to complete a scan and hence either the flow has to be interrupted or the analysis has to be done off line. In some commercial instruments the sample is collected on a demountable disc which

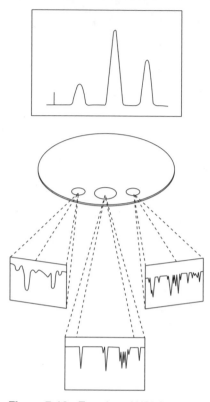

Figure 7.13 Trapping of HPLC output on IR transparent disc. Reprinted with permission of Laboratory Connections Inc., Marlborough, USA

is then taken into the spectrometer at a later date. The disc can be used as a data storage device until all manipulation has been completed. If the collection disc is made of an IR transparent material, e.g. germanium, as indicated in Figure 7.13, then transmission spectra are possible as well as diffuse reflectance methods. Systems can be made almost real time by depositing the eluate on the rim of the disc and removing excess solvent while the disc rotates into the measurement beam position.

7.7 Inductively coupled plasma spectroscopies

Two systems are commonly used with inductively coupled plasma (ICP) sources: mass spectroscopy (ICP–MS) and atomic emission spectroscopy (ICP–AES). Essentially these are only extensions of normal MS and AES which include an unusual source: the ICP. There is a difference between the systems required for MS and AES examination of the plasma in that some of the chemical products from the plasma source have to be introduced into the MS for the measurement but, for spectroscopic measurements, only the light output from the plasma is introduced into the instrument and it is thus easier to handle the plasma source in this case.

7.7.1   Inductively coupled plasma sources

The basic physics of the ICP source is the production of an electrically hot ionised region in an inert gas stream by the passage of a high current through the gas stream. Details are given in Section 5.4.2.2. The ionisation produced by the plasma is equivalent to several thousand kelvins, much hotter than conventional gas-burning or acetylene–nitrous oxide flames, without the problem of hot electrical elements. The inert carrier gas stream is necessary to reduce to a minimum ion–ion reactions within the plasma, which would interfere with the analysis. Other gases will undergo chemical reactions in the plasma and hence cannot be used as carrier gases. Often small quantities of material are added to the rare gas stream as seeds to improve the ionisation process but these must be carefully selected and used only at very low concentrations to avoid these undesirable side reactions. Light output from the plasma arises mainly from the carrier gas ions but there is also some output from the excited analyte ions which can then be analysed by conventional emission spectrometers. Passage of light from a hollow cathode lamp through the plasma produces an absorption spectrum similar to the conventional atomic absorption spectrum.

7.7.2   Inductively coupled plasma–atomic emission spectroscopy

In ICP–AES the light output from the plasma is passed into a spectrometer, which can be a simultaneous complete spectrum analyser, a sophisticated diode array detector as used in HPLC or the more sensitive and specific dispersive analyser, which takes some time to complete the scan and, for best results, requires a more stable ICP source. The improved detection limit for the photomultiplier used in the dispersive system compensates for the additional time required for the analysis. Prototype FT instruments with their rapid scan time and high sensitity are starting to appear but, because of the short wavelengths used, the engineering precision must be high and hence the cost of these machines is far higher than that of the other two systems. Dispersive instruments have a higher tolerance level for spectral interferants, but a number of molecular species present in the plasma cause difficulties with the measurements of the atomic emissions. Although fixed emission background corrections from tungsten filament sources and choppers improve the performance best results are only obtained with dynamic background corrections. Species such as OH cause a number of problems with AES particularly at low analyte concentrations, degrading the detection limits. Sometimes changes in the plasma gas conditions for given elemental wavelengths give improvements but often an alternative wavelength gives better results. Ionisation and matrix effects are apparent in ICP–AES but these can be controlled by the judicious choice of additives to the sample and by internal standard additions.

Normally the background level for ICP–AES is much higher than for the ICP–MS instrument, but this is compensated by the ease of automatic sample injection into ICP–AES with a very rapid system clean-up between samples. Method development is simpler for ICP–AES than for ICP–MS. One of the major advantages of AES is the ability to tolerate high levels of total dissolved solids, up to 30% for simple salt solutions and suspended solids.

7.7.3   Inductively coupled plasma–atomic absorption spectroscopy

The plasma can also be used as the absorbing element in place of the normal flame in atomic absorption spectrometers. Control of the plasma conditions is important and requires optimisation for each analyte, since the ease of ionisation of the element dictates the relative proportions of atoms in the excited and the ground states. Atomic absorption spectroscopy (AAS) is less tolerant of solids than AES owing to scattering of the light beam by involatile solids. Residence time in the plasma may be insufficient to vaporise all the solids.

7.7.4   Inductively coupled plasma–mass spectrometer

Since the MS works at low pressure and the ICP works above atmospheric pressure a separator is required between the MS and the ICP similar to the interface between the GC and MS. In this case, because of the need to sample the ions in the plasma, membrane methods are not possible owing to the removal of the ions on the membrane and also the high temperatures involved and the very aggressive environment in the plasma. Essentially the separator is based on focusing of the ions onto a pinhole entrance and removing the bulk of the plasma, which does not strike the orifice, with vacumm pumps. The sensitivity of the system is extremely high but the specificity depends on the nature of the final MS used to detect the ions. Quadrupole spectrometers are relatively cheap and widely used but have only a limited resolution capability, typically 0.8 of a mass unit in a spectral range from 3 to 400 amu (daltons). This capability will not resolve some of the interferences which occur between diferent elements and ion combinations. For example $^{35}Cl^{40}Ar$, a regular product from the reaction of argon ions in the plasma with chloride ions common to many environmental samples, cannot be resolved from ^{75}As with a quadrupole MS. Some of the interferences can be overcome by use of a chromatographic separator, μ column, before injecting the sample into the plasma.

Electrothermal sample vaporisation or a mixed gas plasma can alleviate the problem. Addition of helium to the argon gas stream changes the ion chemistry and the intensity of the $^{35}Cl^{40}Ar$ ion peak will decrease as a function of helium content whereas the ^{75}As signal will remain essentially unchanged. High resolution (0.01 amu) will eliminate most of these interferent problems. The sample solutions are normally prepared in nitric acid to eliminate many of the problems associated with polyatomic cation interferences caused by HCl, $HClO_4$, H_3PO_4 and H_2SO_4. The sample matrix can also have an effect on ICP–MS owing to interference with aerosol production in the plasma forming stage. Simple matrix matching particularly with internal standards overcomes this difficulty. Also important in environmental samples are high levels of alkali and alkaline earth metal ions since they interfere with the ionisation process in the MS source region. Similarly high ion concentrations can produce space charge effects which result in the preferential loss of light ions in the presence of heavy ions. Sample dilution usually shows whether this effect is present. Solid particles in the sample are particularly difficult for the MS since the inlet can be easily blocked. Ion deflection and focusing can improve the solids tolerance but at the expense of further control electronics. Care must be taken that these additional features do not introduce a non-linear and selective response into the analysis.

A recent addition to the range available is a combined ICP–MS–AES instrument. The MS samples the plasma axially by the normal procedure while the AES examines the light output simultaneously from the plasma normal to the plasma axis (Figure 7.14). Each component can be used separately with the MS being isolated from the plasma when not in use, e.g. if high-solid samples are being analysed, either by a physical shutter or by an inert gas curtain in front of the sampling cones and the absence of an applied field to the gating cone device. The offset gating cone feature again improves the solid tolerance. Dispersive optics using a diode array detector examines the light output. Simultaneous measurements reduce the amount of sample required and can eliminate some of the plasma stability problems.

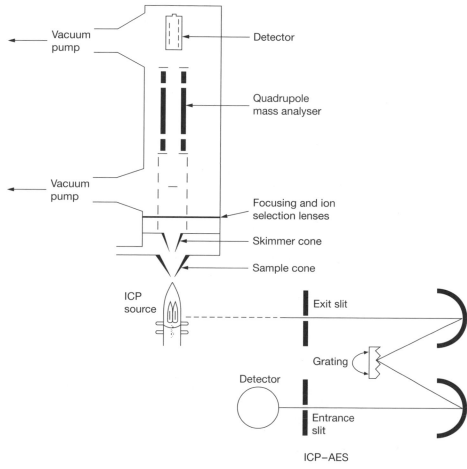

Figure 7.14   Schematic arrangement for combined AES–MS using an ICP source

7.8   Graphite furnace spectroscopy

Graphite furnace atomic absorption spectroscopy has been used as an alternative to the ICP systems largely because of its high solid tolerance. The sample is placed on a graphite rod at the centre of the arrangment and gently dried by passing a limited current through the graphite rod. When the sample is dry the current is increased, raising the temperature to 700 °C in order to char any organic material which may interfere with the light output. After a suitable time interval, large quantities of power are then applied to the graphite rod causing a rapid rise in temperature and vaporising the sample into the optical analysis region. Spectral analysis is different from the normal AAS procedures on conventional instruments owing to the transient nature of the signal and the absence of a steady sample. Although the peak detector output is a function of analyte concentration, the best results are obtained when the signal is integrated with time. This procedure eliminates the effects of variation in the rate of sample vaporisation due to matrix effects and instrumental variations in power delivery to the graphite rod. The detector output must then be captured by a data collecting system, either a chart recorder or digital data capture, at >1 kHz sampling rate. The sensitivity of the graphite furnace AA is an order of magnitude higher than conventional AA.

Because the graphite rod would be oxidised by oxygen in the atmosphere, at the high temperatures used during the sample vaporisation stage, a stream of inert gas, usually nitrogen or preferably argon, is passed over the outer surface and through the hollow centre of the graphite tube to eliminate the oxidation. The graphite furnace, with its curtain of inert gas, is contained in a metal tube fitted with silica end windows to allow the passage of the light through the sample. Since the metal tube comes into contact with the hot gases surrounding the graphite it requires water cooling.

7.9   Bibliography

Barker J 1999 *Mass spectrometry*, Wiley, Chichester

Euiman G A 1994 *Ion mobility spectrometry*, CRC Press, Boca Raton, FL

Kitson F G, Larson B S and McEwen C N 1996 *Gas chromatography and mass spectrometry: a practical guide*, Academic Press, New York

Russell D H 1994 *Experimental mass spectrometry*, Plenum, New York

Yinon J 1995 *Forensic applications of mass spectrometry*, CRC Press, Boca Raton, FL

8

Remote sensing of the environment

8.1 Introduction

8.1.1 Definition and purpose

'Remote sensing' is the aided acquisition of information about the nature and properties of a target that is not in direct contact with the sensor. By definition, therefore, the monitoring probe has to be electromagnetic (EM) radiation, and the target, solid or liquid or gas, should be remote, i.e. outside the confines of the sensor system.

Vision is a perfect example of unaided remote sensing. The advent and use of telescopes, spectral analysers and photodetectors have vastly extended the limitations of unaided eyes, so that slight variations in colour of reflected or scattered radiation can be discerned and the molecular species of the designated target can be identified and measured from long distances.

Recent advances in the state-of-the-art electro-optics technologies have enabled the implementation of diverse and esoteric schemes for remote sensing. Recent developments and improvements in lasers, providing high spectral radiance and purity, very short pulses and tunability over the whole UV–visible and near-IR (NIR) wavelength ranges, have made remote sensing practicable and commercially viable for the surveillance of the environment.

Effective remote sensing and surveillance of the environment are essential for the assessment of environmental changes and pollution loading, preparation and execution of schemes for environmental and resource management, planning and development of future projects and prediction of natural and man-made calamities and pollution incidents. These will allow necessary steps to be taken for the protection of the delicate balance necessary for the well-being and sustenance of life, for disaster preparedness and for public health control.

8.1.2 Environmental parameters

Parameters that need to be monitored by remote sensing relate to the terrain and the atmosphere. The terrain consists of land mass and water bodies. Parameters of these three components of the environment are interdependent and continuously interact with each other.

The land mass can be broadly divided into soil, vegetation and forest. The soil again may be cultivable earth, sand or rock. The erosion, deforestation and salinisation of land are some of the manifestations of important environmental changes. Parameters relevant to vegetation, such as stress factor, yield, type and biomass, are essential for resource management.

Both surface water and underground water directly affect the characteristics of soil and vegetation. Soil moisture, drainage, flood, sewage, ice, etc. control soil erosion and increase vegetation yields. Local and global monitoring of these parameters is needed for the overall resource management and the future planning for this planet, and these can be effectively carried out only by utilising remote sensing technologies.

The properties of the land mass and the water bodies are ultimately governed by the atmosphere. Broadly speaking, the atmosphere, which governs the whole ecosystem on Earth, can be divided into lower atmosphere and upper atmosphere. The weather, climate, aerosol loading and water vapour content are the factors related to the lower atmosphere. The ozone depletion of the upper atmosphere has become a parameter of great concern in recent times.

Because of the rapid industrialisation and urbanisation, the delicate balance in the ecosystem is being rapidly eroded through increasingly high loading of the environment by pollutants at an alarming rate. Natural calamities such as volcanic eruptions and floods and incidents such as the release of radioactivity from the reactor in Chernobyl (Ukraine) and the burning of oil fields in Kuwait (during the Gulf War) have contributed to the rapid increase in the pollution levels of land, water and most importantly the atmosphere. The measurement of many environmental pollution parameters is, therefore, essential for the effective implementation of control and the maintenance of the delicate balance in the global ecosystem.

There are other environmental factors which affect life directly or indirectly. Among these are natural and man-made radioactivity, temperature, pH value of water and soil etc. Technologies are currently available for the aerial monitoring and survey of such parameters. These aspects are beyond the scope of this book.

8.1.3   Interaction of electromagnetic radiation with matter

EM radiation interacts with matter through the electrons associated with its molecular structure. The interactions give rise to radiative and non-radiative effects which can be utilised for remote sensing. The photon energies required to initiate such effects lie in the UV–visible and NIR spectral bands, extending from about 0.3 to 20 μm. The effects are classified as reflection, absorption and scattering and are described below in the context of remote sensing.

8.1.3.1   *Reflection*

Smooth and shiny surfaces give rise to specular reflection which follows Snell's law, i.e. the incident and reflected angles are the same ($\theta_i = \theta_r$). This type of reflection occurs when EM radiation crosses an interface of media with two different refractive indices. The greater this difference, the higher will be the reflection, which is given as

$$R = (n_2 - n_1)^2/(n_2 + n_1)^2$$

where n_1 and n_2 are the refractive indices of the two media. In the case of atmospheric remote sensing the value of n_1 for air at visible wavelengths is taken as unity. For reflection from a shiny surface with air as a medium with $n_1 = 1$ at visible wavelengths, $R \simeq 1$ when the absorption is neglected.

In the case of rough surfaces having surface irregularities larger than the wavelength of the incident radiation, diffuse reflection results. In this, the reflected (scattered) intensity is uniformly distributed over the 2π sr solid angle (hemisphere) from the point of incidence.

Although most topographic solid targets are not perfectly diffuse reflectors, for remote sensing purposes these may be considered to reflect isotropically in the UV–visible wavelengths. The efficiency with which a target reflects is called reflectance, $R = P_r/P_i$, where P_i and P_r represent total incident power and reflected power respectively and the reflectance is generally quoted for a specific wavelength. The average reflectance of a target expressed as a percentage over the visible wavelength band is sometimes referred to as the 'albedo'. It is to be noted that, for a remote target, the strength of the reflected signal at the detector is related to the power density at the target and the detection solid angle.

8.1.3.2 *Black-body radiation*

The radiation incident upon a target is partly reflected (scattered) and partly absorbed. For most solid topographic targets, scattering and absorption are primarily surface phenomena, whereas for liquid and gases these are bulk phenomena. While some of the absorbed energy is re-radiated with characteristic spectral signature of the target (inelastic scattering), some of it is converted to heat energy. Hot bodies, even at room temperature and below, emit EM radiation in the IR wavelengths. In the case of solids the interatomic or intermolecular interactions are so strong that they invariably radiate over a continuous band of wavelength. Theoretical studies on the emission properties of solids at different temperatures are based on an ideal emitter which does not 'reflect', i.e. totally absorbs and re-radiates EM radiation of all wavelengths. Such an object is called a 'black body'. An incandescent metal or a condensed gas such as that in the Sun can be considered to be black bodies. For a temperature, T (kelvins), the emission takes place over a broad wavelength band, having a maximum at λ_m, given by Wien's displacement law as

$$\lambda_m (\mu m) \simeq 2.9 \times 10^3/T \tag{8.1}$$

The spectral radiance of a black body at different temperatures has been described mathematically by Planck and a series of theoretical spectral emission curves for different temperatures are shown in Figure 8.1.

8.1.3.3 *Absorption of radiation*

The transmittance, T, of EM radiation at a wavelength λ through semi-transparent media such as water, glass and gas is given by Beer and Lambert's law as

$$T = I_t/I_0 = e^{-\mu x}$$

where I_t and I_0 are the transmitted and incident intensities (power densities) respectively, x is the distance traversed through the medium and μ is the attenuation coefficient of the medium at λ. If the designated absorbers are uniformly distributed in the medium at a concentration c, then

$$T = e^{-\mu x} e^{-\alpha c x} \tag{8.2}$$

where α is the molar absorptivity or the absorption cross-section per molecule at λ.

In remote sensing applications, the wavelength of the probe radiation is chosen such that μ is minimum and α is maximum. For example, in the remote sensing of airborne pollutants, the wavelength of the probe beam is chosen to fall within an 'atmospheric window', that is the wavelength range over which the atmospheric attenuation of the probe beam is minimum. The radiation at this wavelength should also be significantly and selectively absorbed by the target species (optimum α) to allow remote sensing by absorption methods.

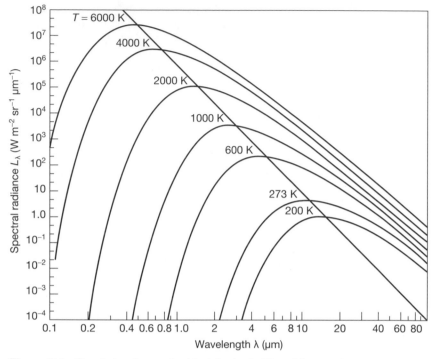

Figure 8.1   Spectral radiance of a black body at different temperatures

8.1.3.4   *Elastic scattering*

For discrete scatterers such as those found dispersed in water and air, the EM radiation is scattered over a 4π sr solid angle (compared with a 2π sr solid angle in the case of reflection from continuous solid targets). In elastic scattering process, the incident wavelength, λ_i, remains unchanged, i.e. $\lambda_i = \lambda_s$, the latter being the wavelength of the scattered radiation. The efficiency of the scattering process, however, is dependent on the dimension of the scatterer (a) in relation to the wavelength of the incident radiation and also on the angle of observation.

The scattering efficiency of a molecule (or an atom) is generally called the cross-section, σ, and is defined as

$$\sigma\,(\mathrm{cm}^2) = P_s / I_i$$

where P_s is the total scattered power and $I_i = P_i / a_m$ is the power density incident onto a molecule having an effective cross-sectional area a_m.

Most molecules of interest have sizes (diameter $\sim 10^{-10}$ m) much smaller than the wavelengths of practical probe radiation, and the scattering is therefore governed by Rayleigh's law, i.e. $\sigma_R \propto \lambda_i^{-4}$, and the intensity is generally spherically uniformly distributed. For particles with $a_m \gg \lambda_i$, the scattering is practically independent of the wavelength and is strongest at the forward and backward directions with respect of the direction of the incident beam. This type of scattering is known as Mie scattering.

8.1.3.5 *Non-elastic scattering*

In this scattering process, the scattered radiation consists of spectral components different from those of the incident radiation. The scattering manifests itself as photoluminescence (fluorescence and phosphorescence) and the Raman effect. Strictly speaking the former cannot be called a scattering process as it involves transitions to real states and therefore is not instantaneous. The efficiencies of the phosphorescence process are considerably lower than that of fluorescence and it is of little practical significance in remote sensing.

Fluorescence The fluorescence efficiency of a target species is quantified by a parameter called the quantum yield, η_f, defined as the ratio of the scattered intensity to the absorbed intensity. Observed fluorescence from practical targets is always quenched, i.e. fluorescence is reduced primarily as a result of non-radiative energy loss by collisions. The quenching is, therefore, governed by the environment of the fluorescing species and the prevailing conditions of temperature and pressure. In the wavelength region where the absorption band of the species is an increasing continuum with decreasing excitation wavelength, the fluorescence quantum yield is generally a strong function of excitation wavelength ($\eta_f \propto \lambda^{-4}$).

For most cases, fluorescence appears as a broad spectrum, not only because of the crowding of vibrational energy levels of the ground electronic state and the de-phasing due to temperature- and pressure-dependent collisions but also because of the overlapping fluorescence contributions from many molecular species in real-world samples. Thus the utilisation of fluorescence in remote sensing is only possible if the target has a stronger fluorescence signal within a unique band than that emitted by its surroundings, which is generally spread over a broad spectrum.

Raman scattering The efficiency of the Raman scattering, expressed in terms of a cross-section, is much lower than that of other scattering processes: however, unique, isolated and extremely narrow (spectrally) lines are emitted in this process. Signals can be detected with high spectral and temporal discrimination. The Raman scattering is characterised by the shift, $\Delta v' = 1/\lambda_i - 1/\lambda_r$ where λ_i and λ_r are excitation wavelength and the Raman scattered wavelength respectively. Normally, the intensity of the Raman scattered signal depends on the inverse fourth power of the excitation wavelength ($I_R \propto \lambda_i^{-4}$). However, at excitation wavelengths near an isolated absorption line of the target species, drastic enhancement (2–3 orders of magnitude) of the Raman scattering cross-section (σ_R) can be obtained for some species. This is known as resonance Raman scattering.

8.2 Basic concepts in remote sensing

From operational considerations, remote sensing can be classified under two broad headings passive and active. Both of these methods are applicable to remote sensing of all three components of the environment – land, water and atmosphere. In passive remote sensing, either the reflected solar radiation or the spontaneously emitted thermal IR radiation from terrestrial targets is monitored. The spectral properties of the reflected or absorbed solar radiation provide identifiable signatures of targets. Passive remote sensing does not provide any range information of the target and cannot generally be used for quantitative measurements. Despite these drawbacks, passive remote sensing has become the most extensively used and highly developed branch of technology, encompassing remote photography, radiometry and thermal imaging and Earth-orbiting satellite technologies. Passive remote sensing is essential for the meso-scale surveillance of the terrain and the monitoring of large-scale pollution incidents in the atmosphere. These aspects are discussed in the next section. In the following, basic schemes in operational active remote sensing are described.

8.2.1 Light detection and ranging

In active remote sensing, EM radiation is used to excite a response (e.g. scattering, fluorescence, absorption) in a remote target. Either the absorption of the probe radiation by the target is measured or the response of the target as re-radiated EM radiation is collected, spectrally analysed and converted to electrical signals prior to analysis. Active remote sensing is universally known as 'lidar', which stands for 'light detection and ranging' (similar to radar in microwave technology). The collection optics (telescope), wavelength selector, transducer (photodetector), etc. constitute the receiver part of a lidar, whereas the system which generates, collimates and transmits the probe beam is called the transmitter. Because of unparallel spectral power density, spectral purity, low beam divergence and wide wavelength tunability, lasers have now become (Chapter 5) an indispensable component of a lidar transmitter and the remote sensing community sometimes refers to this as ladar (laser detection and ranging) instead of lidar.

When the receiver and the transmitter are at the same location and are practically collinear, the system is called the monostatic lidar. In this configuration, the back-scattered response from the target species allows detection and identification or in some applications the back-scattered signal from a retro reflector or distributed scatterers (e.g. particulate in air) provides absorption data of the target species distributed along the laser path. Currently operational active remote sensing systems have been fashioned in many different ways for different applications. The basic principles on which these operate are summarised below.

8.2.1.1 *Long-path absorption*

The scheme for remote sensing of atmospheric pollutants by long-path absorption relies on choosing a radiation source having the smallest possible wavelength spread (narrow line), coincident with the peak of an absorption line of the target species. The probe radiation is reflected by a retro reflector or a topographical target and detected with a receiver at the same location as the transmitter. The received power, P_r, and the incident laser power, P_i, are related through Beer's law:

$$P_r = P_i \rho \exp[-2R(\alpha_a + \alpha_m)] \tag{8.3}$$

where ρ is the reflectance of the retro reflector (or the topographical target) at λ_i, R is the range of the reflector over which the absorbers are assumed to be uniformly distributed, α_a and α_m are the attenuation coefficients of the beam at λ_i due to the absorbers and the medium (air) as a whole respectively. λ_i is chosen so that $\alpha_a(\lambda_i) \gg \alpha_m(\lambda_i)$. Also, if the receiver signal is normalised with respect to the signal (P_r') when the distributed absorbers are absent, $P_r/P_r' = \exp(-R\alpha_a)$ and, therefore, the instrumental and the atmospheric parameters are eliminated. Since $\alpha_a = \sigma_a N$, σ_a being the absorption cross-section (per molecule) and N the concentration of the species, the average concentration of the species, distributed over the range R, can be calculated from the *a priori* knowledge of σ_a and the measured value of α_a.

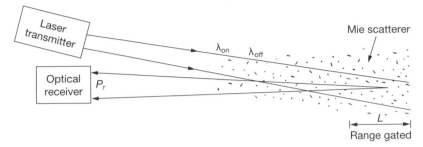

Figure 8.2 Schematic of remote sensing arrangement by differential absorption and scattering technique

The long-path absorption method is very sensitive and could be effectively applied for continuous monitoring of pollution emission from landfill sites, bogs and also from industrial chimneys and other not easily accessible sources. However, it does not provide any range information and is susceptible to changes in atmospheric conditions and relies on locating and locking a suitable wavelength for a particular species.

8.2.1.2 *Differential absorption and scattering*

This technique is commonly known as DIAL (or DIAS) which stands for 'differential absorption lidar' (or 'differential absorption and scattering'). DIAL is currently by far the most widely used active remote sensing technique for atmospheric pollution monitoring. A scheme for a DIAL system is illustrated in Figure 8.2. In this, the transmitter alternately sends two laser pulses, one at λ_{on} which coincides with the peak of the absorption line of the target species and the other at λ_{off}, which is just beyond the absorption edge. Since the atmosphere is never devoid of aerosols, the radiation back-scattered from these distributed Mie scatterers is detected at these two wavelengths. The ratio of scattered power at these two wavelengths is then given as

$$P_r(\lambda_{on})/P_r(\lambda_{off}) \simeq \exp(-2RN\sigma) \tag{8.4}$$

where σ is the absorption cross-section of the target species at λ_{on} and it is assumed that the absorption cross-section of the species at λ_{off} is negligible ($\sigma_{on} \gg \sigma_{off}$) and the wavelengths λ_{on} and λ_{off} are close enough to justify the assumption that the atmospheric attenuation and other wavelength-dependent experimental parameters are practically the same at these wavelengths and can be cancelled in the ratio. Under the above considerations, the minimum detectable concentration of the species distributed over the range R is limited by the noise power, P_N, contributed by the detection system, so that for signal detection the condition $P(\lambda_{on}) - P(\lambda_{off}) > P_N$ must be satisfied. From this, the minimum detectable concentration is given as

$$N_{min} = (\tfrac{1}{2}\sigma R)\ln[(1 - P_N/P_{off})^{-1}] \tag{8.5}$$

In a practical DIAL instrument, the transmitter is repetitively pulsed to enable averaging over many differential signals $P(\lambda_{on})/P(\lambda_{off})$, and the detection is electronically gated so that signals from a specific gate period, i.e. range segment ΔR, are detected (time-of-flight dependent).

8.2.1.3 *Photoacoustic emission*

The generation of sound (acoustic) signals by irradiating substances with IR light pulses is a well-known phenomenon. This takes place when the wavelength of the light corresponds to a strong absorption line of the molecular species of the target. The increased vibrational energy of the absorbing molecules is quickly converted to translational energies through collisions, thus causing an increase in temperature. The heated gas expands, or, in the case of a solid target, the buffer gas (an inert gas) in contact with it expands. If the light is chopped (modulated), the pressure will alternately rise and fall, thus generating an alternating acoustic signal. If the power of the incident beam is P_0, the magnitude of the pressure pulse, p, is given as

$$p = K(\gamma - 1)NP_0/f \qquad (8.6)$$

where $\gamma = C_p/C_v$, C_p and C_v being the heat capacities of the gas in a cell at constant pressure and constant volume respectively, N is the gas concentration, f is the modulation frequency (i.e. pulse repetition frequency) and K is the constant which accounts for the experimental parameters.

Detection of species by the photoacoustic technique is extremely sensitive as, unlike the previously mentioned absorption technique, it is directly proportional to the absorbed power and not to the difference in absorbed and transmitted powers. In laboratory conditions where the target species are confined to a photoacoustic cell, detection at parts per billion (ppb) level of concentrations (i.e. measurement of $\alpha \sim 10^{-11}\,\text{cm}^{-1}$) is possible. Because of this and the prospect of achieving much higher range resolution compared with that possible with a DIAL system, the photoacoustic technique has found important but limited applications in remote sensing of atmospheric species. Techniques based on this principle are now called padar in comparison with lidar.

The principle of remote sensing by padar is quite simple. In this, the pressure pulses are generated by irradiating a target (mainly gaseous) by high-power laser pulses, tuned to an absorption line of the target species. Detailed analysis of the dynamics of the absorption of a laser pulse and the evolution and subsequent propagation of the pressure pulse has been carried out. This suggests that the amplitude of the pressure pulse generated by a localised concentration of gases at a range given by $R = vt$, where v is the velocity of the sound and t is the signal arrival time at a parabolic microphone detector near the transmitter, decreases in proportion to R^{-1}. Additionally, these pulses are substantially attenuated by the air along the range, R.

The padar technique offers the potential for use in short-range ($\sim$100 m), sensitive detection of some gas species with high range resolution. For example, a microphone with a frequency response of 25 kHz could resolve distances of less than 20 mm. For effective utilisation of this technique, two laser lines, λ_{on} and λ_{off}, like those in a DIAL system, need to be used. This technique is therefore useful for detection of leaks of hazardous gases. Although very simple in operation, the need for high-power tuned laser pulses in the IR is expected to render this technique generally unsuitable for practical applications in terms of cost and convenience.

8.2.1.4 *Photothermal deflection*

The principle of photothermal detection is explained with reference to Figure 8.3. When a tuned laser beam, having a Gaussian intensity profile, passes through an absorbing gas, the absorption causes a temperature rise and a consequent change in the refractive index of the

Figure 8.3 Illustration of photothermal deflection technique for measuring gas concentration

medium. A non-interacting laser beam (probe) passing collinearly with the pump beam will respond to this change by producing a deflection. The angle of deflection, ϕ, is related to the refractive index gradient:

$$\phi = \frac{1}{n_0} \frac{dn}{dT} \frac{dT(r,t)}{dr} \tag{8.7}$$

where n_0 is the mean refractive index of the gas at the ambient temperature, dn/dT is the rate of change of refractive index with temperature and the last term is the transverse (spatial) temperature gradient which has been theoretically evaluated for various dynamic situations. In the case where the thermal diffusion length is much larger than the radius of the pump beam and at low frequencies the deflection is practically independent of the modulation frequency of the pump beam (the low-frequency approximation) and is given as

$$\phi = (dn/dT)(P_0/K\pi^2 x_0)\,[1 - \exp(-\alpha l)]\,[1 - \exp(-\alpha_0^2/a^2)] \tag{8.8}$$

where K is the thermal conductivity of the medium, P_0 is the incident laser power and α is the absorption coefficient of the gas of extent l. The deflection is maximum when $x_0 = a$, a being the distance from the centre of the Gaussian profile to where the intensity has dropped to $1/e$ of its maximum value.

Although the use of this technique has so far been confined to sensitive measurements of concentration of pollutants in the laboratory condition, it has the potential to be applied in remote sensing. Like photoacoustic detection, the signal intensity in the photothermal detection technique is also directly proportional to the incident power. A scheme for its application in range-resolved remote detection of pollutant species can be implemented by using a retro reflector or a topographic target to reflect the probe beam, which could be a small He–Ne gas laser. The pump beam (usually an IR laser) has to be chosen to give output at a wavelength which corresponds to a strong absorption peak of the target species. The sampled volume and its range can be varied by varying the angle between the probe and the pump beams.

8.2.1.5 *Back-scattering lidar equation*

Remote sensing based on the detection of elastically and non-elastically scattered radiation is carried out in many different ways. The principles of such remote sensing methodologies are basically the same and depend on how the response parameter is defined. In elastic scattering, the scattered radiation has the same wavelength as the incident wavelength ($\lambda_i = \lambda_s$) and is useful only for the remote sensing of high-density pollution clouds or major atmospheric constituents. Non-elastically scattered radiation such as Raman-scattered signals and fluorescence is species specific and bears characteristic spectral signatures of the scattering or fluorescing species and, therefore, techniques based on these interactions have great potential for being used in remote sensing of environmental species and parameters.

Scattering lidar   The lidar equation for the detection of back-scattered Raman or Rayleigh signals is based on the definition of the response parameter called the differential scattering cross-section, σ', defined as the cross-section, σ, per unit of solid angle, Ω, i.e.

$$\sigma' \equiv d\sigma/d\Omega = P_s/I_i\Omega$$

In other words the differential Raman scattering cross-section is defined as the total power scattered per unit of incident power density on the effective area of a molecule per unit of detection solid angle (Ω), in a specific mode (line). The lidar equation for the power received, P_r, at the receiver telescope of area A_r at a range R from the target is given as

$$P_r = \sigma'(P_0/A_T)(NA_T c\tau_1/2)(A_r/R^2)\eta_t T_{at} \tag{8.9}$$

where N is the concentration (number density) of the target species, A_T is the area of the laser beam at the target and τ_1 is the pulse length of the laser (detector integration time $\tau_d < \tau_1$) so that $c\tau_1/2$ becomes the length over which the signal is sampled at any instant of time during the propagation of the laser pulse and gives the range resolution. The term in the second pair of parentheses, therefore, gives the number of molecules sampled at any instant of time. The term in the third pair of parentheses is the solid angle of detection of a receiver of an effective area of A_T, located at a distance (range) R. The overall transmission efficiency of the optical components including the monochromator (optical filter) is accounted for by the total efficiency parameter, η_t. The atmospheric attenuation is given by the transmittance factor, $T_{at} \simeq \exp(-2\alpha R)$. In the above derivation it is assumed that the atmospheric attenuation coefficients at λ_i and λ_r are practically the same, i.e. $\alpha(\lambda_i) = \alpha(\lambda_s) = \alpha$, and the field of view of the telescope covers the irradiated target area (i.e. complete overlap).

Rearranging the terms in the lidar equation, the minimum detectable concentration in the above scheme is

$$N_{min} = P_{r,min}R^2(KP_i \sigma' T_{at})^{-1} \tag{8.10}$$

where the constant K accounts for all the wavelength-independent experimental parameters and $P_{r,min}$ is the minimum detectable power. The latter is limited either by the photodetector dark current or by the statistics of the signal photoelectrons (shot noise). The sensitivity of shot noise limited detection can be improved either by adequate signal averaging or by lowering the bandwidth of the detection electronics.

For a Raman lidar system, the detection sensitivity (detection of lowest concentration, N_{min}) can be improved by increasing the Raman cross-section. This can be achieved by exciting at shorter wavelengths, as $\sigma' \propto \lambda^{-4}$, and, in some cases, by excitation at a near-resonance wavelength, i.e. at a wavelength near an absorption line of the target species (resonance enhancement). However, the trade-off in increased signal power at short-wavelength excitation is the increased loss of signal strength due to higher atmospheric attenuation.

8.2.1.6 *Fluorescence lidar*

The analysis of remote sensing data obtained by fluorescence lidar is more complex than that by scattering lidar. This is primarily because of the finite lifetime of the excited molecules before they radiate in the fluorescence process. This allows convolution of signals from the target species distributed along the range, R, with the detector's integration time. The interdependence between the laser pulse length, τ_1, the fluorescence lifetime, τ_f, and the detector's integration time, τ_d, plays a major part in this analysis. For most practical purposes and under some simplifying assumptions, the fluorescence lidar equation can be set in a similar fashion to that for scattering lidar:

$$P_r = (P_i N c \tau_d / 2) \, (A_r / R^2) \, [\sigma_f(\lambda_f)/4\pi] \eta_t T_{at} \tag{8.11}$$

The parameters, as before, refer to the range R and the excitation wavelength λ_i. The spectrally integrated total fluorescence cross-section at the excitation wavelength λ_i is defined as

$$\sigma_f(\lambda) = \sigma_A(\lambda_i) \, \tau_f / \tau_{rad}$$

where $\sigma_A(\lambda_i)$ is the absorption cross-section (per molecule) and τ_f and τ_{rad} are the observed fluorescence lifetime and the unquenched radiative lifetime of the excited states. In the above analysis the assumptions made are as follows: (i) the fluorescence is radially uniform; (ii) a fraction within $\Delta\lambda$ at λ_f is selected by a filter having a transmission efficiency factor included in η_f; (iii) the fluorescence lifetime is smaller than the laser pulse duration; (iv) there is no sharp boundary of the pollution cloud at the target location.

8.3 Propagation of electromagnetic radiation through the atmosphere

8.3.1 Introduction

Both active and passive remote sensing rely on the efficient propagation of EM radiation through the atmosphere. The transmission is governed by the wavelength-selective scattering and absorption of the EM radiation by the species which constitute the atmosphere. In addition to major constituents such as N_2 (~78%), O_2 (~21%), Ar (~0.9%) and CO_2 (0.3%), the atmosphere also contains many pollutant species in trace concentrations. Some of these species have very high absorption cross-section and limit the wavelength range for effective atmospheric transmission. For example ozone, even at sub-ppm level, limits the transmission of EM radiation to wavelengths above 200 nm (UV-A) whereas CO_2 cuts off transmission above about 20 μm (mid-IR).

The atmosphere is never free from aerosol particles and water vapour. These scatter and absorb EM radiation and affect atmospheric transmission. The atmospheric transmission is strongly dependent not only on the wavelength of the radiation but also on the variable characteristics of the atmosphere. In general, the transmission decreases exponentially with the decreasing wavelength. The response of a target on EM irradiation is also dependent on the excitation wavelength, generally increasing with decreasing wavelength. In all remote sensing applications, therefore, a trade-off between atmospheric attenuation and the optimum response from the target species must be considered in choosing the optimum wavelength of the probe beam.

8.3.2 Atmospheric transmittance

All atmospheric species, distributed along the path of propagation, absorb and scatter EM radiation and cause its attenuation. Assuming that the species are uniformly distributed along a range, R, the atmospheric transmittance factor in the lidar equation can be expressed as

$$T(R, \lambda_i, \lambda_s) = \exp\left\{-R \sum_n \left[(\alpha_a + \alpha_s)_{\lambda_i} + (\alpha_a + \alpha_s)_{\lambda_s}\right]\right\}$$

where α_a and α_s are attenuation coefficients due to absorption and scattering respectively. The subscripts λ_i and λ_s refer to the sum of the coefficients at the incident wavelength and the scattered wavelength respectively and the summation is over the number of species, n, involved in the attenuation process. Therefore, the transmittance due to all processes by all major atmospheric constituent species can be separated as

$$T_{total} = [T(\alpha_a)T(\alpha_s)]_m [T(\alpha_a) T(\alpha_s)]_p$$

where the subscripts m and p refer to molecular and particulate contributions respectively.

The attenuation due to molecular absorption in the UV–visible wavelengths below 400 nm are dominated by oxygen and below 250 nm by ozone. In the IR wavelengths, the molecular absorption corresponds to discrete lines. Evaluation of the contribution of molecular absorption must therefore be carried out using absorption data available in the literature and an atmospheric model pertaining to ambient conditions.

Contributions from the molecular scattering can be quantified from Cabannes–Rayleigh scattering formula, so that

$$T_m(\alpha_s) \simeq \exp[-R \times 1.17 \times 10^{-24}(\lambda_i^{-4} + \lambda_s^{-4})]$$

The distinction between aerosol and water vapour attenuations is arbitrary. For practical purposes the attenuation due to particulate absorption and scattering can be parametrised in terms of a visibility factor according to an empirical relation of the following form:

$$[T(\alpha_a)T(\alpha_s)]_p \simeq \exp\{-2R \times 3.9 \times 10^{-5}[(1 - 2.9 \times 10^{-3}V)/V](5.5 \times 10^3/\lambda_i)^q\}$$

where the exponent factor $q \simeq 0.58V^{1/3}$, V being the visibility factor, i.e. meteorological range (in km) and is approximately given (in LOWTRAN computer code) as $V \simeq 3.9/\beta$, β being the extinction coefficient (in km^{-1}) measured at 550 nm wavelength. In practice, the meteorological range is estimated from subjective measures of the horizontal naked-eye observation limit of a dark object against the horizon sky in daylight.

In the above derivation, it is assumed that the attenuation coefficients pertaining to particulate absorption and scattering are practically the same at λ_i and λ_s.

8.3.3   Database for atmospheric transmittance

Computer software has been developed to calculate atmospheric transmittance for various standard atmospheric models and by adopting modifications based on prevailing conditions. In the UV–visible wavelengths the atmospheric attenuation is primarily governed by Rayleigh scattering and oxygen–ozone absorption, giving an increasing attenuation with decreasing wavelength, as shown in Figure 8.4, for a standard clear atmospheric path at sea level.

The spectral attenuation characteristic changes quite dramatically with the atmospheric conditions as shown in Figure 8.5. The conditions are described by the sea-level visibility parameter, i.e. meteorological range. In the IR wavelength region the transmission is strongly influenced by the water vapour continuum. Under humid tropical conditions, the continuum absorption can be up to 80% for a transatmospheric path of 1 km or so. In normal clear atmosphere conditions, the atmospheric absorption spectrum exhibits some highly trans-mitting wavelength bands (windows) as shown in Figure 8.6. For effective remote sensing it is, therefore, necessary to locate an appropriate probe beam whose wavelength falls within one of the atmospheric transmission windows.

Many computer programs for modelling atmospheric transmission involving line spectra of absorbing species are now commercially available (e.g. Emission Spectra-ES, a JPL code; LOWTRAN-6) (Kneizys et al., 1983). The database in these has five major elements:

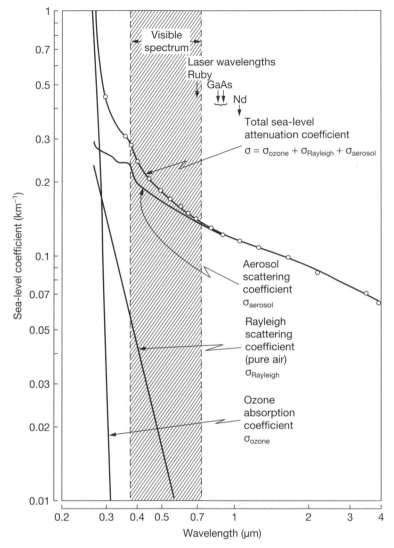

Figure 8.4   Calculated atmospheric attenuation coefficients for horizontal transmission at sea level in a model clear standard atmosphere (absorption by H_2O and CO_2 is omitted; it may be appreciable at wavelengths other than those at plotted points) (from RCA 1974 *Electro-optics handbook*, RCA, Harrison, NJ)

1. Atmosphere model – such as Mediterranean, summer or winter, based on temperature, pressure and the total number density of the constituents (air mass) as functions of altitude.
2. Volume mixing ratio – some 40 known constituents are incorporated for this parameter with the option of incorporating user-specified constituents.
3. Line lists – all relevant parameters of the absorption lines of the relevant species (included in element 2) are incorporated and further line parameters can be included according to prevailing conditions.

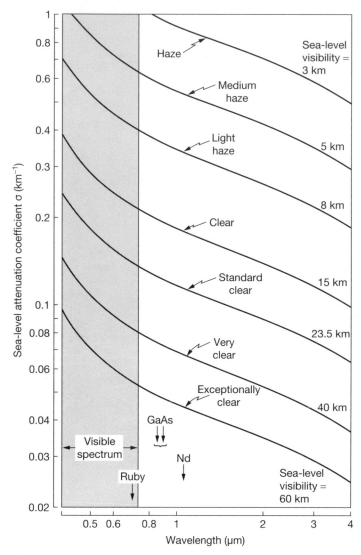

Figure 8.5 Approximate variation of attenuation with wavelength for various atmospheric conditions (from RCA 1974 *Electro-optics handbook*, RCA, Harrison, NJ)

4. Absorption coefficient tables – these are primarily H_2O and N_2 continuum absorption data and data for some trace species for which no line absorption data are available.
5. Solar flux tables – these are based on experimental measurements of solar spectra. However, the Sun itself is not a well-calibrated source and other atmospheric factors usually introduce large errors in atmospheric transmission calculation through these parameters.

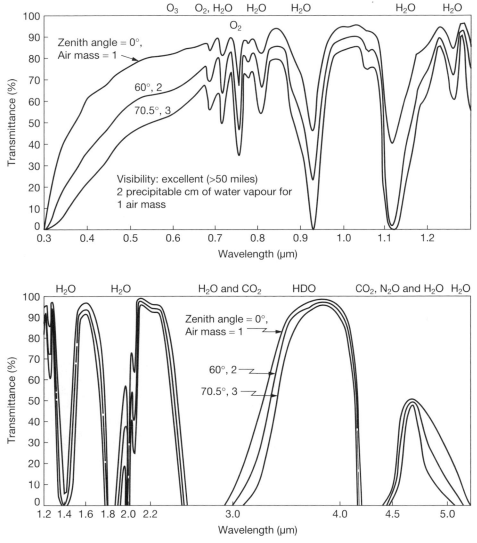

Figure 8.6   Spectral transmission of Earth's atmosphere for varying optical air masses (from RCA 1974 *Electro-optics handbook*, RCA, Harrison, NJ)

8.4   Remote sensing of atmosphere

The remote measurement of atmospheric pollutants and parameters such as temperature, pressure, humidity, wind speed and direction is mainly carried out by active lidar systems. The basic components of such a system are the transmitter, optical receiver, wavelength selector, photodetector, data processing electronics, computer for instrumental parameter control and data management and finally a device for obtaining hard copy. A typical lidar

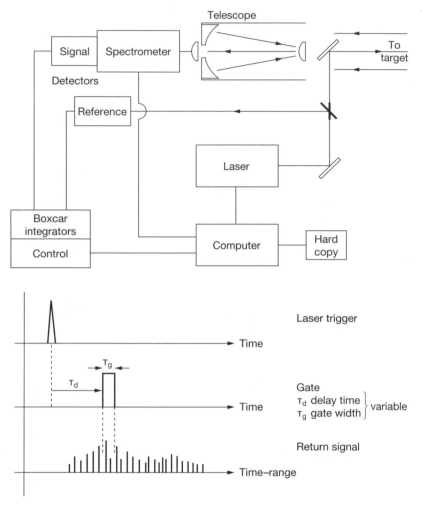

Figure 8.7   Basic layout of a scattering lidar for atmospheric remote sensing and illustration of time (space) gated measurement

set-up for atmospheric measurements is shown schematically in Figure 8.7 and is briefly described below.

8.4.1   Optical transmitter

Although a high-pressure Xe lamp in combination with appropriate optical filters can be used in long-path absorption lidars, albeit with limited range capabilities, lasers are now universally used as the lidar transmitter. The type of laser to be selected depends on the mode of lidar operation, e.g. DIAL or Raman lidar, and also on the specific remote sensing application. The complexity, cost, reliability, tunability, beam power and quality, portability and commercial availability of lasers dictate their application in a specific lidar system.

Table 8.1 *Absorption properties of some pollutant gases relevant to CO$_2$ laser^a (9.174 µm–11.61 µm) monitoring*

Molecule	Peak absorption wavelength (µm)	Absorption^b cross-section (10^{-22} m^2)
Ozone (O$_3$)	9.505	0.45
Fluorocarbon-113 (C$_2$Cl$_3$F$_3$)	9.604	0.77
Chloroprene (C$_4$H$_5$Cl)	10.261	0.34
Ammonia (NH$_3$)	10.333 (9.22)	1.0 (3.6)
Ethylene (C$_2$H$_4$)	10.533	1.19
Sulphur hexafluoride (SF$_6$)	10.551	30.3
Trichloroethylene (C$_2$HCl$_3$)	10.591	0.49
Fluorocarbon-12 (CCl$_2$F$_2$)	10.719	1.13
1-Butene (C$_4$H$_8$)	10.787	0.13
Fluorocarbon-11 (CCl$_3$F)	11.806	4.4

^{a} In addition to the listed species, major absorption lines of carbon monoxide (4.709 µm), nitric oxide (5.263 µm), etc. could be accessed by frequency-doubled CO$_2$ laser lines.
^{b} The values are quoted from the literature.

For the DIAL system, the lasers should provide two outputs, one at a wavelength to match up with the peak of an isolated absorption line of the target species and the other at a wavelength close to the absorption peak at the pedestal level. Moreover, these wavelengths must fall within a good atmospheric transmission band (window).

From the above considerations, CO$_2$ lasers, operating in the 9.174 µm–11.61 µm wavelength band, have become the most widely used transmitters in DIAL systems. CO$_2$ lasers can be tuned to deliver output selectively at about 80 spectral lines and their operating wavelength band can be extended up to 12.16 µm with the use of isotopic CO$_2$ gas. A large number of molecules which are potential atmospheric pollutants have absorption lines accessible by the emission lines of a CO$_2$ laser transmitter and some examples are given in Table 8.1.

For Earth-orbiting lidars, high-energy CO$_2$ lasers, based on the state-of-the-art electro-optic technologies, have recently been developed by NASA under the Laser Atmospheric Wind Sounder programme. These lasers, developed by Textron Defence system (USA), can deliver 10 J pulses at 10 Hz and at 10.6 µm wavelength. These are expected to deliver more than one billion pulses during a three-year period over which the orbiting lidar will monitor wind and cloud parameters.

Dye lasers are also extensively being used as transmitters in DIAL systems operating in the UV–visible wavelength range. These are pumped by flash lamps or by laser diodes or by the output of an excimer laser. The excimer-laser-pumped dye lasers are rarely used in mobile lidar applications because of the complexity and cost of such a transmitter system. The dye lasers are normally tunable over the whole visible wavelength band (using different dyes). Tunable outputs in the UV wavelengths could be obtained by a frequency doubling technique, albeit at only 10–12% efficiencies. Some examples of pollutant gases detectable by the DIAL system using dye lasers are given in Table 8.2.

Table 8.2 *Absorption properties of some pollutant gases relevant to dye laser monitoring*

Molecule	Peak absorption wavelength (μm)	Absorption cross-section (10^{-22} m^2)
Nitric oxide (NO)	226.8	4.6
	253.6	11.3
	289.4	1.5
Benzene (C_6H_6)	250	1.3
Mercury (Hg)	253.7	5.6×10^4
Sulphur dioxide (SO_2)	300.0	1.3
Chlorine (Cl_2)	330.0	0.26
Nitrogen dioxide (NO_2)	448.1	0.69

It is to be noted that as long as the ratio of the aerosol-scattered optical power of on- and off-resonance excitations is detectable against the noise, the power of the transmitted beam needs only to be sufficient to overcome transmission losses. The detection of signal and the evaluation of concentration will ultimately depend on achieving a signal-to-noise ratio of greater than unity ($S/N > 1$).

High output power at UV wavelengths is most desirable in scattering lidars. Excimer lasers provide high output power at 0.193 μm, 0.248 μm, 0.308 μm and 0.351 μm. Nitrogen lasers operating at 0.337 μm are also potential transmitters in these lidar systems. For ease of operation, diode- or flash-lamp-pumped Nd–YAG or Nd–glass lasers, operating at about 0.266 μm (frequency quadrupled output of the fundamental line), are quite suitable as transmitters in mobile or airborne Raman or fluorescence lidars. Another solid state laser which is a potential candidate as a transmitter in a scattering lidar is the titanium–sapphire laser. These lasers can be operated at a very high repetition rate and can provide high average power at wavelengths between 0.35 μm and 0.45 μm after frequency doubling the fundamental modes.

The beam divergence of lasers varies from a fraction of a milliradian to a few tens of milliradians. For long-range lidar operations it may be necessary to reduce the beam divergence by using some collimating optics, so that the beam is confined to the designated target area and, therefore, delivers optimum energy to the target. This ensures optimum scattering from the target species and minimum background scattering noise from the surrounding.

8.4.2 Optical receiver

The lidar receiver comprises signal collection optics, often referred to as the telescope. Essentially the lidar telescope is either a refractive or a reflective lens having a large aperture (effective diameter, D). $F^{\#}$ (f number) defined as $F^{\#} = f/D$, where f is the focal length of the lens, needs to match up with that of the aperture of either the photodetector or the monochromator. A lens system with the longest possible focal length and the largest practicable diameter will produce a highly resolved image of the illuminated target at the detector plane and collect the maximum amount of scattered light. Lenses of large diameters are extremely difficult to fabricate. Often their performances are degraded by

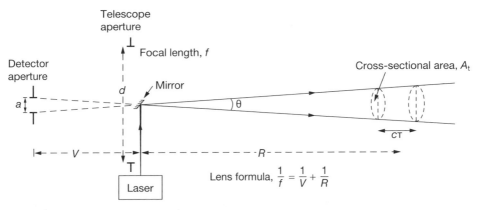

Figure 8.8 Imaging configuration for a telescope in a lidar system

spherical and chromatic aberrations and high losses, particularly in the UV wavelengths, due to impurities in the lens materials. In addition to these, increasingly high dispersion at shorter wavelengths makes refractive lenses unsuitable in lidar applications. Telescopes constructed of all reflective surfaces are therefore almost always used in lidar systems. Most commonly used telescopes are of the Cassegrain type which consists of a large concave mirror, known as the collecting mirror, with a small central hole to transmit the signal (see Figure 8.7). This intercepts (collects) a fraction $\pi r^2/4\pi R^2$ of a spherically expanding signal from the target at a range R. A small subsidiary convex mirror, known as the transfer mirror, is placed at an appropriate position along the axis of the large mirror. This is necessary to transmit the converging optical signal through the central hole and to focus it to the entrance slit of a monochromator or directly onto a detector through an appropriate interference filter.

The field of view of a telescope or a lens is defined by the circular area in a target plane which is completely detected by a circular detector of radius r_D. For the optimum condition there should be a complete overlap between the laser-illuminated area at the target plane and the field of view of the telescope. This can be achieved for moderate to long ranges in a coaxial lidar configuration as shown in Figure 8.8. In this, the telescope is represented by a lens of focal length f and radius r_t ($= d/2$) and the laser beam divergence angle, θ, and the telescope opening angle, ϕ, are kept approximately equal, i.e. $\theta = \phi$. The radius of the field of view at the target plane is given as $r_{FV} = r_t + r_D R/2f$ whereas the radius of the illuminated target area is approximately $R\theta/2$ for moderate to high ranges. It is understood from the above equation that, for a particular range, the field of view of a lidar receiver depends not only on the effective area of the collecting mirror but also on the dimension of the active area of the photodetector (or an aperture in front of it).

The subsidiary mirror at a position along the axis of the collecting mirror reduces the effective area of the collecting mirror. From detailed analysis it is found that for ranges in excess of 300 m, and for $\phi \geq \theta$, the effective area of the telescope reaches the maximum value $A_t^c = \pi(r_t^2 - r_s^2)$, where r_t is the radius of the collecting mirror and r_s is the radius of the subsidiary mirror. At shorter ranges the effective area of the collecting mirror is drastically reduced owing to the shadowing effect of the subsidiary mirror. It is to be noted that an additional shadowing effect due to the support structure of the mirror is unavoidable.

The collecting mirror and the subsidiary mirror (transfer mirror) collectively image the source in accordance with the conventional lens formula $1/V + 1/R = 1/f$, where R and V

are the object and image distances from the lens (telescope). The object distance, R, is essentially the range of the target location. For large ranges, the received beam may be considered to be near parallel and $V \simeq f$, i.e. the image is tightly focused at the focal position of the lens. For small ranges, the size of the image at $V \simeq f$ will be enlarged (diffused) and a detector with a relatively small aperture may not intercept the whole image so that only a fraction of the collected signal will be detected.

The shadowing effect and the image confusion effect mentioned above, jointly prevents the amount of scattered laser power at the photodetector from decreasing in accordance with R^{-2} dependence prescribed in the lidar equation (8.9). These geometric signal compression effects are more drastic at shorter ranges and need to be carefully considered in the design and operation of lidars involving a large dynamic range, including short ranges (e.g. 100 m–10 km).

For the optimum performance, the lidar receiver should have the largest practicable front-silvered collecting mirror having a long focal length, smallest possible secondary mirror and a photodetector with smallest possible effective area. Although a photodetector with a large active area will intercept more of the image confusion region and generally have a higher sensitivity than a detector of relatively small active area, the latter has many advantages in lidar applications as described in Section 8.4.4.

8.4.3 Wavelength selector

The simplest wavelength selectors are colour filters (absorption filters). These have broad transmission bands, low transmittance (throughput) at the peak transmission wavelengths and very poor stray light rejection capabilities. These are also restricted to the visible wavelength band only. Therefore, absorption filters have very limited use in spectroscopy or remote sensing.

For remote sensing applications, requiring EM signal detection over a narrow wavelength band with high stray light (noise) rejection capability, interference filters with customised transmission characteristics are most commonly used. These filters function on the principle of interference between the incident and reflected beams at the interfaces of successive layers of two transparent dielectric materials (e.g. calcium fluoride and magnesium fluoride). The interference filters usable throughout the UV and visible regions and up to 14 μm in the IR are commercially available. These usually have peak transmittance of 80–85% and an effective bandwidth of about 1.5% of the wavelength of peak transmittance. In comparison, the transmission filters usually have transmittance in the region of 20% and a bandpass of about 10% of the peak transmission wavelength (i.e. about 50 nm at visible wavelengths).

For high-resolution spectral measurements, the use of monochromators or spectrometers with reflection gratings or prisms as dispersive elements, sometimes in conjunction with interference filters, becomes necessary. The types and mode of their operations are described in Chapter 5. It is to be noted that high throughput, good resolution and adequate background rejection can be achieved by a spectrometer at the expense of cost, weight and complexity and, in mobile lidars, this should be avoided if possible.

8.4.4 Photodetector

For efficient lidar operation, the photodetector, sometimes called the transducer, should be small and highly sensitive and have broad spectral response, minimum dark noise at room temperature and fast response. A small detector has the advantage of a small inherent noise level. It also provides a small field of view which allows good spatial resolution and a match with the laser-illuminated area at the target plane. Therefore, a well-matched small detector reduces the background radiation noise arising from beyond the lateral extent of the laser-irradiated area. A small detector also inherently provides faster response than a detector with larger active area.

A judicious choice of detector size can lead to a trade-off between the signal power diminution and the enhanced signal detection efficiency with increasing range. In this, a compression of the lidar signal dynamic range is achieved with almost no loss of sensitivity for measurements involving a large dynamic range.

Although photodiodes are the best choice as lidar detectors, in some applications their gains are not adequate for sensitive measurements. However, state-of-the-art integrated and intensified diode arrays provide high sensitivity and good temporal resolution, and allow multi-spectral detection. However, these need sophisticated and dedicated computers (optical multi-channel analysers) to run and are inherently complex, cumbersome and costly. Photomultipliers with internal gain in excess of 10^8 are commonly used as detectors in active lidars. Unlike photodiodes these need a stabilised high-voltage power supply and sometimes need cooling. These difficulties have now been overcome by the commercial development of small compact low-voltage photomultipliers (OPTO Hamamatsu) and micro-channel plate detectors.

8.4.4.1 *Signal-to-noise considerations*

The noise associated with the lidar signal can be categorised into external and internal with respect to its origin. The external noise originates from the passive background (e.g. solar radiation) and active background (e.g. Rayleigh-scattered radiation). The effectiveness of a lidar depends on the extent to which the external noise is reduced by spatial, temporal and spectral filtering to improve signal-to-noise ratio (S/N). This can also be achieved electronically by signal averaging.

The internal noise generated in the photodetector by the 'thermionic' emission, ohmic leakage etc. is known as dark current. This is the average direct current that flows in a photodetector at a certain applied voltage in the absence of any optical signal or background radiation. The dark current of a photomultiplier depends on the ambient temperature and the applied voltage. Modern photomultipliers have a few nanoamps (10^{-9} A) of dark current at typical operating conditions (25 °C and 2 kV). The photomultiplier dark current noise can be made negligible compared with other internal noise at typical operating conditions by cooling the photomultipliers.

Thermal agitation of electrons and other charge carriers in electronic components associated with the photomultiplier circuitry gives rise to a net current. This appears at the output as a noise signal. The value of the root-mean-square voltage, V_{rms}, arising from the fluctuating thermal noise is given as $V_{rms} = (4kTR\,\Delta f)^{1/2}$, where k is the Boltzmann constant ($1.38 \times 10^{-23}\,\mathrm{J\,K^{-1}}$), T is the temperature in kelvins, R is the resistance in ohms of the resistive element in the circuit and Δf is the frequency bandwidth of the circuitry. The rise time of the electronic circuit, i.e. the time required for the circuit to respond between 10% and 90% of the pulse amplitude, is related to the bandwidth: $\Delta f = 1/\tau_c$, where τ_c is the rise time of the circuit. The thermal noise, also known as Johnson noise (or white noise), can be minimised by reducing the bandwidth of the detection system, albeit at the expense of increased rise time of the circuit and the consequent deterioration of the reliability in the faithful reproduction of a fast-rising signal.

The random statistical nature of charge movements in photodetectors of all types gives rise to what is known as shot noise. The noise current due to such a statistical nature of charge carrier movements is given as $I_{rms} = (2Ie\Delta f)^{1/2}$, where I_{rms} is the root-mean-square current fluctuation associated with the average current I and e is the electronic charge. This noise can be minimised only by reducing bandwidth or by signal averaging. In the photon counting mode of signal detection, the signal-to-noise ratio limited by shot noise is given by Poisson's statistics as $S/N = \sqrt{S}$ where S and N are signal and noise counts respectively.

8.4.5 Data processing

The data processing method applicable to remote sensing depends on the mode of laser operation (continuous wave, pulsed or chopped), response time of the electronics, available data averaging time, the required signal-to-noise ratio and the signal strength. Various data processing methodologies are described in Chapter 5. The most effective method of data processing is, however, offered by digital filtering such as boxcar averaging. Digital filtering can also be carried out by Fourier transform procedures. In this case, the signal in the time domain is converted into the frequency domain signal. The processing of data and signal retrieval are carried out by computers and specially developed software.

The range capability or the detection sensitivity (minimum detectable concentration) of an active lidar is dictated by the minimum detectable signal strength. When the detection is shot noise limited, the single photon counting detection technique is often the preferred method. However, this requires much sampling and data averaging, often taking an unacceptably long time to build up the required photon statistic. It is also wasteful in the sense that only one photon per sampling interval is detected, although the signal may consist of many thousands of photons.

8.5 Remote sensing of terrain

8.5.1 Introduction

The survey and analysis of terrain are most often carried out by passive remote sensing instruments on board aircraft, balloon or Earth-orbiting satellite. In passive remote sensing, solar heat and light re-radiated by terrestrial objects is detected.

The spectral characteristics of the solar radiation reaching the Earth's surface depend on the location, time of the day and month of the year and the ambient atmospheric condition. The atmosphere is a dynamic system having variable quantities of water vapour (H_2O), ozone (O_3), aerosol particles and trace contaminants, in addition to its major stable constituents. These absorb solar radiation and reduce the sea-level irradiance. Some of these species, principally H_2O, selectively absorb radiation and produce sharp absorption lines in the solar spectrum at sea level as shown in Figure 8.9.

The solar irradiance on the Earth's surface is maximum at about 470 nm. The UV radiation reaching the Earth is strongly attenuated by the oxygen and ozone and is practically cut off below 200 nm. In the IR wavelengths, up to about 2.5 μm, the atmospheric windows allow radiation to reach the Earth's surface. Above the atmosphere, the solar spectrum resembles that of a black body at 5000 K.

Different objects absorb, retain and re-radiate solar energy differently. In the UV–visible wavelengths, the absorption is negligible and the incident solar radiation is mostly reflected. This, of course, is the basis of remote sensing of terrain by conventional photography. The solar radiation in the IR wavelengths is mostly absorbed by terrestrial

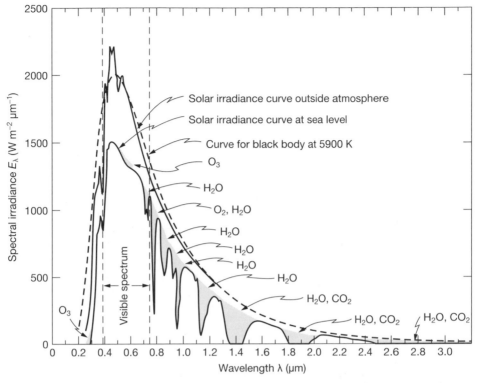

Figure 8.9 Spectral radiance of the Sun (shaded area indicates absorption at sea level due to the atmospheric constituents) (from RCA 1974 *Electro-optics handbook*, RCA, Harrison, NJ)

objects. This causes a temperature rise and emission of IR radiation at wavelengths governed by the black-body radiation law. This is the basis of remote sensing by thermal imager (TI).

The use of photography and thermal imaging in remote sensing of terrain is restricted to good weather conditions and, in the case of photography, daytime recording only. To overcome these problems, active remote sensing, based on laser-induced fluorescence and reflectance and microwave-aided radiometry, is being used in recent times. The techniques, applications, advantages and disadvantages of different systems for the remote sensing of terrain are given below.

8.5.2   Passive methods and devices

8.5.2.1   *Camera*

Remote photography by the conventional camera with a telephoto lens is extensively used for studies in forestry, agriculture and climatology, in addition to its widespread application in map-making and military operations. The state of the art in camera technology has greatly improved in recent years; so has the achievable resolution of photographic films. The scope of photography has now been extended to include IR-sensitive films and cameras.

Qualities of aerial photography of terrain depend primarily on the resolutions of the lens and films. Other external factors, such as the optical qualities of filters or windows placed in front of the lens, and the flatness and location of the focal plane, also affect photographic quality. The usefulness of aerial photographs is also determined by the roll, pitch and vibration of the airborne platform.

The analysis and interpretation of aerial photographs are carried out manually by trained personnel. The recognition of targets and identification of features are primarily based on the shape, size, pattern, site, shadow and most importantly on the tone, texture and colour of the image recorded in the photographic film. The variations in colour and tone in the image fields allow discrimination of many spatial variables, for example crop or vegetation types on land or water bodies of contrasting depths or temperatures.

The cameras for normal colour photography operate in the 0.3–0.7 µm wavelength band, whereas IR-transmitting (UV–visible-blocking) filters and IR-sensitive photographic films are used to record aerial photographs of terrain in the 0.7–0.9 µm band. The latter, known as IR photography, has some advantages over UV–visible photography. IR photographs provide better discrimination between the healthy and stressed vegetation, clear and silty water, dry and wet soils and excellent contrast between land and water bodies.

The photographic method for remote sensing of terrain (and in some instances clouds and storms) offers high ground resolution, relatively high sensitivity, high reliability and, importantly, large information storage capability. The main disadvantages of this method are that this is only applicable at daytime and good weather conditions, does not have real-time data transmission capability and does not give quantitative information.

8.5.2.2   *Image tube*

The image tube, also known as image intensifier, overcomes the disadvantages of the photographic camera and enables real-time transmission of images in the 0.35–1.1 µm wavelength band with a high resolution and sensitivity.

The operational principle of image tubes is explained with reference to Figure 8.10. Basically, this consists of a photocathode upon which an image is focused, an electron lens

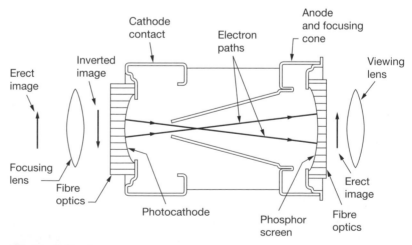

Figure 8.10   Schematic of an electrostatic-type image tube

system to manipulate the ensuing photoelectrons from the cathode and a phosphor screen upon which the image is reconstructed by the photoelectrons.

The image intensification results when the photoelectrons are accelerated by a high applied voltage. When two or more image tubes are coupled together, maximum luminance gains of up to 10^6 can be obtained.

There are many variations in the operation of image tubes mainly based on the electro-optical focusing mechanism. The resolution of magnetically and electrostatically focused single-stage image tubes is comparable with that of photographic films.

8.5.2.3   *Vidicon camera*

Another class of image tubes are generally known as television camera tubes. Among these, the most widely used tube type is the Vidicon. More sensitive and intricate versions are known as the image orthicon and image isocon. The Vidicon utilises an electron beam to scan a target. A transparent photoconductive layer applied to the front side of the photo-conductor serves as the cathode. The back side of the photoconductor is initially charged to the cathode potential (near zero) by the scanning electron beam. When the image is focused onto the photocathode, its conductivity increases in the illuminated area and the rear side charges to more positive values. The electron beam deposits electrons on the positively charged point locations and thereby provides capacitively coupled trains of signals at the cathode side. These video signals are converted to visual images on a television screen or recorded on a video tape.

The sophistication, sensitivity, resolution and range of operation of solid state imaging devices, known as charge-coupled devices (CCDs), have improved drastically over the last two decades. Although CCD cameras are now primarily used for low light level detection in laboratories (e.g. Raman spectroscopy), these are expected to find important applications in terrain reconnaissance from satellites because of their small size, ruggedness, reliability and high sensitivity.

The active element of a CCD camera is an array of gates laid over a suitably doped silicon substrate. Incident light creates electron–hole pairs and the electrons are collected into the potential wells created by the gate electrodes. These two-dimensional arrays of potential wells, known as pixels, constitute the parallel registers. By switching the voltages on these electrodes, the charge collected in the wells can be 'shuffled' sequentially to a serial register and then to the input of a low-noise pre-amplifier circuit. These signals are then used to recreate the image on a television screen.

8.5.2.4   *Thermal imager*

This is a device which detects IR radiation ($3–14\,\mu m$ band) from terrestrial objects and creates thermal profiles of objects on a screen through a scanning process. Temperature differences between a target and its background enable detection and recognition of the target from the visual facsimile of the thermal scene.

Since the spectra of the IR radiation from almost all terrestrial targets are quite broad (black-body radiation) the wavelength band at which the intervening atmosphere is reason-ably transparent to the radiation can be selected (by a filter) for thermal imaging. The detectors and optics in these devices need to be matched and have maximum response and throughput respectively at this wavelength band.

Basically, the TI consists of either a single detector or a multi-element detection system such as a CCD. IR detectors are inherently noisy and need cooling. The radiation emanating

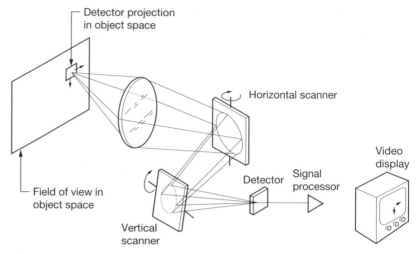

Detector projection
in object space

Horizontal scanner

Video
display

Detector

Signal
processor

Field of view in
object space

Vertical
scanner

Figure 8.11 Schematic layout of a scanning thermal imaging system

from a scene is collected by IR optics and through appropriate filters as shown in Figure 8.11. The optics provide a field of view at the target plane where a detector of small dimension is projected through an opto-mechanical scanning mechanism. The scanning is done horizontally and vertically in steps across the field of view and synchronously with a video processing system. The video screen reconstructs a visual facsimile of the thermal scene. The recognition of the target and the changes in its thermal properties are then interpreted by trained personnel.

Although the major application of real-time thermal IR imaging is still in the field of military intelligence gathering, its application in remote sensing of the environment is increasing. This is finding important applications in thermal pollution surveys, detection of fissures and loose materials in coal mines and in Earth resource surveys. The main advantages of thermal imaging are its all-weather and both daytime and nighttime operation capabilities.

8.5.2.5 *Infrared radiometer*

Radiometers are non-imaging devices which record reflected and scattered solar radiation or self-emission (thermal IR) from terrestrial (or atmospheric) components as electrical signals. Spectral information is obtained by dispersing the optical radiation by filters and detecting by several detectors (or a detector array) at different wavelength bands. The use of specific IR bands over the wavelength range 1–14 μm is dictated by the available atmospheric transmission windows in this region.

For long-term stability and the reduction of noise, IR photodetectors for remote radiometric applications need to operate at low temperatures. Most often cooling down to 100–77 K is required.

Airborne radiometers are usually equipped with a raster form of scanning system similar to that used in thermal imaging.

8.5.2.6 *Microwave radiometer*

Remote radiometry in the microwave region (1–100 mm) has the important advantage that measurements can be carried out under all weather conditions. The large penetration depth of microwaves (0.5–20 wavelengths) allows profiling of the surface layers of rocks, sands or soils by multi-wavelength measurements.

The microwave radiometers usually have comparatively low spatial resolution (~1 km). The resolution in these systems is governed by the size of the aerial and the altitude of the orbiting platform. Since the emissivity of most topographic targets is comparatively very weak in the microwave band, the spatial resolution is therefore limited by the requirement of large aerials and measurement at low altitudes to achieve acceptable signal-to-noise ratio.

Microwave radiometry is now routinely used for Earth resource surveys by Earth-orbiting satellites. Some of the parameters monitored are the water temperature (ocean, sea or lakes), apportionment of water surface, ice–water boundaries and soil moisture content. The technique is also used for monitoring oil films on surface waters.

8.5.3 Active methods (recent developments)

8.5.3.1 *Reflectance*

Conventional flash photography is an example of active remote sensing method based on the spectral reflectance properties of terrain. Non-imaging detection of spectral signatures, induced by a powerful artificial light, extends the scope and reduces the limitations of remote sensing of terrain offered by imaging methods.

Ground truth data on the spectral reflectance of most terrestrial targets bear their characteristic fingerprints. Rocks and soil provide data on their characteristic colours and shades in addition to data on their CO_2 and H_2O contents. Water has a series of strong absorption bands in the IR region between about 0.8 μm and 8 μm. The low reflectance of grass at about 1.4 μm is due to absorption by water. Therefore by comparing the remote sensing spectral data with those of the ground truth and adopting appropriate corrections and scaling factors, it is possible to identify vegetation types, to survey biomass and, most importantly, to analyse the physiological conditions of plants, e.g. stress and disease.

Most promising is the simultaneous multi-spectral scanning of terrain from air platforms. Such measurements of spectral signatures of terrain are now providing valuable data for global resource management.

The effectiveness of both imaging and non-imaging remote sensing relies heavily on the analysis and interpretation of data. These are now carried out by digitisation of analogue data and applying computer-aided image and data processing techniques. The science of data and image processing plays an important role in the fast-expanding field of remote sensing technology.

8.5.3.2 *Fluorescence*

Most organic substances and some inorganic compounds exhibit fluorescence under short-wavelength, narrowband optical illumination. Fluorescence is, however, a very weak interaction process compared with reflectance. Therefore, the application of fluorescence spectroscopy has so far been limited to laboratory studies for obtaining ground truth data on vegetation and water bodies.

Recent advances in electro-optics and laser technologies have prompted increasing research and development activities in the use of laser-induced fluorescence (LIF) detection for remote sensing. In particular, the strength and spectral properties of fluorescence from oil slicks in seas and the vegetation canopy on land have now been measured and analysed from airborne platforms. Aircraft-mounted fluorosensor systems, currently in use, have the capability of a very intensive sampling rate and can collect data along transects covering thousands of kilometres.

Some of the optical energy in the UV–visible wavelengths absorbed by the chlorophyll in plants is used for photosynthesis. The remaining energy is reflected and re-radiated as fluorescence. Therefore, the more efficient the photosynthetic process, the lower is the fluorescence return. This property allows fluorescence to be used as a measure of plant vigour. In leaves, there are chromophoric constituents other than chlorophyll a which exhibit characteristic spectra superimposed on that due to chlorophyll a. Concentrations of these constituents and their oxidation states affect the overall spectral shape and magnitude of the fluorescence. These properties enable identification of the plant and provide information on the environmental stress factors causing damage to the plant (acid rain, dehydration, etc.).

The oceanographic lidar systems developed in Germany, USA and China have demonstrated the feasibility of not only surveying the extent of oil spills in sea waters but also identifying the type of oils, from airborne platforms, by comparing the spectral signatures with those record in the laboratory.

8.5.3.3 *Midar*

Sometimes this is referred to as microwave radar. In this, pulsed microwave radiation in the 300 MHz–100 GHz waveband is used to illuminate the terrain. Reflected signals, known as radar echoes, are recorded by a receiver, electronically processed and fed into a cathode ray tube (CRT). The scanning of the incident microwaves and the time base of the CRT are synchronised, so that the CRT glows in response to the strength of each echo and produces a shadow-effect image of the scanned terrain. In a side-looking radar system, the forward motion of the platform provides the 'alongside' scanning and the 'across-track' scanning is carried out electronically.

In most, so-called 'real aperture', side-looking systems, the resolution across the CRT (or films) is much better than that achieved in the 'along-track' direction. The resolution in the along-track direction can only be improved by increasing the size of the antenna. However, there is a limit to the size of an airborne antenna. This problem has been overcome by using the motion of the platform to effectively create a long 'synthetic' aerial.

Remote sensing by microwave radar imaging has some advantages over that by IR radiometry. It has all-weather capability, can measure surface roughness and dielectric properties and can probe these parameters at certain depths. However, it provides relatively poor resolution and produces image distortion and shadowing in areas of pronounced relief. The modern 'side-looking synthetic aperture' radars can record terrain from spaceborne platforms with a few metres of ground resolution.

8.6   Remote sensing platforms

The platforms for remote sensing systems could be static, mobile or continuously moving. These are categorised into ground-based observation posts, airborne platforms and satellite-mounted systems.

8.6.1   Ground-based posts

The remote sensing observation posts could be static, such as dismantleable metal masts or permanent concrete towers, or mobile bases, such as van-mounted 'cherry-picker' type platforms. The van-mounted platforms can only be used for roadside measurements of terrain parameters and therefore are limited in their applicability. Recording instruments can be mounted on a cross-country vehicle to extend the scope of measurement of crop or soil conditions.

Cameras and sensors can also be mounted in an adjustable mast attached to a van. However, these and dismantleable static platforms often suffer from mechanical instability in windy conditions and are limited to local measurements.

8.6.2   Airborne platforms

Airborne platforms can be categorised into vertical profilers such as balloons and horizontal profilers such as drones, aircraft and helicopters.

Balloons are quite extensively used to carry radio-transmitting instrumented packages for meteorological measurements. They typically carry aloft a compact combination of temperature, pressure and humidity sensors. Radiosondes on board balloons are also used to transmit data on wind speed and direction, and cloud conditions at different heights. The height ceiling for the radiosondes is about 30 km. Modern balloons are equipped with other terrain surveillance devices. The radiosonde balloons are used all over the world and are currently providing atmospheric measurement data from over 20% of the land surface.

Research balloons are usually made of rubber, polyethylene or neoprene materials and are inflated by a 'lighter-than-air' gas (e.g. hydrogen, helium, ammonia, methane). To maintain a balloon to a predetermined air density level, it is fitted with an automatic pressure valve. At night, the gas cools down and the balloon descends to the ground unless ballast is released by remote control.

Instruments carried in balloons vary from electronic thermometer, hygrometer, barometer, etc. to imaging cameras, multi-spectral photometers and radiometers. They also carry power supply units and remote control apparatus. Such balloons are returned to Earth by a remotely operated parachute.

Aircraft and helicopters are routinely used for both atmospheric and terrain surveys. These allow economical means of performance evaluation of sensors under development. Although continuous measurements of terrain parameters and atmospheric conditions by aircraft-mounted instruments could be costly and limited in extent, these have provided validation of ground truth data. Specially designed high-altitude aircraft have attained heights of the order of 15 km, recording images of $100-400\,km^2$ terrain area per frame.

Although aircraft can carry a host of remote sensing equipment, the data received from such platforms are not adequate for global surveillance. A project called Global Horizontal Sonding Technique (GHOST) has currently been undertaken in the USA for extending the scope of both balloon and airborne platforms. Under this programme data from Earth-orbiting satellites and the network of free-floating instrumented balloons will be collectively analysed for more accurate weather forecasting, disaster predictions and resource management.

8.6.3   Satellite platforms

The first artificial satellite (Sputnik) was launched by the then Soviet Union as early as in 1957. This was followed by the race in satellite launching between USSR and USA; the European Space Agency (ESA) joined the race in 1983.

Satellites are normally used to carrying passive sensors for routine measurements of terrain resources and climatic conditions worldwide. Some of these have also been equipped with active sensors (e.g. lidar, radar) to measure ionisation of the upper atmosphere, density changes in the ozone layer, cosmic radiation, number density and the size distribution of the micrometeorites and the strength and direction of geomagnetic fields.

Artificial satellites are powered by solar cells, by batteries and in some cases by nuclear reactors. These are equipped with radio transmitters to relay data to the ground station from onboard instruments, which are, in turn, remotely controlled from the ground observation posts. These satellites (shuttles) can be retrieved and re-used as and when necessary.

In recent years, launching of satellites for obtaining computerised Earth imagery data has become highly commercialised. Many companies have mushroomed, selling such data worldwide. It has been estimated that satellite-based imagery is more cost-effective than that obtained from airborne platforms.

Earth-orbiting satellites are designed and equipped with specific objectives. Communication satellites such as COSMOS (USA), MOLNIYA (Russia) and recently by many other developing countries are designed for the transmission of data to command posts or simply for international TV and radio communications. Meteorological satellites such as those launched by NASA (USA) and NOAA are primarily weather watchers. These are also designed for long-term studies and research on the global climate change, on the depletion of ozone layers and on the interactions between land, sea and atmosphere. As a part of NASA's short-term objective named 'Mission to Planet Earth', these satellites are also equipped to monitor parameters in terrain for agriculture and fishery industries. Satellites launched by ESA (ERS-1 and ERS-2) circle the Earth every 100 min and cover the entire planet in three days, monitoring greenhouse gases, surface winds, temperature of the sea surface, etc. Utilising an onboard synthetic aperture radar (10 cm ground resolution), these satellites can produce three-dimensional images of terrain which reveal distinction in soil moisture, thereby identifying excessive rain or drought. These satellites also carry both passive and active remote sensing systems to map ozone layers at the upper atmosphere and monitor trace gases in the Earth's atmosphere. Satellite-borne geographical surveys which include monitoring of drought, flood, wasteland, biomass, mineral (including oil) prospects and ground movements undertaken by the world's remote sensing community are, currently, providing invaluable data for many diverse applications. The operational multi-purpose satellite systems – Système Probataire d'Observation de la Terre (SPOT) launched by a European commercial consortium, Landsat (USA) and ERS – are now playing vital roles in the remote sensing of the Earth and the atmosphere for global resource management, disaster prediction and climatic modelling.

8.7   Outlook

The remote sensing of the environment is, probably, the most rapidly advancing technology in recent times. This is primarily due to the urgent need for maintaining the delicate balance in the environment and the transfer of the state-of-the-art military technologies in reconnaissance to civilian needs, following the end of the Cold War.

In active remote sensing systems (lidars), the vital component is the laser source. These lasers not only need to provide high-energy outputs at appropriate wavelengths but also need to be portable, reliable, durable and rugged. These criteria are more important for airborne and satellite-borne lidars than those operated from vans or fixed ground stations.

The dye lasers and CO_2 lasers, currently in use in static or van-portable DIAL systems at UV–visible and IR wavelengths, do not adequately meet the above criteria. The excimer lasers, nitrogen lasers, frequency-doubled Nd–YAG (or glass) lasers and Ti–sapphire lasers can provide suitable outputs for applications in fluorescence or Raman lidars, but, still, these are not suitable for cost-effective versatile lidar applications. The pressing need, therefore, is to look for new laser sources.

Research is underway in some establishments such as NASA on the development of a new generation of CO_2 lasers specifically for satellite- or aircraft-borne lidar applications. These lasers are radio frequency pumped, injection seeded and have master-oscillator power amplifiers for high stability, efficiency, reliability and compactness.

Research has shown good prospects for a new generation of high-power diode-laser-pumped solid state lasers. For example, Ti–sapphire, alexandrite and thulium- or holmium-doped laser systems have been found to be capable of providing high output energy tunable over small wavelength ranges in the UV and visible.

The pressing need for resource management, mineral prospecting and disaster prediction worldwide has made satellite imagery a big and thriving business. Many multi-national and multi-disciplinary projects are currently underway and some are planned for future to improve the quality of the images and make the imagery more cost-effective. The trade-off restrictions that are currently being addressed are the spatial, radiometric and spectral resolutions. These trade-offs are weighted in accordance with particular applications.

The future objectives of the major spaceborne remote sensing programmes such as those under SPOT (French), COSMO (European consortium) and ERS (European) are to extend the scope and reduce the limitations of their imaging performances. In particular, immediate objectives of all these projects are to launch multi-orbital satellites carrying high-resolution spectrometers, sensitive detection systems including synthetic aperture radars and high-speed data processors to obtain imagery with ground resolution of 2 m or less and temporal resolution of 3–24 h depending on the latitude and height. It is expected that these data will be used for global weather forecasting and resource evaluation and also for regional and specific applications such as locating sewer and water lines in urban planning, solving traffic problems, management of forest fires, disaster management, treating pollution incidents and enabling efficient use of seeds, fertiliser and pesticides in farming.

8.8 Reference

Kneizys F X, Skettle E P, Gallery W O, Chetwynd J H, Abrea L W, Selby J E A, Clough S A and Fenn R W 1983 *Atmospheric transmittance/radiance computer code LOWTRAN-6*, Air Force Geophysics Laboratory Report AFGL-TR-83-0188

8.9 Bibliography

Asrar G (ed) 1989 *Theory and applications of optical remote sensing*, Wiley, New York
Barret E C and Curtis L F 1976 *Introduction to environmental remote sensing*, Wiley, New York
Danson F M 1995 *Advances in environmental remote sensing*, Wiley, New York
Kondratyev K Y 1996 *Global change and remote sensing*, Wiley, New York
Lillesand T M and Kiefer R W 1994 *Remote sensing and image interpretation*, 3rd edn, Wiley, New York

Methods for environmental monitoring: biological methods

9.1 Introduction

This chapter covers assays and tests which make use of biological properties as diagnostic tools, largely for the detection of contaminants in the environment. The chapter does not cover the use of aquatic community structure monitoring for assessment of water quality or the use of plants and animals for air quality monitoring. The methods described in this chapter have been divided into cellular and molecular. The cellular methods make use of whole cells, e.g. bacteria, or whole multicellular organisms including fish, crustaceans and plants. The response of the organism to pollutants in the sample can be used to assess the toxicity of the sample or the ability of biological processes to degrade the pollutant and render it harmless. In contrast, molecular methods use the specific properties of isolated components or products of cells. In this chapter the exploitation of the specific recognition properties of antibodies and the use of nucleic acid sequences specific to certain organisms as diagnostic tools are discussed.

9.2 Cellular methods

The cellular methods covered in this section focus on tests and assays where the key element is the use of whole cells or intact organisms to perform the test.

9.2.1 Bioassay and toxicity testing methods

Environmental monitoring of the air, water and land for compounds that may pose a risk to human health or the ecosystem is of increasing importance to regulatory authorities. Substances such as heavy metals, organic and inorganic toxic chemicals must be kept below a safe threshold level. However, the quantity and diversity of industrial, agricultural, pharmaceutical and other chemicals produced by humanity are continually increasing. Therefore a range of whole-organism bioassays have been developed to assess the impact of chemical pollutants on the environment. The general principle of these methods is to expose the test organisms to various doses of the pollutant and to monitor their biological integrity as a measure of toxicity. Living organisms respond to the components in chemical waste and integrate the effects of chemical impacts and environmental conditions experienced during their growth. Such tests are important in providing useful information critical in completing an overall risk assessment. Since it is not feasible to determine the specific toxicity of each of the toxic compounds in an environmental sample, whole-sample toxicity testing using organisms is a simple, cost-effective and relevant means of determining sample toxicity. Bioassays can provide a more direct measure of relevant toxicity than chemical and physical analyses alone which are not sufficient to asses the potential effects on biota. Toxicity testing is often carried out in conjunction with biodegradability testing (Section 9.2.2), initially to determine the non-inhibitory concentration of a test substance and also where a substance has been found to be resistant to biodegradation and its environmental impact must be assessed. Toxicity testing is related to biodegradability testing in that both use biological indicators, very frequently bacterial cells.

A battery of bioassays rather than single-species assays is usually used in toxicity evaluation to represent the range of sensitivities of field organisms (Keddy *et al.*, 1995). This is due to the fact that plant and animal communities are diverse in composition and in their sensitivity to toxicants. The exposure of organisms to toxic levels of pollutants can cause disease, contamination of animal flesh, death and also stop or damage the biological processes occurring in the environment including long-term inhibition of growth, reproduction and migration. The application of biological toxicity tests for environmental samples is increasing rapidly and the need for fast, simple, sensitive and cost-effective field analytical technologies that can increase the number of analyses and decrease the time required to perform them has influenced the research and development of new tests.

Many types of bioassays are available and tests can be conducted in the laboratory or the field and monitored manually or automatically. The test organisms incorporated in these assays include representatives from four groups, namely invertebrates, fish, plants and micro-organisms (Novotny and Olem, 1994). The diversity of contaminated matrices contributes to the overall environmental monitoring challenge. Typical examples of environmental matrices include air, drinking water, ground water, soil and sludge. Freshwater and marine species are widely used as test organisms in toxicity tests for water samples. Soil leachates or elutriates have often been used to assess soil toxicity via water quality tests with aquatic organisms. However, soil quality testing should also concern soil-dependent organisms. A number of terms are utilised in expressing toxicity test results. These are acute toxicity, chronic toxicity, lethal toxicity, sublethal toxicity and cumulative toxicity.

9.2.1.1 *Invertebrate tests*

Toxicity tests with aquatic organisms have been used extensively in water pollution investigations and also for sediment screening. Invertebrate chronic toxicity tests are widely applied in the evaluation of toxic hazards and species which have been widely used in toxicity tests are listed in Table 9.1 (US EPA, 1991). The two main freshwater toxicity tests with invertebrates, which are routinely used, are the 21 day *Daphnia* (water flea) and the 7 day *Ceriodaphnia* survival and reproduction tests (Persoone and Janssen, 1993). Acute lethality tests with *Daphnia* are well established and being standardised internationally (ISO, 1982, 1989). Tests are usually conducted by exposing a number of the test organisms to the toxins under control conditions. Living (mobile) daphnids are counted after the required incubation period. A 24 h and 48 h screening test with daphnids has also been reported. The use of daphnids has many advantages important for routine toxicity testing, such as sensitivity to toxicants, short reproductive cycle and parthenogenetic reproduction (Buikema *et al.*, 1980). However, these test organisms represent only one of the three main groups of pelagic invertebrates in freshwater communities. The other two groups are the copepods and the rotifers, which are not used in routine testing.

Recently, however, the potential use of the freshwater rotifer *Brachionus calyciflorus* as test organism for toxicity tests has been examined (Janssen *et al.*, 1994). The main

Table 9.1   *Test organisms widely used in toxicity tests*

Invertebrates		Vertebrates	
Cold water	Warm water	Cold water	Warm water
Stoneflies (*Pteronarcys* spp.)	Midges (*Chironomus* spp.)	Brook trout (*Salvelinus fontinalis*)	Bluegill (*Lepomis macrochirus*)
Mayflies (*Baetis* spp., *Ephemerella* spp.)	Mayflies (*Hexagenia limbata, Hexagenia bilineata*)	Coho salmon (*Oncorhyncus kisutch*)	Channel catfish (*Ictalurus punctatus*)
Crayfish (*Pacifastacus leniusculus*)	Crayfish (*Orconectes* spp., *Cambarus* spp., *Procambarus* spp.)	Rainbow trout (*Salmo gairdneri*)	Fathead minnow (*Pimephales promelas*)
	Amphipods (*Hyalella* spp., *Gammarus lacustris, Gammarus fasciatus, Gammarus pseudolimnaeus*)		
	Cladocera (*Daphnia magna, Daphnia pulex, Ceriodaphnia* spp.)		

advantage of using this species in toxicity tests is the ability of *Brachionus* to produce cysts (resting eggs) which can be stored for months and used on demand. This observation has led to the development of a 24 h test with the freshwater rotifer *B. calyciflorus* and its marine counterpart *B. plicatilis* (Snell *et al.*, 1991a,b). Survival and reproduction data are obtained by exposing isolated rotifer neonates to the toxicant and recording their life history characteristics at regular intervals. Toxicity tests are conducted in sterile, 24-well polystyrene plates and started by introducing one neonate into each well containing food and test solution and incubating the plates at 25 °C, in darkness. The rotifers are checked every 12 h and the number of attached eggs, offspring and mortality recorded through the 3–4 day test. The assay sensitivity for copper (Cu) and pentachlorophenol (PCP) compared favourably with sensitivity of the chronic toxicity tests with *D. magna*, but the test proved rather insensitive for 3,4-dichloroaniline (DCA) and lindane.

Other tests based on the same concept have also been developed, such as with the crustaceans *Streptocephalus proboscideus* (Centeno *et al.*, 1993) for freshwater testing and *Artemia salina* (Van Steertegem and Persoone, 1993) for marine toxicity testing. The shrimp tests are based on survival, growth and reproduction during a 7 day exposure to a range of effluent concentrations. The brine shrimp (*A. salina*) assay has been used as a rapid screening method for potentially hepatotoxic and also neurotoxic cyanobacterial blooms. Suspensions containing 15–25 brine shrimp larvae in a growth medium were added to each microtitre plate well. Test samples are then added to these wells and the plates incubated for 18 h at 25 °C under cool fluorescent lights. The percentage mortality relative to controls is determined.

Bioassays using earthworm survival and reproduction (*Eisenia foetida*, *E. andrei*) are used for soil toxicity testing (ISO, 1991a,b). Surface and subsurface sediment-dependent organisms are used for freshwater sediment quality assessment. Tests such as amphipod survival and reproduction (*Hyalella azteca*), midge survival and reproduction (*Chironomus tentans*, *C. riparius*) and mayfly survival and growth (*Hexagenia* spp.) are used in sediment toxicity testing (Keddy *et al.*, 1995).

With respect to their labour demand and general requirement, tests with invertebrates are practical and feasible. However, operator skills are required to culture and maintain test organisms. Assays using invertebrates are usually based on results of visual or microscopic examination. Tests are being developed and assessed to produce more rapid assays through attempts at automation and the use of advanced equipment for image recognition.

9.2.1.2 *Fish tests*

Fish bioassays have been employed in toxicity assessment for decades. The knowledge that fish show distinct physiological and behavioural responses to low levels of pollutants has been exploited in the development of fish monitors to act as indicators of water quality. The fish tests are usually based on larval growth and survival, where newly hatched fish are exposed to a range of effluents for 1–2 days or up to 7 days. The acute lethality test with fish measures the concentration of a chemical that is lethal to 50% of the exposed population after 96 h (LC_{50}). Fish assays based on the measurement of ATP as the biochemical indicator of energy stress in white muscle tissue are also carried out. Species such as rainbow trout (*Oncorhynchus mykiss*) and fathead minnow (*Pimephales promelas*) are commonly used for acute lethality tests. Salmonids (such as *Salmo gairdneri*) have been used and continue to be a popular choice of fish for various assays to assess toxicity of wastewater.

The commercially available fish activity monitor (Water Research Centre (WRC) Mark III fish monitor) developed by the WRC (UK) is an on-line automated field-based bioassay. This system uses rainbow trout placed in a series of tanks through which the test water is pumped. Electrodes in the tanks measure the minute electrical signals generated from the fish by the general muscular activity associated with gill movements and swimming. The data are then processed automatically with a microcomputer to give an alarm when the monitored signal varies significantly from that produced during control activity. Two aspects of the signal are monitored: amplitude of the potentials, to establish whether the fish are alive or dead, and frequency of the signal which is related to the ventilation or breathing rate. Fish breathing rate is known to increase, decrease or become erratic when the fish are distressed, e.g. subject to toxicants. The response time and the detection limit of the WRC fish monitor are in the ranges 17–60 min and 0.003–100 ppm respectively, depending on the type of pollutant tested. The use of fish tanks lined with photocells to track the movement of individual fish has also been reported. In these tanks the measured parameters included velocity of movement, turning behaviour, distance travelled and the time spent in various areas of the tank.

Fish bioassays are generally more sensitive than microbial tests and permit real-time analysis, but they suffer from a number of disadvantages such as difficulties in maintenance and standardisation problems since variations in the test organism occur as a result of species diversity, age and culture conditions. During recent years, however, research has been initiated to reduce or replace acute fish tests with *in vitro* assays using cultured fish cell lines. Cytotoxicity studies with fish cell lines hitherto have utilised fairly few cell lines, mainly RTG-2 fibroblasts derived from gonadal tissue of rainbow trout, BF-2 fibroblasts from the caudal trunk of bluegill sunfish fry, FHM epitheloid cells derived from tissue posterior to the anus of the fathead minnow and BB fibroblasts derived from the posterior trunk tissue of brown bullhead catfish. For determination of cytotoxicity, different concentrations of the test compounds are placed in individual wells of a 96-well microtitre plate and 5×10^4 cells are added to each well and incubated for 24 h at 18 °C. At the end of the incubation period the cells are stained with crystal violet and the absorbance is measured in a microplate reader at 590 nm. Cultured fish cells are a potential alternative to fish bioassays for monitoring wastewater effluent, but recent studies have classified them as being less sensitive than fish tests. For a broad spectrum of test compounds, including heavy metals and organic chemicals, there was good correlation of *in vivo* lethality and *in vitro* cytotoxicity. However, of 21 chemicals, $HgCl_2$, $CuSO_4$, 2,4-dichlorophenol, 4-chloroaniline, chlorobenzene and phenol were toxic to fish at concentrations 10 times lower than those found to be toxic in the cell culture assay. Research is being carried out to improve these tests.

9.2.1.3 *Plant and algal tests*

Bioassay methods based on different characteristics of various test plants or algae have been developed. Most of the older methods are considered unsuitable for routine work. Bioassays based on growth responses of plant are potentially sensitive but require a long time for growth to occur, e.g. 4–6 days for length measurement of root and shoot of plants, 14–30 days for fresh or dry weight measurement and 21 days for germination scores.

Several plant bioassays have shown potential as effective monitors in detecting genotoxic pollutants in environmental media. The *Tradescantia* micronucleus (Trad-MCN) bioassay, which was developed for detecting gaseous agents, has also been adapted for waste water and drinking water and soil quality testing. The assay uses the *Tradescantia* clone 4430 which is a hybrid between *T. hirsutiflora* and *T. subacaulis*. A principal advantage of this clone is its inability to reproduce sexually, so, in the absence of mutagenic changes, genetic homogeneity from plant to plant and experiment to experiment is ensured. For *in situ* monitoring plant cuttings bearing young inflorescences are suspended in the polluted water or exposed to the test material for 3–24 h. At the completion of the exposure regimen, inflorescences are fixed in 1:3 acetone–ethanol solution and the buds stored in 70% ethanol after 24 h fixation. Slides can be prepared from the buds for cytological studies. An automated image analysis system for scoring MCN was developed recently and can be used to score a large number of slides with high efficiency. This assay has a broad database and has been reported to be a simple, cost-effective short-term bioassay which can yield comparable test results within independent laboratories. The meristematic mitotic cells of plant roots have been used in developing bioassays for *in situ* monitoring of water contaminants. In these fast-dividing cells the endpoint indicator of cytological damage is micronucleus formation. The *Allium cepa* and *Vicia faba* root meristem MCN assays have been used as bioassays in the past two decades; an improved and more efficient test using these bioassays has been developed by Ma *et al.* (1995). According to the authors, these assays are quick, easy and inexpensive. The effect of toxic substances on seed germination can also be used as a bioassay. The lettuce and radish seedling emergence test (120 h test) exposes the seed to total available toxic constituents in the soil, while the root elongation test exposes the seed to the water-soluble constituents eluted from the soil. The seedling emergence test is, therefore, likely to demonstrate greater sensitivity than the root elongation test if non-water-soluble toxic constituents are present in the soil (Keddy *et al.*, 1995).

Bioassays involving the direct measurement of oxygen evolution from aquatic flowering plants have also been developed and have been shown to be sensitive to photosynthetic inhibitors. Simple and rapid bioassays for the measurement of the herbicide simazine in water using the aquatic flowering plants *Ceratophyllum oryzetorum, Ranunculus trichophyllus* and *Alisma plantago-aquatica* have been reported. These tests are based on the effect of simazine on the amount of oxygen produced by photosynthesis which is measured directly using a Clark-type oxygen electrode. The method is capable of measuring a simazine concentration of $0.02\,mg\,l^{-1}$ within 10 min of exposure and maintenance of stock cultures of the plants is simpler and more easily conducted than for methods using algae.

The use of algae in bioassays has proved useful in detecting metals, herbicides, pesticides and crude oil compounds (McFeters *et al.*, 1983). Test species such as *Chlorella fusca* (unicellular green alga) and *Selensastrum capricornutum* are utilised in these tests as indicator species. This test has been used for testing wastewater, soil leachates and elutriates. Structural damage observed in algal cells and percentage algal growth are taken as toxicity measures in these assays. These tests are performed by inoculating the test water with the test strain to give a density of 1×10^4 cells ml^{-1} and incubating the algal growth test under the 5000 lx continuous fluorescent light for 3 days at 24 °C. Cell density is then measured by a Coulter counter. The use of electron microscopy to allow the examination of structural changes in algal cells as a result of toxic stress has been applied to effluent water studies. In tests such as the red algae test, female and male plants are exposed to a range of effluent concentrations for two days and then incubated in clean seawater. The number of cystocarps (reproduction products) is counted and compared with the numbers for controls. Bioassays using algae, with measurement based on direct oxygen evolution or fluorescence emission of photosynthetic processes, are usually more sensitive and rapid than those based on growth of plants. However, the culturing and preparation of the algal suspension is a lengthy process and, in addition, it is difficult to maintain an identical culture of algae each time the bioassay is conducted, which can affect the results.

9.2.1.4   *Microbial tests*

The use of micro-organisms in toxicity testing has received increasing attention during recent years. Micro-organisms possess a number of features which encourage their use as indicators of pollution in bioassays over higher organisms since the latter tests require specialised equipment and skills and are time consuming. An extensive range of microbial techniques have been developed and are now used as toxicity screening procedures. Studies of effects on microbial function or activity constitute a more direct, rapid and sensitive approach to measure chemical stress. Several laboratory-based bioassays utilising micro-organisms have been described in the literature. These assays can be categorised according to the measuring principle used: (a) monitoring transformation of carbon, sulphur or nitrogen; (b) determination of the activity of a microbial enzymes such as dehydrogenases and adenosine triphosphatases (ATPases); (c) measurement of growth, mortality or photosynthesis; (d) determination of glucose uptake activity using radioisotopes; (e) measurement of oxygen consumption using a dissolved-oxygen electrode or respirometric techniques (respirometer); (f) measurement of luminescence output using a photometer. Ideally, soil toxicity to bacteria is examined using a representative soil bacterium or freshwater bacterium and is conducted in the soil.

Bioassays based on microbial transformations Most examples of this type of assay are still in their research phase. An example of these tests is the use of an NH_3 ion selective electrode for quantifying bacterial nitrification rate. Stroud and Jones (1975) developed a biomonitor incorporating nitrifying bacteria growing on rocks in a column through which test water is pumped. The depletion of ammonia in solution was monitored with an NH_3-sensitive electrode. A method using *Nitrobacter* has been developed specifically to determine effects on nitrification (ISO, 1986).

***In vivo* microbial enzyme bioassays** Bioassays based on the determination of the activity of microbial enzymes have been developed. The dehydrogenase enzymes used in this type of assay are involved in catalysing the oxidation of substrates by the transfer of electrons through microbial electron transport chains (ETCs). Specific dyes can be used as indicators of ETC activity where they act as artificial hydrogen acceptors, changing colour once they are in the reduced state, a process which can be followed with a spectrophotometer. Resazurin, methylene blue, triphenyl tetrazolium chloride (TTC) and 2-(p-indophenyl)-3-(p-nitrophenyl)-5-phenyl tetrazolium chloride (INT) are the most commonly used dyes in these assays. Bioassay tests such as the resazurin reduction (RR) method are performed with mixed bacterial cultures obtained from activated sludge or with a pure *Escherichia coli* (ATCC 10536) culture. This test is based on the reduction of resazurin by microbial dehydrogenases and involves incubating the bacteria in the presence of resazurin and toxicant. After an incubation period of 90 min the absorbance of the supernatant is read at 610 nm (the absorption maximum of the oxidised resazurin). The assay has proved useful in the determination of EC_{50} (effective concentration which causes 50% inhibition of dehydrogenase activity) of sodium arsenate (100 ppm), sodium arsenite (25 ppm) and mercuric chloride (<1 ppm). The dehydrogenase test has been used in soils, and the performance of sewage treatment plants following toxicant input. TTC and INT are the most widely used redox dyes in ecological toxicity tests and are reduced to formazan, an insoluble red precipitate which is monitored at 480–485 nm using a spectrophotometer. The SOS chromotest (Environmental Biodetection Products, Inc.) is a semiquantitative genotoxicity test based on a colorimetric reaction measurement of microbial enzymatic activity of the bacterial tester strain *E. coli*. Other tests such as the MetPAD heavy metal toxicity test kit (Group 206 Technologies, Inc.) is based on enzyme inhibition of an *E. coli* strain. Response is observed on an assay pad by purple colour development (a lower colour intensity corresponds to higher toxicity).

The majority of ATPases are membrane-bound enzymes and are involved in high-energy metabolic transformations. The inhibition of these enzymes affects cellular functions which enables them to be used as toxicity indicators. The activities of enzymes such as acid and alkaline phosphatase, glucose-6-phosphatase, urease and lipase are all reduced in fish exposed to mercuric chloride. The inhibition of some of these enzymes in micro-organisms can also be used as pollution indicators.

Tests based on growth and mortality Bacterial growth and mortality can be used as indicators for wastewater toxicity testing. The assimilable organic carbon (AOC) bioassay uses a defined bacterial inoculum of *Pseudomonas fluorescens* P-17 and *Spirillum* sp. strain NOX to measure the potential for bacterial growth in drinking water. The bacteria used can oxidise reduced carbon compounds for energy and also require these organic molecules as a source of carbon building blocks. The assay involves the growth of a stationary-phase inoculum of the bioassay organisms to a maximum density when introduced into 600 ml of pasteurised test water. Bacterial density is enumerated with spread plates on

Oxoid Lab-Lemco nutrient agar and the yields of the bioassay organisms on model compounds are used as to calculate AOC concentrations. Despite a growing need within the water industry to measure biodegradable organic matter concentrations, widespread use of the AOC bioassay has been hindered by the fact that the procedure is not easy for routine use and it remains as a research tool.

Tests such as the *Spirillum volutans* motility inhibition test and *P. fluorescens* growth inhibition are also used. The *S. volutans* assay is a short-term toxicity test which has received much attention in the last few years and employs the motility patterns of this large aquatic bacterium as the test end point. The procedure involves mixing a defined test medium with the wastewater sample and healthy (with >90% motility) bacteria. Slide preparations from the mixture are examined using dark field microscopy after up to 120 min incubation. Relative toxicity is measured as MEC 90 which is the minimum effective concentration of a toxicant that causes loss of reverse or forward motility in more than 90% of cells after exposure to the toxicant for different time periods. This test is both rapid and sensitive, but cultures of *S. volutans* are fragile and have a short life-span of 3–5 days.

Another toxicity test based on bacterial growth is the ECHA dipstick test (ECHA Microbiology Limited). This test has been used for direct sediment testing and it contains a small absorbent pad impregnated with *Bacillus* sp. and a growth indicator dye. Bacterial growth is detected by incubating the dipstick (35–37 °C, 18–24 h) after contact with the samples. No colour development indicates a very toxic sample and a red colour indicates a non-toxic sample. The agar plate method is a similar rapid screening method for chemical toxicity and is based on the growth inhibition of *Bacillus cereus*. Evaluation of toxicity is scored by measuring the inhibition zones after overnight incubation.

Glucose mineralisation assay The principle of this test is the evaluation of the inhibitory effects of toxicants on the rate of labelled CO_2 released by *E. coli* in the presence of glucose U-^{14}C. The assay is performed in flasks equipped with a central well containing phenylethylamine as the CO_2 absorbent. Glucose is added to the bacterial suspension (50×10^5 bacteria ml^{-1}) and the mineralisation reaction is carried out for 12 min, at 20 °C in a shaking incubator. The reaction is stopped by the injection of 0.05 mol l^{-1} H_2SO_4 into the test medium and the flasks are shaken for further 4 h to permit complete absorption of CO_2, before the radioactivity of the phenylethylamine is measured using a scintillation counter. Glucose mineralisation is determined from the kinetics of released $^{14}CO_2$, established in toxic and control media simultaneously. In this type of test *E. coli* was found to be specifically sensitive to zinc and copper, but unaffected by other tested ions such as cadmium and chromium at 100 mg l^{-1}. Other bacterial strains can be used in the glucose mineralisation assay that are more sensitive to other specific compounds than *E. coli*.

Respiration inhibition test Biomonitors based on measuring bacterial oxygen consumption (respiration) have been developed and these incorporate mixed bacterial populations derived from activated sludge and synthetic sewage. The 'Respiration' test is based on studying the metabolic criteria of a mixed bacterial population. The inhibition of the oxygen consumption by the addition of toxic compound is compared with the oxygen consumption of a control test. The respirometric activity is evaluated by recording the kinetics for 3 h of dissolved oxygen reduction in the medium by means of an oxygen electrode (ISO, 1986). The percentage of inhibition is calculated from the reduction in respiratory activity relative to the control sample. A plot of percentage of inhibition versus logarithm of concentration is then constructed. The concentration of chemical required to reduce the respiration rate (mg O_2 l^{-1} min^{-1}) by 50% (EC_{50}) is calculated with reference to the activity of

231

a toxicant-free control. This test has been reported to be well adapted to toxicity screening of effluents, easy to use and can provide reproducible and repeatable results. Since the test is conducted on a microbial community, it has the advantage of being more representative of the microflora in rivers. Wastewater treatment plants can be subjected to sudden surcharges of toxicants, a condition known as shock loading. Respirometric tests can be useful for estimating the effect of shock loading at a particular treatment plant. Since these plants were not designed to receive such hazardous chemicals, plant upsets caused by influent toxicity may lead to costly process down-time and low water quality.

The respiration test is conducted on a mixed bacterial community, which simulates the natural conditions of activated sludge and can allow the effect of industrial effluents on the activated sludge process treatment plant to be tested. This test is also well suited to the evaluation of the treatability of individual wastes (Elnabarawy *et al.*, 1988). However, varying results are often reported when using municipal or industrial activated sludge substrate in toxicity assessments. The lower sensitivity of respiration tests compared with the 'Microtox' (see following section) test has been related to the activated sludge inoculum. The inoculum is known to absorb toxicants or modify the tested compounds, resulting in a decrease of their effective concentration in the test medium.

Earlier respirometers were large, heavy and unreliable and their use was time consuming and they had no facilities for data handling. The new respirometers overcome most of the previous shortcomings. Laboratory respirometers can be used on line or off line in testing industrial wastewaters for toxicity prior to their introduction to a biological treatment process. This is becoming common practice at industrial and municipal sewage treatment plants. There are several respirometers reported by different authors and manufacturers such as the RODOX system, the N-CON COMPUT-OX wastewater respirometer, model WB-1000 (N-CON System Company Inc., Larchmont, NY 10538, USA) and the Gilson respirometer. However, some of these respirometers are not commercially available. A new respirometer ('Merit 20') has recently been developed by Yorkshire Water plc and E. R. Addingtons. The Merit 20 is an automated manometric electrolytic respirometer which is capable of assessing the kinetic effects of toxicant.

The 'Polytox' toxicity kit is a commercial preparation produced by Polybac Corporation (Polybac Corporation, 1986, Polytox rapid toxicity test procedure, Polybac Application Procedure, Allentown, PA). It utilises a specialised blend of bacterial cultures. This mixed bacterial culture is placed in standard biochemical oxygen demand (BOD) bottles. By introducing the wastewater sample to these bottles, the percentage reduction in respiratory activity due to the presence of toxicant is calculated relative to the control sample. All dissolved oxygen measurements are made with oxygen probes. The toxicity of wastewater and chemicals to biological treatment systems can be determined in 30 min with the Polytox toxicity procedure. The major advantage of using a standard bacterial preparation as in the Polytox kit in place of activated sludge culture is that inter-laboratory comparison studies can be established easily when using standard materials.

Luminescent microbial tests   Luminescent micro-organisms have been used to produce several toxicity test systems. The 'Microtox' test or 'LUMIStox' was developed by Beckman Instruments, Inc., as a rapid toxicity screening test with the Beckman Microtox analyser. The test uses the bioluminescence of the marine bacterium *Vibrio fischeri*, often referred to as *Photobacterium phosphoreum*, as a measure of its activity. The degree of toxicity is measured using a luminometer by monitoring the reduction in light output by luminescent bacteria on contact with toxic substances. The bioassay is based on the production of light per unit time by living luminescent bacteria, which is a reflection of the rate at which a complex set of energy-producing reactions is operating. Chemical inhibition of any of the enzymes involved will alter this rate and subsequently the amount of light produced. Toxicity is expressed as EC_{50} which is the effective concentration of a toxicant causing 50% reduction of light output during the designated time intervals at 15 °C. This test has the advantage of being rapid, simple, sensitive and reproducible. However, it may be unrepresentative of freshwater ecosystems since it uses a marine bacterium. This test is valuable in toxicity screening of effluents and leachates but its suitability for predicting the possible toxic effects of wastewaters on activated sludge from sewage treatment plants is in question.

The Microtox assay system has been used extensively to assess the toxicity of a wide range of aquatic and terrestrial pollutants. This assay appears to be the best available method with high sensitivity for field monitoring and toxicity screening of industrial effluents. The Microtox Toxicity Test System (Microbics (UK) Ltd, Hitchin, Herts, UK) collects and processes the test data automatically and prints a detailed report within 30 min of test initiation. This test has become a standard method for assessment of ecotoxicological effects of water samples. However, test development is still being carried out to improve the relevance of this bioassay to freshwater ecosystems.

The Mutatox Test (Microbics (UK) Ltd, Hitchin, Herts, UK) is an analytical test system which is used to detect the presence of genotoxic agents. This test uses a dark variant of *V. fischeri* as the test organism. When grown in the appropriate concentration of genotoxic agents, the Mutatox test strain produces light after an incubation period of 16–24 h. Suspected genotoxic agents are those samples that induce increased light levels of at least two times the average control reading. This test is used for characterising the genotoxicity of water, wastewater and hazardous wastewater materials.

Anaerobic toxicity assays Several anaerobic toxicity assay procedures have been developed to test the treatability of industrial effluents by the anaerobic sludge digesters. Changes in methanogenic activity have been used with a gas pressure lock syringe, manometrically or with a pressure transducer. On-line analysis of dissolved methane and hydrogen is carried out using membrane mass spectrometry. Some batch bioassay methods already exist for toxicity testing using municipal digester sludge as inocula source (Department of the Environment, 1978). A new assay was reported by Erasin *et al.* (1994), which uses the change in methanogenic activity through methane production rate as the monitored process parameter and compares the performance of intoxicated inocula to activity prior to adding test compound. Suspended and fixed film biomass assays were compared and their performances with chlorinated solvents and heavy metals were tested. Results showed that the suspended growth was found to be five times more sensitive to trichloroethane, but there was no clear inhibition with the heavy metals even at high concentrations (750 mg Cu l^{-1}). The Microtox test was found to be more sensitive than this test.

Whole-cell biosensors Microbial biosensors, a combination of whole-cell organisms with transducers, have opened up new and exiting field in toxicity assessment. Some of the old bioassay methods are now being supplemented by a range of sensors which can detect toxic materials by their effect on living cells or organelles. Increasing interest in environmental analysis has created a need for rapid methods and simple devices that can detect toxic substances in the aquatic environment. Biosensors are now being developed to fulfil this role. Biosensors are analytical devices, constructed by combining a biological sensing element (e.g. enzymes, antibodies, micro-organisms or DNA) with a transducer (e.g. electrochemical, optical, calorimetric or piezoelectric transducers) to obtain a measurable signal. A live-cell biosensor contains living organisms, usually bacteria, and a method of interrogating its general metabolic status. This usually involves detecting oxygen or substrate consumption, the production of carbon dioxide or metabolites, detection of bacterial luminescence, or direct electrochemical sampling of the electron transport chain.

Whole-cell based sensors for specific aspects of toxicity, such as BOD, have been developed. BOD is one of the most widely used tests in the measurement of organic pollution in wastewaters, effluents and polluted waters. Conventional methodology for assessing the amount of biodegradable substances in water takes 5 days, is expensive to perform and is subject to a high degree of fluctuation. A new BOD biosensor system has been developed to provide a fast accurate way of measuring BOD of wastewater. The BOD biosensor is based on the use of micro-organisms immobilised behind a membrane at a Clark oxygen electrode. On introducing a wastewater sample containing readily biodegradable substances into the sensor, the respiratory activity of the micro-organisms, and therefore their oxygen consumption, increases immediately. The decrease in oxygen is then correlated with the organic material present in the water sample.

Various BOD sensors using micro-organisms such as *Trichosporon cutaneum*, *Hansenula anomala*, *Pseudomonas* sp., *E. coli* and *Bacillus subtilis* have been developed. The 'BOD module' biosensor produced by Medigen (Prüfgeräte-Werk Medigen GmbH, Germany) has a BOD measuring range of 2–22 mg l^{-1} with a fast response time of less than 1 min. Other sensors such as the 'ARAS BOD Biosensor System' from Dr Lange (Dr Bruno Lange GmbH, Willstatterstr., Düsseldorf, Germany) are based on the same principle as above. This sensor has a measuring range of 6–600 mg l^{-1} BOD with response time of 60 s.

A whole-cell biosensor for the on-line detection of herbicides in drinking water has been described. The biosensor uses redox mediators to monitor the photosynthetic activity of the cyanobacterium *Synechococcus* electrochemically by transporting electrons from the cells to an electrode, resulting in the flow of current which is measured in an external circuit. This sensor is reported to be capable of detecting herbicides from the nitrile, urea, anilide and triazine families at 1–3 ppm concentrations. A mediated amperometric eubacterial biosensor system for pollution monitoring and surface water intake protection has also been reported. This biosensor uses a chemical mediator to divert electrons from the respiratory systems of the physically immobilised *E. coli* bacteria to an amperometric carbon electrode poised at 550 mV with respect to a silver/silver chloride reference/counter electrode. The sensor was reported to be capable of detecting a range of aquatic pollutants down to ppb levels within minutes of exposure.

A new immobilised cell biosensor uses the changes in the UV absorption of bacteria during metabolism as an assay for toxic compounds in water. In this sensor bacterial cells are immobilised in an agarose membrane in a flow cell. When the microbial cells are stressed so that their energy metabolism is reduced, their UV spectra change, and specifically their absorbance at 200 nm drops within 15 s of exposure to the toxin. This type of sensor has potential for the rapid and continuous monitoring of toxic contamination in water samples. Another biosensor device for the detection of metabolic activity of cells is based on a conducting polymer (polypyrrole) coated with an agarose layer containing immobilised yeast cells (*Saccharomyces cerevisiae*). When their metabolism is inhibited by a toxic substance the response declines; this makes the sensor suitable as a general monitor of toxic compounds in water.

Whole-cell biosensors for the detection of specific pollutants in water are also being developed, e.g. detection of the anionic surfactants linear alkylbenzene sulphonates, 2,4-dichlorophenoxyacetic acid, formaldehyde, sulphide and chlorinated phenols. Biosensors with immobilised active cells can show only short-term viability and stability. However, methods for improving cell stability in these types of sensors are being investigated.

Biosensor devices are suitable for mass production for rapid field tests or for on-line monitoring owing to simple manufacturing procedures and the uncomplicated measuring principle they employ. Therefore interest in these types of devices for water monitoring is increasing.

9.2.2 Biodegradability tests

Biodegradability and treatability tests are a range of test methods which aim to assess how quickly and completely a substance will be degraded once it is released into the environment and becomes subject to the action of natural microbial populations. To do this these tests use a pure culture or, more commonly, a mixed microbial population (often derived from sewage effluent or sludge flocs) to simulate the degradation of a substance in the environment. For most substances there are now standard test methods and regulations which the substance must pass before it can be marketed. For example, the problem of detergent foams in natural watercourses, which was widespread only a few years ago, has been largely eliminated by the introduction of biodegradability standards which detergents must meet (OECD, 1971). Biodegradability tests are also of value in determining discharge consent limits for industry disposing of waste to sewage treatment plants and the tests can help determine the impact of the waste on the activated sludge process. Limits can then be set as to amounts, concentrations and frequency of discharge.

9.2.2.1 *Considerations for use of biodegradability tests*

Most of the commonly used standard test methods consider aerobic biodegradation as being most representative of conditions which the test substance will encounter in the environment. However, anaerobic test methods also exist and these may be appropriate in particular circumstances, e.g. in considering biodegradation in anaerobic layers in sediments (Kameya *et al.*, 1995). All of the tests are time related and a 'pass' is given as a certain percentage degradation within a certain number of days (Table 9.2). The different test methods may also give different results for the same test substance and there is, therefore, a temptation for a manufacturer to use the test giving the most favourable results for their particular compound. A number of factors can affect the results of the test.

Table 9.2 *Proposed pass levels for biodegradability tests*

Test type	% of theoretical value or % removal	Method
Oxygen uptake	>60	Respirometry
CO_2 production	>60	Sturm test
Specific substance	>80	MOST, die-away, Bunch–Chambers, SCAS, activated sludge simulation
DOC, COD	>70	MOST, die-away
SCAS	>70 – ultimate biodegradation >20 – inherent biodegradation	SCAS

Adapted from UK SCA (1981)

The nature of the microbial inoculum used   As mentioned above, sewage bacteria are often used to provide the inoculum; however, the species present and their numbers may vary depending on source and time of sampling. This inevitably introduces a source of great variability into the test. For some standard tests the numbers or organisms which must be present are stipulated. In some cases it may be necessary to develop acclimatised cultures, by a suitable enrichment techniques, to ensure that organisms capable of degrading the substance under test are present.

The concentration of the test substance   Many substances which are biodegradable at low concentrations may be inhibitory at higher concentrations. The biodegradability test must be carried out at a concentration at which toxic effects on the inoculum are minimised. It is therefore helpful to carry out a respiration rate toxicity test to assess the inhibitory concentration. The concentration which is expected to occur in the environment when the substance is discharged also needs to be taken into consideration to ensure as realistic a test as possible.

The presence of other substances in the medium   The presence or absence of other substances in the medium can affect the ability of the micro-organisms to degrade the test substance. Many micro-organisms have a requirement for nutrients such as N, P, K and trace elements such as Fe, Zn. They may also require organic substances such as vitamins. If these are limiting in the medium then optimum growth cannot occur. In addition, while a test substance may be biodegradable it may not be sufficient as a sole source of C; another C source may have to be supplied so that the test substance can be degraded by a process of co-metabolism. Finally, consideration should be given as to whether the medium may contain other inhibitory substances.

Physical factors   The temperature, pH and dO_2 concentration of the medium must be compatible with good microbial growth. Generally this means $18-25\,°C$, pH $6.5-8.5$ and $dO_2 > 2\,mg\,l^{-1}$, although in some cases lower temperatures are used to more closely simulate the conditions found in the environment.

Definition of biodegradation   Biodegradation can be defined differently and tests can be chosen depending on the definition chosen. Primary (or functional) biodegradability measures simply the disappearance of the test substance; this may mean its conversion to another substance. However, since this product may also have unwanted environmental impacts most definitions of biodegradability focus on complete mineralisation of the substance to CO_2 and water (ultimate biodegradability). In the first case the fall in concentration of the test substance is monitored by whatever means is appropriate, in the second the loss of organic carbon is measured, either directly by chemical oxygen demand (COD) or dissolved organic carbon (DOC) or indirectly by oxygen uptake or CO_2 production.

9.2.2.2   *Biodegradability test methods*

This section will concentrate on those assays which have been adopted as standard test methods by the various regulatory bodies. However, there are many other methods described in the literature which may be appropriate in particular circumstances. A comprehensive review is given by Kilroy and Gray (1995). The methods can be classified as indirect (i.e. those where CO_2 production or O_2 is measured) or direct (i.e. the disappearance of the test substance or DOC is measured).

Indirect methods

The ratio of the 5 day biochemical oxygen demand (BOD_5) to COD, total organic carbon (TOC) or theoretical oxygen demand (TOD)   Standard methods exist for determination of BOD, COD and TOC (see Appendix 6). TOD is calculated from the formula of the test substance.

Oxygen uptake· or respiration rate   A number of different methods can be used to measure the demand for oxygen. Manometric tests involve the placing of a measured volume of the test substance together with the inoculum and medium into a closed, partially filled bottle which is connected to a manometer, e.g. the Hach manometric method. The consumption of oxygen is measured by observing the change of pressure. CO_2 evolved must be absorbed by alkali held in a small cup within the headspace of the bottle. A control bottle is set up without test substance for comparison. Adopted methods include the UK Standing Committee of Analysts (SCA) (1982), OECD (1992) and the Japanese Ministry of Trade and Industry (MITI) tests. A further method of measuring the use of oxygen is by the use of an oxygen electrode immersed in the test solution which measures the fall in dO_2. Aerated samples are introduced into a closed vessel and the rate of fall in dO_2 is measured with time and used to calculate a respiration rate. The toxicity of a test substance can also be assessed by determining the concentration which causes a 50% inhibition of respiration rate (EC_{50}). Standard methods include OECD (1971), EC (1981), UK SCA (1982) and OECD (1984, 1992). Electrolytic respirometry measures oxygen uptake by re-oxygenation of the test vessel. The test solution is sealed into the vessel with a CO_2 absorber included; as pressure in the headspace falls a sensing device causes a current to flow in an electrolytic cell also contained within the headspace. This replenishes the O_2 and the amount required is measured. This method has the advantage over the previously described methods in that the O_2 concentration remains approximately constant and perhaps more closely simulates conditions in the environment. Several manufacturers market electrolytic respirometers and their use is widespread in biodegradability testing.

CO_2 production   As an alternative to O_2 consumption the production of CO_2 can be measured. Again standard methods for this measurement exist. The Sturm test, adopted by OECD (1992), is widely used. A chemically defined liquid medium without other C sources and containing a relatively large inoculum is seeded with the test substance and aerated at a constant temperature. The CO_2 evolved is captured by a barium carbonate scrubber or alternatively an infrared analyser may be used. The CO_2 production of the test vessel is compared with that of a control. The use of ^{14}C-labelled test substance and measurement of radioactivity released can also be used but is, understandably, a less widespread method.

Direct methods

Die-away, or screening, tests   These are tests designed to screen compounds for ready degradability and are assessed by fall in DOC or COD. The modified OECD screening test (MOST) (OECD, 1992) uses a relatively low concentration of inoculum and incubation for 28 days. DOC is determined using the UK SCA (1979) method. This test is designed to identify compounds which are readily degradable; therefore failing this test does not mean that a compound is not biodegradable, merely that further testing is required. The DOC die-away test (based on a method of the UK Standing Technical Committee on Synthetic Detergents, 1966) is a similar test but using a much higher concentration of micro-organisms in the inoculum. This identifies compounds which may not pass the MOST test but which will normally be removed by sewage treatment works. The test can be used for primary degradation, in which case specific assays are required, or ultimate biodegradation, in which case DOC is followed (ISO, 1984; OECD, 1992).

Bunch–Chambers test (Bunch and Chambers, 1967)   This test is a modified die-away type test which takes into account the fact that some compounds can only be degraded when they are co-metabolised with another source of carbon. Since other carbon compounds are present measurement of primary degradation must be made using an appropriate assay. The test substance is placed in a medium containing yeast extract as the additional carbon source and inoculated with raw sewage. Weekly subcultures are made and after each 7 day incubation the level of test substance remaining is analysed. A parallel control containing a substance of known biodegradability and, preferably, a similar chemical structure to the test substance is also monitored.

Semi-continuous and continuous tests   These types of tests are used where compounds have not passed tests of ready biodegradability. The modified semi-continuous activated sludge (SCAS) procedure is adapted from a method developed by the American Soap and Detergent Association and has been adopted by the OECD (1981). In this test a semi-continuous activated sludge unit with a known concentration of the test substance is dosed with sewage at intervals and the DOC in the effluent is compared with that of a control unit without test substance. The difference in DOC is assumed to be due to residual test substance. The test can be prolonged as necessary to determine biodegradability. The Zahn–Wellens test (OECD, 1981, 1992) is a similar type of test, widely used in Germany and Switzerland, in which the test substance is mixed with activated sludge and incubated with agitation and aeration for 28 days. A high concentration of test substance allows for greater analytical reliability but does not allow for acclimatisation of the inoculum to the test substance as with the SCAS test.

Continuous simulation (activated sludge) tests are used for substances which have failed the biodegradability tests described earlier or where more information on the effect of a compound on the activated sludge process is required. They are time consuming and more expensive assays to perform and are therefore only used where absolutely necessary. These tests are intended to simulate the activated sludge process and various methods have been described (OECD, 1971, 1981; EC, 1982, WRC, 1978). The international standard method is the Husmann apparatus, a small-scale activated sludge unit, and in the UK the porous pot method (WRC, 1978) is also accepted; these differ in that they do not contain a settlement vessel. In both these types of tests there is a danger that adsorption of the test substance may be a source of error, although generally this occurs early on in the test and as it proceeds an equilibrium is reached. Biodegradation can be followed either by DOC or by analysis of the test substance. A synthetic sewage medium is generally used, for reasons of hygiene and to avoid variability in the test.

9.2.2.3   *Strategies for biodegradability testing*

In order to keep the cost of biodegradability testing as low as possible a hierarchy of tests is used (Gotvajn and Zagorc-Koncan, 1996; Nyholm, 1996). In some cases a simple determination of BOD_5:COD may be sufficient, where a value of 0.5 would indicate a readily biodegradable substance. Otherwise the zero level test is one of ready biodegradability, e.g. respirometry, Sturm test or die-away test. If the substance fails this then a test of inherent biodegradability, e.g. SCAS test, is applied. In choosing the test and conditions consideration must always be given to whether the level of test substance is inhibitory, whether the initial inoculum is sufficiently concentrated or requires acclimatisation and whether the test substance can only be degraded by co-metabolism (Bunch–Chambers test). Finally, if the substance is not satisfactorily degraded in any of these tests, a longer-term simulation test is performed. A strategy for biodegradability testing (from UK SCA, 1981a) is shown in Figure 9.1.

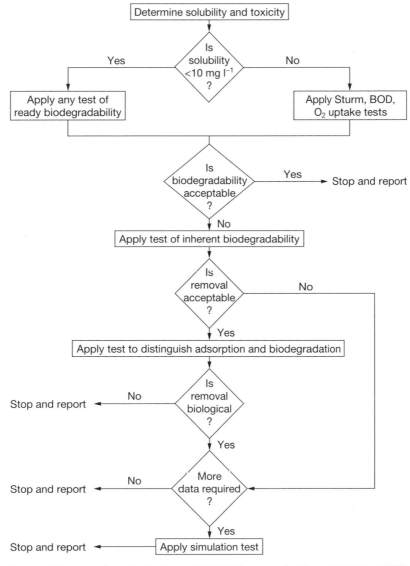

Figure 9.1 A strategy for biodegradability testing (adapted from UK SCA, HMSO, 1981. Crown copyright is reproduced with the permission of the Controller of Her Majesty's Stationery Office.)

9.3 Molecular methods

This section focuses on assays where the key biological element of the assay is an isolated cellular component or product e.g. antibodies (used in immunoassays) and nucleic acids.

9.3.1 Immunoassays

The potential of immunoassay as an emerging technique for environmental analysis is discussed in this section. The impact of immunoassays on the environmental field is evident in the extensive variety of kits which are available for the detection of trace contaminants, including pesticides, industrial residues and their degradation products. Many enzyme-linked immunosorbent assay (ELISA) kits are designed for screening tests; the result demonstrates the presence or absence of a particular analyte. For more detailed information on immunoassays in general the reader is referred to current literature (Hammock and Gee, 1995; Despande, 1996; Gould, 1996; Hermanson, 1996). Detection of environmental pollution using immunoassay techniques has grown over recent years. The presence of pesticides, insecticides, herbicides, drugs and other environmental contaminant residues in water, soil and food is an escalating problem that has aroused public concern over potential health hazards. This concern has prompted more stringent legislation related to the concentration of these residues, which made the development of a fast and inexpensive screening methods essential. Most of these analytes are generally detected using conventional methods of analysis, such as high-performance liquid chromatography (HPLC), gas chromatography–mass spectrometry (GC–MS), IR spectroscopy and nuclear magnetic resonance (NMR) which can give high sensitivity at a parts per billion (ppb or $\mu g \, l^{-1}$) range. However, these are complicated and expensive, require highly trained personnel and many entail a multi-stage cumbersome and time-consuming measurement procedure. Development of technologies which provide simpler, less expensive tests is creating a greater demand for more pollution testing, with new immunoassay formats for testing water, soil and food samples reducing the cost from approximately \$100–\$200 per test to \$3–\$20 per test. The number of rapid field-based analyses carried out is likely to increase. The impact of immunoassays on the environmental field is evident in the extensive variety of kits which are available for the detection of trace contaminants, including pesticides, industrial residues and their degradation products. There has been a trend of small start-up companies developing rapid, portable immunoassay test kits for environmental monitoring, but these have tended to be acquired by larger companies who then market the products. The immunoassay kits which are being sold are largely based on conventional colorimetric enzyme (ELISA) formats. Most of the commercial kits have been developed specifically for soil and water analysis. Although some kits specify that they are suitable for food analysis, others may also be applicable but would require an appropriate sample preparation step. Immunoassay kits, however, have not, as yet, played a significant part in environmental diagnosis. It would seem that the market for these rapid tests has not yet reached its full potential – the easier the tests are to use and the cheaper they become, the more people will use them.

Over the past decade immunoassay methods have been adapted and validated for environmental monitoring (Marco *et al.*, 1995), and were found to provide a relatively inexpensive and rapid screening methods. Immunoassays have been developed for a wide range of analytes. They are analytical tests based on the selective and sensitive antibody (Ab)–antigen (Ag) reaction and exploit the immune system's ability to produce antibodies in response to any organic molecule. The portability and ease of use have meant that immunoassay is ideal for field evaluations.

9.3.1.1 *The immune response*

The principal function of the immune system is to protect animals from infectious organisms and their toxic products. This system has evolved a powerful range of mechanisms to locate foreign cells, viruses, or macromolecules, to neutralise these invaders and to eliminate them from the body. The immune response in vertebrates can be divided into two functional responses:

1. **Non-adaptive immunity.**
   This system acts as a first line of defence, by a fixed manner through producing internal defensive cells (phagocytes, lysozyme complement and interferon). The non-adaptive system does not improve with repeated exposure to foreign molecules.
2. **Adaptive immunity.**
   This system is called on when the first line of defence is compromised and is enhanced by repeated exposure. It is characterised by two forms of response, i.e. cellular and humoral defences. The cellular immune response is effective against parasites, viral cells and cancer, while the humoral response is effective against extracellular phases of bacterial and viral infection. When a foreign molecule or antigen enters the tissue of a vertebrate, humoral immunity is responsible for producing B lymphocyte cells that are stimulated, divide and differentiate. This results in the production of plasma cells that secrete proteins called antibodies, the antibodies are highly specific for the antigen that caused the initial B lymphocyte response.

The produced antibodies can recognise and attach to the antigen to form a complex. This causes precipitation, neutralisation or death by various mechanisms including phagocytosis, followed by removal from the circulating body fluid. It is the highly specific nature of the antibody–antigen reactions that make them fundamental reagent components in immunochemical techniques.

9.3.1.2 *Antibody structure and function*

Antibodies or immunoglobulins (Ig) are a group of glycoproteins which are divided into several classes. Five classes are recognised in vertebrates, namely IgA, IgG, IgD, IgE and IgM (designated α, γ, δ, ε and μ respectively). These differ from each other in function, size, charge, amino acid composition and carbohydrate content. The IgG class is the major class of immunoglobulin in normal vertebrate serum and accounts for about 70–80% of the total immunoglobulin pool and it is utilised mainly in immunoassays. The IgG molecule is a monomeric protein with a molecular weight of 164 000 Da.

The structures of all immunoglobulins are based on a four-polypeptide chain sequence, two heavy chains and two light chains. The smaller light (L) chain has a molecular weight of 25 000 Da and is common to all classes of immunoglobulins. The larger heavy (H) chain has a molecular weight of 50 000–70 000 Da and is structurally different for each class. A schematic illustration of the IgG molecule is shown in Figure 9.2 (Harlow and Lane, 1988). Both the light and the heavy chains contain different regions. Constant regions (C_L and C_H) do not vary much between antibodies of the same class. The variable regions (V_L and V_H) are different between antibody classes. The chains are arranged to form a flexible Y shape stabilised by an intermolecular disulphide bonds, with a hinge region which gives the antibody molecule its flexibility. Variations in the amino acid residues in the N-terminal portion (variable region of Fab) of the molecule result in the unique topography of the antigen binding site and account for the specificity of each antibody. The stalk fragment of

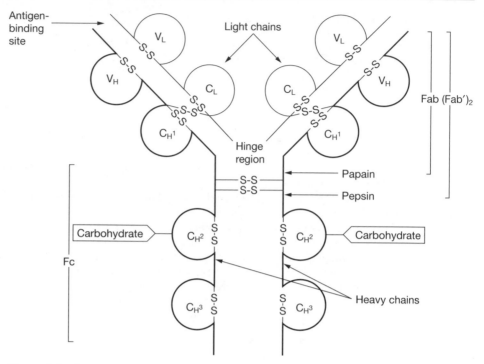

Figure 9.2 A schematic of the structure of IgG (adapted from Harlow and Lane, 1988)

the antibody (Fc region) contains the site where the antibody binds to the cell and plays no part in antigen binding. A fundamental function of all antibodies is the recognition and binding of target antigens. Antibodies typically have two or more identical antigen-binding sites. Each of these sites is formed by an interaction between a variable domain located at the N terminus of a heavy-chain polypeptide and a variable domain located at the N terminus of a light-chain polypeptide.

Antibody fragments that contain the variable domains often retain full antigen-binding activity. Such fragments can be produced either by proteolytic cleavage of whole antibodies or by recombinant DNA methods.

9.3.1.3 *Immunogens*

The ability of a molecule to induce an immune response is determined by:

1. the chemical structure of the molecule;
2. the host animal.

For a compound to elicit antibody production, it must contain an epitope that can bind to the cell-surface antibody of a virgin B cell and promote cell-to-cell communication between B cells and helper T cells. Binding site imposes a minimum size limit on an immunogen, and molecules of less than about 3000–5000 Da are generally not good immunogens. Proteins, peptides, carbohydrates, nucleic acids and many other naturally occurring or synthetic compounds can act as successful immunogens. However, most organic pollutants of current interest are small molecules and have a molecular weight of less than 1000 Da. These small chemicals can be used to raise antibodies, if they are coupled to larger protein molecules. The small compounds are known as 'haptens', while the proteins to which they are coupled to are called 'carriers'. Bovine serum albumin (BSA) and keyhole limpet haemacyanin (KLH) are both popular carriers. However, the carrier protein must be from a different species to the host animal if a strong immune response is to be induced. Often, the target compound contains a functional group which can be covalently coupled to the carrier molecule. If not a suitable group (e.g. $-NH_2$, $-COOH$, $-OH$, CO) can be introduced (Sherry, 1992). The coupling mechanism will vary with each hapten. The design and preparation of the immunogen can have a major influence on an assay's characteristics. The immunogen is then purified and characterised and used to induce the production of polyclonal antibodies (PAbs) in a suitable host animal.

9.3.1.4 *Production of antibodies*

An antibody of the desired affinity and specificity is vital to the process of developing immunoassay technology. In order to produce the antibody, the preparation of a suitable immunogen of the analyte of interest is required. Once the immunogen has been prepared as described above, it is introduced into the host animal to induce the production of antibodies.

Polyclonal antibodies   Most of the antibodies used in immunochemistry are raised or induced by injection of a solution or suspension of the appropriate antigen into a rabbit. A wide range of vertebrate species can be used for the production of antisera: mice, rats, hamsters, guinea-pigs. For large-scale production, sheep, goats, pigs, donkeys and horses are commonly used. Inoculating an animal with a single injection of a strongly antigenic compound will result in the production of specific antibodies that are detectable in its serum after about 10 days, reaching a maximum after 15–20 days. Following a series of inoculations, blood is taken from the animal and the serum is separated from it. The resulting liquid is termed antiserum and is a complex mix of proteins and specific antibodies directed against the inoculated immunogen. Because of the multiplicity of antigenic sites on any single immunogen or of impurities in the immunogen inoculated, a heterogeneous mixture of different antibodies of varying specificity and affinity are produced, i.e. PAbs. Purification of immunogen helps to narrow the range of affinities and specificities of the induced PAbs.

Monoclonal antibodies   When antibodies of monospecificity are required, antibodies can be derived from single cell lines using hybridoma technology. Monoclonal antibodies (MAbs) are produced by inoculating mice with an antigen. When the mouse is producing antibodies of a high titre the spleen is removed and a cell suspension prepared. The suspended cells are fused with myeloma cells and the fused cells (hybridomas) are purified, cloned and screened for ability to produce high-affinity Abs of the required specificity. Selected clones are used to produce Mabs in cell culture. The result is a clone of cells that is derived from a single progenitor, which both is immortal and produces a single species (monoclonal) antibodies.

Hybridoma technology is not problem free since MAbs are costly to produce, the chromosome complement of the hybrids can be unstable and assay specificity can be too narrow for some screening tasks.

Recombinant antibodies Hybridoma technology had a significant impact on antibody production over the last 10 years. The use of recombinant antibodies is having the same effect today. By using antibody engineering, antibodies can be produced at a fraction of the cost of the production of poly- or monoclonal systems. However, because of the difficulty in cloning, assembling and expression of antibody molecules, this task is complex. There are two main types of recombinant antibodies: produced by a cloning system (using an existing monoclonal cell line) and produced *in vitro* (bypassing the animal entirely).

9.3.1.5 *Antibody–antigen interaction*

The interaction of an antibody with an antigen forms the basis of all immunochemical techniques. Antibodies contain in their structure recognition–binding sites for specific molecular structure of the antigen. The interaction is highly specific and follows the basic thermodynamic principles of any reversible biomolecular interaction:

$$Ab + Ag \rightleftharpoons Ab\text{--}Ag$$

The affinity constant $K_A = [Ab\text{--}Ag]/([Ab] + [Ag])$, where $[Ab]$ is the molar concentration of unoccupied Ab binding sites, $[Ag]$ is the molar concentration of unoccupied Ag binding sites and $[Ab\text{--}Ag]$ is the molar concentration of the Ab–Ag complex.

The interaction is divided into three sections:

1. the structure of the Ab–Ag bonds;
2. the strength of these interactions, a characteristic known as affinity;
3. the overall stability of immune complexes, a property called avidity.

The region of an antigen that interacts with an antibody is called an 'epitope'. Because antibodies can recognise relatively small regions of antigens, occasionally they can find similar epitopes on other molecules. This forms the molecular basis for cross-reaction.

The binding of the Ab–Ag is entirely dependant on reversible non-covalent interactions:

1. van der Waals forces;
2. electrostatic bonds;
3. hydrogen bonds;
4. hydrophobic interactions.

The resulting complex is in equilibrium with the free compounds. The immune complex is stabilised by the combination of the above weak interactions that depend on the precise alignment of the antigen and antibody. These stringent binding requirements between the Ab and Ag make immunoassays so selective. Small changes in antigen structure can affect profoundly the strength of the interaction. Also, changes in the epitope structure can prevent antigen recognition. Therefore antibodies have been isolated that will differentiate between conformations of protein antigens, detect single amino acid substitution or act as weak enzymes by stabilising transition forms.

9.3.1.6 *Labels used in immunoassays*

Labels are employed in immunoassays to detect the immunological reaction. There are different types of labels used in immunochemical techniques to monitor the antibody–antigen binding reaction including latex particles (blue latex), radioisotopes (^{125}I, ^{3}H), metal and dye sols (colloidal gold, fluorescent chromophore), enzymes (horseradish peroxidase, alkaline phosphatase and β-D-galactosidase), substrates and cofactors. Although fluorescent and chemiluminescent labels have been gaining popularity in the last few years, enzyme labels are still the most popular in environmental immunoassays. Enzymes are extremely useful as labels as their catalytic properties allow the detection and quantitation of low levels of analytes and their signal can be easily detected through colour formation. It is also possible to amplify the signal by several orders of magnitude in an enzyme immunoassay by the use of enzyme amplification systems. Depending on the assay format, the label is incorporated into either the Abs (primary or secondary) or the analyte (antigen or hapten).

9.3.1.7 *Classification of immunoassays*

The first major distinction between immunoassays is:

(a) **Homogeneous immunoassay.** A homogeneous system does not require separation of free and bound antigen: the assay relies on the alteration of the properties or function of the label on formation of the antibody–antigen complex. For example the Ab–Ag interaction will either inhibit or enhance the enzyme label used in the assay. The assays are simple and easy to automate and therefore are commonly used in the diagnostic industry, such as the enzyme multiplied immunoassay technique (EMIT), enzyme channelling immunoassay (ECIA) and enzyme modulator-mediated immunoassay (EMMIA).

(b) **Heterogeneous immunoassay.** In a heterogeneous system there is a separation step to remove the unbound reagents before the tracer is determined. This assay format is a more sensitive approach, is less prone to interference and is most commonly employed in test kits.

Detection techniques in immunoassays can be divided into two groups:

(a) **Direct detection.** The antigen-specific antibody is labelled and used to bind to the antigen.

(b) **Indirect detection.** The antigen-specific antibody is unlabelled and its binding to the antigen is detected by a secondary reagent, such as labelled anti-immunoglobulin antibodies.

The choice of the direct or indirect method depends on the test. The use of directly labelled antibodies in an immunoassay involves fewer steps and is less prone to background problems, but it is less sensitive than indirect methods and requires a new labelling step for every analyte to be tested. In contrast, indirect methods offer the advantages of widely available labelled reagents which can be used to detect a large range of antigens and are available commercially. Since the primary antibody is not modified by the label the loss of activity is also avoided.

In immunoassays either Abs or Ags are immobilised on a solid phase and the most-used configurations are shown in Figure 9.3 and are listed below.

Competitive immunoassay   Competitive assays are usually used for small molecular weight compounds. Analytes of environmental importance including pesticides are too small to allow binding of two Abs simultaneously and therefore competitive assays are usually used in the diagnosis of these compounds. In an Ab coating format, these tests work on the principle that there is competition for a limited number of binding sites on an antibody between the analyte and a labelled form of analyte. An equilibrium will be established between the Ab bound to the solid surface, the analyte, and the analyte–enzyme

Direct competitive ELISA

Indirect competitive ELISA

Sandwich ELISA

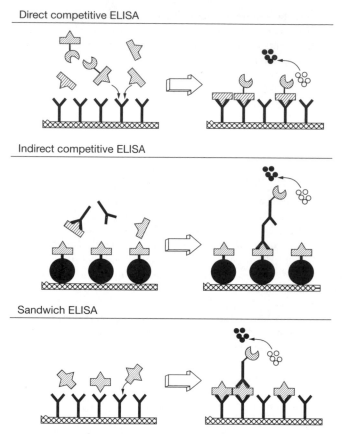

Figure 9.3 Most commonly used configurations for immunoassays (adapted from Marco M P, Gee S and Hammock B C 1995 *Trends Anal. Chem.*, **14**; 341)

tracers which are in solution. After the incubation period the unbound reagents are washed away and the amount of enzyme bound to the solid phase by the Ab is measured. A reduction in the enzyme activity is directly proportional to the amount of analyte present. In an Ag coated format, a competition between the immobilised Ag and the analyte for a fixed amount of antibody–enzyme tracer will take place.

Sandwich immunoassay   This assay differs from the competitive in several ways: it utilises two antibodies, the antibodies are present in excess not limiting concentrations and one of the antibodies is labelled with the enzyme tracer. There are many variants in the assay format but the most common uses one antibody immobilised to a solid surface, e.g. well or tube. This antibody recognises one site on the analyte and binds after the sample is added. The second antibody is added which binds to the antigen and so forms the sandwich. The signal generated is proportional to the amount of analyte. This type of format is used to detect analytes with a molecular weight that can allow the binding of two Abs simultaneously.

Displacement immunoassay   In this assay the antibody is immobilised on a solid phase and the antigen is labelled. At the start of the assay all of the available binding sites on the immobilised antibodies are occupied by labelled antigen. On addition of the unlabelled antigen there is a displacement of labelled material and under appropriate conditions the extent of this displacement will be dependent on the amount of analyte. This format is similar to competitive assays.

9.3.1.8 *Immunoassays for environmental analysis*

Immunoassay techniques, particularly ELISA, have several attributes that make them ideal for environmental monitoring. These include their selectivity, sensitivity, rapid turnover time (<2 h), simplicity and portability. These kits can be tailored to target specific analyte or a class of analyte at very low concentrations in samples containing many other compounds. The robust simplicity of immunoassays can generate on-the-spot data that can be used to make quick decisions. Manufacturers of environmental immunoassay kits have invested in research that aims to correlate the results of immunoassays with conventional and accepted laboratory methods such as HPLC and GC–MS. The aim being to encourage regulatory bodies such as the environmental agencies to use them for routine monitoring. The results of several studies have shown that ELISA kits compare favourably with conventional methods (Selim *et al.*, 1997).

9.3.2   Nucleic-acid-based tests

Detection of nucleic acids has been widely used in molecular biology laboratories over the last few decades and is now beginning to make an impact on diagnostics for use in medical, food and environmental applications as the assays are developed and packaged in such a way as to make them usable by the non-specialist.

9.3.2.1 *Nucleic acids*

Nucleic acids as the genetic material   All organisms contain nucleic acids as the genetic material, that is the heritable material that is used to accurately pass information from one generation to the next. In the majority of organisms the genetic material is DNA (deoxyribonucleic acid) but in some viruses it is RNA (ribonucleic acid). In eukaryotic organisms the DNA is present as several linear chromosomes and in prokaryotes it is present as a single, circular chromosome. Further, extrachromosomal DNA is present as circular plasmids and also in organelles, such as mitochondria and chloroplasts. The genes determining an organism's characteristics are sequences of DNA, within the chromosome, which encode proteins. In eukaryote organisms there is also much DNA which does not have a protein encoding function, some of which may have other functions, e.g. in stability, inheritance and replication of the chromosomes. Human beings are believed to have about 50 000 genes arranged on 46 chromosomes in a diploid cell, i.e. a haploid genome of 23 chromosomes. There are about 3 pg of DNA in the human haploid genome (1 pg of DNA is approximately 1×10^9 base pairs) compared with 0.004 pg in the genome of the bacterium *E. coli*.

Structure and replication of DNA   The structure of DNA is the famous double helix elucidated by Watson, Crick and Wilkins, using the X-ray crystallography results of Franklin, in 1953 (Figure 9.4). The helix consists of two sugar phosphate backbones, running antiparallel to one another; each sugar residue carries one of four nucleotide bases

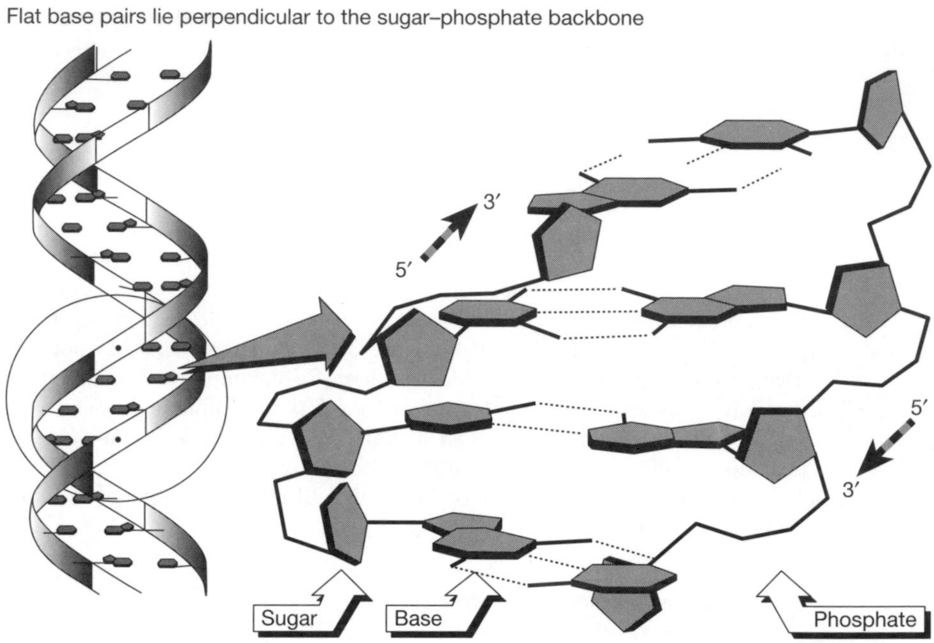

Flat base pairs lie perpendicular to the sugar–phosphate backbone

Figure 9.4 The double helix structure of DNA (Reprinted with permission from Lewin B 1994 *Genes V*, Oxford University Press, Oxford)

(the purines – adenine and guanine; the pyrimidines – cytosine and thymine). It is the hydrogen bonds between bases on opposite strands which hold the helix together and form the base pairs. For steric reasons this hydrogen bonding can only occur between defined base pairs; an adenine on one strand will always pair with a thymine on the opposite strand and cytosine will always pair with guanine (Figure 9.5). It is this specificity of base pairing which makes DNA ideal as the genetic material (the DNA can be replicated accurately as cells divide – see below) and which also makes it such a powerful diagnostic tool. DNA is replicated in a semi-conservative fashion, that is to say that during replication the helix is unwound and the hydrogen bonds between the base pairs are separated and each single strand acts as a template for the synthesis of a new, complementary strand. Synthesis of DNA proceeds in a 5′ to 3′ direction with fresh nucleotides, carrying the appropriate base to complement the next base on the template strand, added by the action of DNA polymerase enzymes to the 3′ end. The fact that synthesis only proceeds in the 5′ to 3′ direction gives rise to a leading strand and a lagging strand during synthesis (Figure 9.6). DNA polymerase requires an RNA 'primer' to provide the 3′ end on which to start addition of nucleotides. This is synthesised by an RNA polymerase which does not require a primer. On the leading strand once priming has occurred synthesis of new DNA can proceed continuously but on the lagging strand as the helix unwinds to expose new template new primers are synthesised and DNA synthesis proceeds in a discontinuous manner. The RNA primers are removed by the 5′ to 3′ exonuclease activity of DNA polymerase and the ends of the new DNA fragments are joined by the action of DNA ligase. Thus, from a single molecule of DNA two new identical molecules are formed and this allows the organism to copy the genetic information intact when cells divide or gametes are produced.

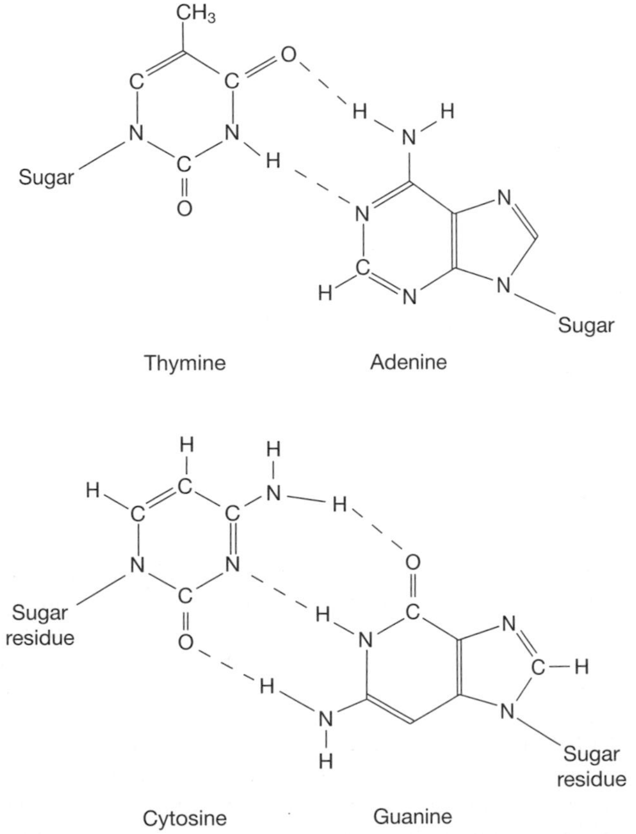

Figure 9.5 Hydrogen bonding between complementary base pairs. Two for the Thymine/Adenine pair and three for the Cytosine/Guanine pair

Nucleic acids as diagnostic tools   The specificity of the base pairing in nucleic acids means that they can be used as diagnostic tools. Where DNA sequence information from a particular gene or organism is known then a complementary probe sequence can be designed and produced, either by chemical DNA synthesis, for which fully automated machines are now available, or from a cloned sequence, typically maintained as a plasmid in a host bacterium such as *E. coli*. These probes can be used for identification of species at several levels – sequences can be chosen such that they can be used to identify one specific organism or they can be chosen such that they will recognise DNA from many organisms, for example a gene that is conserved throughout a particular group of bacteria. Nucleic-acid-based diagnostics have been particularly applied in microbiology, where identification of species may be otherwise based on long-winded culturing and biochemical test techniques, and also in forensic science. Environmental applications of nucleic-acid-based tests include identification of micro-organisms, detection of specific gene functions in a sample, e.g. mercury resistance or PCB-degrading enzymes, and following the fate of genetically manipulated organisms, or genes from those organisms, after release.

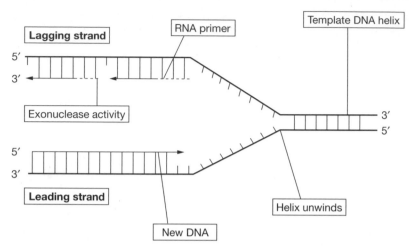

Figure 9.6 The principle of DNA replication

Two techniques are widely used in nucleic-acid-based assays, namely hybridisation and the polymerase chain reaction (PCR). Hybridisation relies on the ability of a given sequence of DNA to find and bind to, by base pairing, its complementary sequence. If the probe sequence is labelled by some means then the fact that this binding has occurred can be detected by detection of the label, be it a radionucleotide, fluorescent group or an enzyme capable of giving a colorimetric or chemiluminescent reaction (see following sections). The PCR is a means of amplifying a given DNA fragment from a small amount of starting (template) material or where you wish to amplify a particular fragment from a background of unrelated DNA. The reaction relies on the user supplying two primer sequences, which hybridise to the complementary sequence in the template DNA and the action of DNA polymerase to extend from the 3′ end of each primer adding bases complementary to the template strand of DNA.

9.3.2.2 *Hybridisation-based methods*

As already stated the use of nucleic acid hybridisation as a diagnostic tool relies on the complementary base pairing (A + T; G + C) between the probe and the sample to be tested. Most commonly the probe is a known sequence of DNA, since it is more stable and easily manipulated than RNA, and the sample, or target, may be DNA or RNA. The stringency of the hybridisation reaction is controlled not only by the probe sequence but also by the temperature and the salt concentration of the buffer used. At low temperatures and higher salt concentrations some mismatch between the probe and target may be tolerated, whereas as the conditions become more stringent the match between probe and target must be more exact in order for the probe to remain bound to the target. Any hybridisation protocol shares the same basic principles and processes: the probe is first labelled, the target is denatured, by heating or by alkali treatment, the probe is hybridised to the target, excess probe is removed and, finally, the bound probe is detected.

Preparation and labelling of probes The most common type of probe is double-stranded DNA (dsDNA). These probes have the advantages of being easy to prepare from cloned fragments or whole plasmids and by cloning random fragments (i.e. making gene libraries) probes can be made from organisms where DNA sequence information is not available. Labelling of dsDNA is straightforward and kits using various different strategies are commercially available. Oligonucleotides (i.e. short DNA sequences, usually less than 40 base pairs) can be chemically synthesised and used as probes. These have the advantage that they can be produced in large quantities and are useful where highly specific assays are required, for example in determining point mutations. Oligonucleotides can also be used as primers in PCR reactions. Single-stranded DNA and RNA can also be used as probes; although not so easy to use they do tend to give increased sensitivity as there is no re-annealing of the complementary strands of the probe. Single-stranded DNA probes are produced by cloning dsDNA into a plasmid vector containing a promoter, e.g. M13, which allows production of single-stranded DNA. RNA probes are made by cloning DNA into a vector with a suitable promoter, e.g. the SP6 promoter using SP6 polymerase to make an RNA transcript.

In order to determine whether hybridisation has occurred between the target and probe DNA the probe is labelled. The majority of hybridisation assays can be described as heterogeneous, meaning that after hybridisation there is a washing step to separate bound from unbound probe. When DNA hybridisation protocols were first developed the only available labels were radiolabelled nucleotides containing either ^{32}P or ^{35}S which are detected by exposing X-ray film (autoradiography). For reasons of safety and waste disposal problems these are now being superseded by non-radioactive methods. All of the following methods are available in the form of labelling kits from manufacturers such as Amersham/Pharmacia and Boehringer Mannheim:

- Biotin; incorporation of biotin dUTP into the probe is detected using a conjugate of streptavidin + a reporter group (enzyme, fluorescent group).
- Enzymes; direct attachment of enzymes, e.g. alkaline phosphatase, horseradish peroxidase, to probes. These enzymes can catalyse colorimetric or chemiluminescent reactions when supplied with appropriate substrates.
- Fluorescent groups; e.g. fluorescein, rhodamine.

There are many companies in the market of synthesising oligonucleotides and if the probe to be used is an oligonucleotide this can be ordered with the appropriate label already attached. Otherwise there are three methods of incorporating labelled nucleotides into a probe. The method is chosen according the type of label and the length of the probe to be labelled:

- **Nick translation.** Nicks are introduced into the double-stranded probe DNA using Dnase I digestion. The nicked DNA is incubated with DNA polymerase and labelled nucleotides are incorporated in the reaction mix. The polymerase synthesises new DNA from the exposed 3' ends at the nicks and the enzyme's 5' to 3' exonuclease activity degrades the existing strand, thus replacing it with a labelled strand. This labelling method generally works best with larger probes (>1000 base pairs) and the labelled products generated are around 600 base pairs.
- **Random priming.** The dsDNA probe is denatured to give single strands and hexamers of random sequence are annealed to the single strands. The 3' ends of these hexamers act as priming sites for DNA polymerase which synthesises new, complementary strands of DNA and incorporates the labelled nucleotides. Again this method is used for larger probes and gives labelled products of 100–500 base pairs.

- **End labelling.** This method is only suitable for short probes, e.g. oligonucleotides, as label is only incorporated at one end of the probe. Labels can be added at the 3′ end using the enzyme terminal transferase or at the 5′ end using T4 kinase.

As an alternative to labelling by incorporation of labelled nucleotides probes can be covalently linked to enzyme labels. This is achieved using various 'linker' molecules, and oligonucleotides with enzyme labels can be ordered from various companies. Amersham (UK) also market the ECL Direct system in which positively charged horseradish peroxidase is bound to the negatively charged DNA molecule and then covalently cross-linked. This method works best for larger probes.

Hybridisation assays In most hybridisation-based assays one of the components, either the target nucleic acid or the probe, is immobilised on a solid support. This has the effect of slowing the hybridisation rate but makes the washing step, to remove unbound probe before detection of the hybridised probe, much easier. It also prevents the target DNA from re-annealing to itself. The common laboratory techniques of Southern and dot blotting involve immobilisation of the previously denatured target DNA onto a membrane (e.g. nitrocellulose or nylon). A range of membranes suitable for different blotting techniques are commercially available. In Southern blotting the DNA is first separated according to size by gel electrophoresis and the DNA is transferred from the gel onto the membrane. In dot blotting spots of the target DNA are applied directly to the membrane. The membrane is then immersed in a hybridisation buffer containing, among other components, the labelled probe, a non-related DNA (commonly salmon sperm or calf thymus) to block non-specific binding of the probe, salts and SDS. The stringency of the hybridisation can be controlled by altering the temperature of incubation and the salt concentration of the buffer. Following the hybridisation the membrane is washed to remove unbound probe. Again, stringency of the wash is controlled by temperature and salt concentration. Finally the probe remaining hybridised to the target DNA is detected in whatever manner is appropriate for the label used.

These laboratory techniques are relatively long winded and require much handling of the membrane. Hybridisation assays are now being developed as kits for detection of specific sequences using novel assay formats and detection methods. The membranes used as the solid phase in blotting can be replaced by latex beads, magnetic beads, microtitre plates and plastic test tubes. In some kits affinity capture of the double-stranded hybrid is used allowing hybridisation to occur in solution and reducing the time required. Affinity capture can be achieved using hydroxylapatite, which only binds double-stranded hybrids, or by using antibodies specific for dsDNA or DNA:RNA hybrids. For example, the Digene Inc. (USA) Hybrid Capture™ system captures the hybrid formed between a DNA target and unlabelled RNA probe using an immobilised anti-DNA:RNA antibody. This is then detected using an antibody–enzyme conjugate and a luminescent substrate; the whole assay is carried out in an ELISA-type format in microtitre plates or in plastic tubes. This assay format is used in the hepatitis B, cytomegalovirus and human papillomavirus detection kits marketed by Murex Diagnostics (UK).

A technique also used in commercial kits is 'sandwich' hybridisation, where the format of the assay is analogous to that of a sandwich ELISA. A 'capture' DNA sequence is immobilised, e.g. on the surface of a microtiter plate, and a mixture of the target DNA and a labelled 'probe' DNA is added. The sequence of the capture and probe DNAs is such that they hybridise to adjacent regions of the target. In this way when the correct target sequence is present it forms a bridge between the immobilised capture and free, labelled

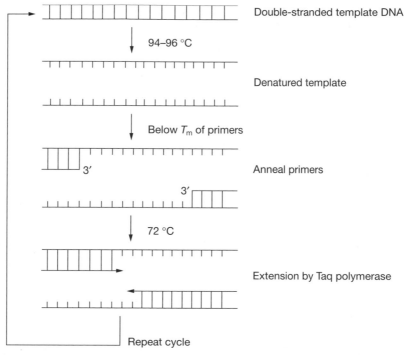

Figure 9.7 The principle of the PCR

probes. Excess probe is washed away and the amount of probe remaining bound to the surface by the sandwich formed is measured. This type of assay has the advantage of achieving very good sensitivity and specificity. The Nunc Inc. (USA, now Nalge Nunc International) Microwell™ assay can achieve attomole detection of target DNA.

There is currently much interest in developing methods of direct detection of nucleic acid hybridisation, that is the double-stranded product itself is detected rather than a label. Such devices have been termed 'genosensors'. Of the techniques being researched the optical approaches of surface plasmon resonance and evanescent wave measurements, where changes occurring as a result of binding events on the surface of a prism can be measured, seem to be the most promising. The Pharmacia (Sweden) Biacore instrument, originally developed for use with antibody:antigen binding assays, can also be used to measure DNA hybridisation occurring on the surface. Hybridisation with a probe is also widely used as a means of detecting the amplified products of the PCR (see the following section) and, again, novel assay formats, replacing the older blotting techniques, are coming onto the market.

9.3.2.3 *Polymerase chain reaction*

The PCR was introduced in 1985 and is a technique for amplifying a specific fragment of DNA either from a background of other, non-related, sequences or from very low amounts of template starting material. The technique in its simplest form uses two oligonucleotide primers (typically 20–30 base pairs) which anneal on opposite strands of the template to be amplified such that the 3′ ends of the two primers face inwards (Figure 9.7). The template

DNA is denatured to yield single strands which anneal to the primers. A DNA polymerase enzyme adds nucleotides at the 3′ ends of each primer using the template DNA and adding the complementary bases. Thus after the first round of replication the number of copies of the target sequence has doubled. Further cycles of denaturation, primer annealing and extension thus give an exponential increase in the number of copies of the target sequence. The cycling is carried out in an automated thermocycler, as the different stages in the cycle require different temperatures. The denaturation of the template DNA is carried out at 94–96 °C; annealing of the primers is carried out at a temperature dependent on the melting temperature (T_m) of the primers, which is calculated from the percentage G + C in their sequence; extension by the polymerase is carried out at 72 °C. The polymerase most commonly used is Taq, from the thermophilic organism *Thermus aquaticus*; this is used as it is able to withstand the high temperatures needed to denature the template DNA. Polymerases from other thermophilic organisms are also available. The use of polymerases from mesophiles would necessitate adding new enzyme at each extension step and would, therefore, make the process very laborious. The conditions of the PCR reaction can be altered such that templates of different lengths can be amplified, from a few tens of base pairs up to several kilobase size fragments.

The specificity of PCR comes from the design of the primers. These are designed from known DNA sequences from the target organism and can be designed such that they recognise particular genes of interest in several organisms or sequences which will only amplify a sequence from a specific organism. As with hybridisation the stringency of primer annealing can be altered by changing the temperature during the annealing stage. The closer the temperature is to T_m of the primers the more stringent the reaction will be as mismatches between primer and template will not be tolerated; as the annealing temperature is lowered some mismatch may be tolerated and allow amplification to occur. This can be exploited, for example if the sequence of a gene is known in one organism and one wishes to look for related genes in other organisms. In designing primers there are a number of factors which need to be considered:

1. The intended size of the product – very short products are less easy to identify while very large products are more difficult to amplify. Generally products between a few hundred and 2000–3000 base pairs are straightforward to amplify and identify.
2. The primers should be at least 20 base pairs long to give a specific product. Very short primers are more likely to anneal at multiple sites on the target DNA.
3. Primers are usually chosen such that their G + C content is approximately 50%. This gives annealing temperatures which are neither too high nor too low. The T_m values of the two primers should be as close to each other as possible.
4. The sequence of the primers should be checked to ensure that they do not have the potential to fold and form hairpin loops with themselves and that there is not complementarity between the primers. If the latter occurs there is potential for the primers to anneal to one another and yield artefactual products, known as primer dimers, in the PCR process.

There are a large number of textbooks and laboratory manuals now available giving details of the design and optimisation of PCR reactions (e.g. White, 1993; Griffin and Griffin, 1994). In addition many of the manufacturers of enzymes and equipment for PCR produce excellent manuals and have technical helplines. The patents on the PCR process are owned by Hoffman-La Roche, manufacturers of thermocyclers, enzymes for PCR and PCR-based kits for detecting specific sequences hold licences from the patent holder and their products are sold with a licence for the purchaser to use PCR in research applications. However, if a commercial application for PCR is developed then a separate licence must be purchased.

Detection of PCR products The most commonly used laboratory technique for detection of PCR products is agarose gel electrophoresis followed by ethidium bromide staining. This allows the size of the amplified product to be determined, by comparison with DNA size markers separated on the same gel, but does not confirm that a specific sequence has been amplified. The detection limit for ethidium bromide staining is typically around 10^4 copies. The Bax™ system from Qualicon (USA) is an automated system for gel electrophoresis and identification of bands of a specific size. Currently the system is available with primers for identification of *Salmonella*, *E.coli* O157:H7 and *Listeria monocytogenes* and is aimed at the food industry for hygiene testing.

In order to confirm amplification of a specific sequence the gel may be Southern blotted and hybridisation with a nucleic acid probe performed as described previously. This increases the sensitivity of detection and provides information on the sequences amplified. This process of hybridisation to identify PCR products may also take the form of a sandwich hybridisation in microtiter wells. Nalge Nunc International market a system for solid phase amplification of DNA allowing rapid detection of the product in an ELISA type system. This system is known as DIAPOPS (detection of immobilised amplified product in a one-phase system); it makes use of the Micro Well plate format with NucleoLink strips, which are coated in such a way that nucleic acids can be covalently bound to their surface. Primer 1 for the PCR reaction is immobilised on this surface and then the rest of the reagents for PCR are added to the well. The difference from a 'normal' PCR is that the ratio of primer 1 to primer 2 in the liquid phase is 1:8 rather than the normal 1:1. As amplification proceeds the amplified product hybridises with the immobilised primer 1 and acts as a template for further amplification. Thus, at the end of the reaction, there are two populations of amplified product, one in liquid phase and one immobilised. The liquid phase is removed and the wells are washed. The remaining immobilised product is then denatured by alkali treatment, leaving single-stranded DNA which can be detected by hybridisation with a suitably labelled probe in an ELISA-type format.

A number of methods of detecting PCR products require the product to be labelled and this is conveniently achieved by using labelled primers. In the Sharp* Signal System Assay (Murex Diagnostics, UK) and the Enviroamp *Legionella* assay (Perkin Elmer, USA) the label used is biotin, enabling the PCR products to be recognised by streptavidin, a protein with a strong affinity for biotin. In the Sharp* system the biotinylated PCR product is immobilised on the surface of a streptavidin-coated microplate and hybridised with an RNA probe. This RNA:DNA product is then detected using an anti-RNA:DNA antibody conjugated to alkaline phosphatase, which is detected by use of a colorimetric substrate. In the Enviroamp kit a probe DNA is immobilised on a test strip and hybridises with the biotinylated amplification product from *Legionella*. The biotin is then recognised by a streptavidin–horseradish peroxidase conjugate and, again, a colorimetric substrate is supplied and a positive result is shown by a coloured spot on the test strip.

In the Captagene-GCN4 system (Pharmacia, Sweden) the two primers used are differently labelled: one is biotinylated and the other contains a sequence recognised by GCN4 (a DNA-binding protein) at its 5′ end. The PCR product is immobilised on the surface of a microtitre plate coated with GCN4 Fusion protein which binds the GCN4 recognition sequence present on one primer. The biotin on the other primer is then detected using a streptavidin–horseradish peroxidase conjugate. Unlike the previous two kits this does not provide confirmation (by hybridisation) that a specific sequence has been amplified. However, the manufacturers recommend that specificity can be ensured by using the technique of 'nested' PCR. In this process an initial PCR reaction using unlabelled primers is performed to increase the amount of desired target present, and then a further PCR using the labelled primers is performed. The second set of primers is designed such that they amplify a fragment internal to the fragment amplified by the original primers.

The Gene Chip™ being developed by Affymetrix Inc. (USA) is a novel approach allowing the screening of PCR products by hybridisation to very large arrays of probe sequences immobilised on the surface of a chip. The user amplifies the target sequence using primers labelled with fluorescent dyes. Hybridisation to the sequences on the chip is carried out automatically by a fluidics station and a scanner using an argon-ion laser to excite the fluorescent dye reads the intensity of hybridisation occurring for each of the probes on the chip. For example the Gene Chip HIV PRT probe array carries over 15,000 different probes and is designed to perform highly accurate sequence analysis on parts of the HIV-1 virus. Affymetrix is currently licensing the Gene Chip technology to companies interested in developing DNA-based diagnostics. These have largely been for medical applications, to date; for example, an agreement has recently been made with bioMérieux Vitek Inc. (USA, France) to develop assays for identification of bacteria and antibiotic resistance analysis. While initially aimed at the medical diagnostics market such assays may also prove useful in food and environmental analysis.

A further novel means of detecting PCR products is the TaqMan™ technology produced by Perkin Elmer Applied Biosystems Inc. (USA). In this assay a short DNA probe, which anneals at a site lying between the forward and reverse PCR primers, is included in the reaction mix and anneals to the denatured template DNA. This probe is labelled with two fluorescent dyes, a reporter and a quencher; while the two dyes are in close proximity to one another on the probe the fluorescence emission from the reporter is quenched. However, as Taq polymerase extends from the PCR primer and reaches the $5'$ end of the annealed probe the $5'$ to $3'$ exonuclease activity of Taq polymerase degrades the probe and releases the reporter dye. Once it is no longer in proximity to the quencher the fluorescence of the reporter is detectable. As the PCR cycling continues more reporter is released and the intensity of fluorescence increases. This system has the advantage that it not only detects formation of the correct PCR product but can be used to quantify the amount of starting template DNA in the reaction. This may be used, for example, to infer the number of bacteria present in a sample.

Normally PCR is difficult to use as a quantitative technique – in order to extrapolate back to the amount of template measurement of product must be made during the exponential phase of the reaction. However, detection techniques such as gel electrophoresis and ethidium bromide staining are not sensitive enough to detect product early in the reaction and multiple samples must be taken through the cycles in order to be certain that measurement in the exponential phase is made. Often a known amount of an internal control (a template amplified by the same primers as the sample DNA but giving a different size product) is included in the reaction for comparison with the amount of sample amplified. This is to take into account the effect of any inhibitory substances which may be present in the PCR mix, e.g. from sample contaminants. The TaqMan system, when used in conjunction with a thermal cycler equipped with a fluorimeter (7700 Sequence Detector, Perkin Elmer Applied Biosystems), can make real-time measurements of the progress of a PCR reaction and from the cycle at which the fluorescence is first detectable the amount of starting template can be inferred. The availability of multiple reporter dyes means that a probe functioning as an internal control can also be included.

Other PCR techniques   The basic PCR technique has been modified and adapted to different applications; nested PCR to improve specificity has already been mentioned in conjunction with the Captagene detection system. Other PCR techniques which may be useful in environmental microbiology include random amplified polymorphic DNA (RAPD) and reverse transcriptase PCR (RT-PCR).

RAPD is a technique for producing a genomic profile or genetic 'fingerprint' from organisms and can be used to characterise different strains within a population. In this technique short primers (usually 10 base pairs) of randomly chosen sequence are used in the PCR reaction. These primers are short enough that they are likely to anneal at multiple sites on the genome, where two primers anneal on opposite strands with 3′ ends facing inwards and a suitable distance apart amplification of the fragment in between occurs. The resulting PCR reaction will contain several such bands and these are separated by size using gel electrophoresis. The organism is characterised by the bands produced by several such primers and differences (polymorphisms) in the bands amplified between strains can be found. The technique is more technically challenging than a normal PCR reaction as it is more prone to primer artefacts and the amount and quality of the starting DNA can influence the amplification of bands. Therefore care must be taken to include adequate replicates and controls.

RT-PCR is a technique for producing an amplified product from an RNA template. This was originally developed for studying gene expression (where messenger RNA is of interest) and RNA viruses. However, its particular interest to environmental studies is that it allows living and dead cells to be distinguished. DNA persists for a surprisingly long time in dead cells whereas RNA is unstable and is rapidly degraded after the death of the cell. Thus when amplification is performed from an RNA template the inference can be made that the source was living cells. For this technique a DNA copy of the extracted RNA is made first using the viral enzyme reverse transcriptase, which is able to make a DNA copy of an RNA template. This cDNA is then used in a PCR reaction in the normal way.

9.3.2.4 *Analysis of ribosomal RNA and ribosomal sequences*

The analysis of ribosomal RNA (rRNA) and the DNA sequences (rDNA) encoding the ribosomal RNA has found widespread application in environmental microbiology as a means of characterising microbial populations. As mentioned above, RNA analysis can be used to distinguish living from dead cells; however, in addition to this there are a wide range of gene probes and PCR primers available for ribosomal sequences. Ribosomal sequences have regions within them that are highly conserved throughout all organisms and also highly variable regions which can be used to identify individual species. Within these two extremes are sequences which can be used to identify particular groups of organisms, for example Archaea from Bacteria, or a particular genus, e.g. *Bacillus* or *Legionella*. This is particularly useful when characterising organisms from previously unstudied environments, for example the populations in deep-sea sediments and hydrothermal vents where many previously unknown microbes have been discovered.

Analysis of ribosomal sequences has been automated by the RiboPrinter® produced by Qualicon Inc. (USA). In this system the bacterial samples are prepared and then an automated system performs lysis of the cells and digestion of the DNA released by restriction enzymes. The fragments are separated by gel electrophoresis and transferred to a membrane before hybridisation with a chemiluminescently labelled DNA probe. A digitising camera captures the light emission image data indicating the fragment sizes hybridising to the probe and this is used to produce a pattern which is compared with patterns in a database for identification of the bacterial strain. While this system is currently being aimed mainly at applications in the food industry it could be used to produce genetic 'fingerprints' for bacteria isolated from any source. A further commercially available kit for identification of ribosomal RNA is the RiboTag (NCIMB, UK). In this assay the cells are cultured overnight and then lysed to release RNA. This RNA is fixed to a membrane and then hybridised with the RiboTag probe. These probes are oligonucleotides coupled to alkaline phosphatase which is detected colorimetrically. Currently RiboTags are available for *Helicobacter pylori*, *Mycobacterium tuberculosis*, *Legionella*, *E. coli*, total coliforms, *Aeromonas salmonicida*, *Renibacterium salmoninarum*, *Vibrio vulnificus*, *Listeria*, *Streptomyces* and sulphate-reducing bacteria.

9.3.2.5 *Preparation of nucleic acids from environmental samples*

Sampling of nucleic acids from environmental samples is generally for the purpose of detection of particular micro-organisms or gene functions, e.g. in studying bioremediation it may be helpful to know whether organisms with the necessary catabolic pathways to degrade contaminants are present. Often the easiest way to obtain sufficient high-quality nucleic acid for analysis is enrichment culture of the sample. There are well-established techniques for enrichment of different groups of bacteria, and specialised enrichment media are commercially available, and laboratory techniques for preparation of nucleic acids and DNA–RNA extraction kits (e.g. from Qiagen, Promega, Amersham etc.) are well known. However, in some cases the organisms of interest may not be known or may not be culturable or the aim may be to quantify the numbers of organisms present in a sample by detection of their nucleic acids. In these cases nucleic acids must be extracted directly from the sample.

For water and air samples micro-organisms may need to be concentrated from large volumes by filtration. However, once collected the extraction of nucleic acids is straightforward and good-quality product is generally obtained using normal DNA–RNA extraction techniques. Extraction from soils and sediments is much more difficult to achieve. Clay and organic fractions have a strong effect on the purity and quality of nucleic acid obtained. Clay tends to bind DNA and make extraction difficult whereas humic acids tend to co-purify with DNA and have an inhibitory effect on PCR. It may be necessary to develop variations to extraction procedures depending on the composition of the sample. Some examples of extraction techniques for soils are given in the bibliography and references; basically they fall into two groups, lysis of the cells *in situ* and subsequent extraction, and separation of the cells from the sample followed by lysis and extraction. The difficulties of sample preparation are probably the biggest obstacle to developing DNA-based diagnostics as a field analytical technique. Nevertheless there is commercial interest in developing hand-held devices for carrying out PCR and nucleic acid analysis outside the laboratory, although these are, probably, still a long way from the marketplace.

9.4 Reference

See Appendices 5 and 6.

9.5 Bibliography

See Appendices 5 and 6.

10

Electroanalysis for environmental applications

10.1 Introduction

Electrochemistry is a powerful analytical tool, which has been in widespread use in analytical and research laboratories for many years. Its origins lie in Galvani's experiments with frog leg muscles as early as the late 18th century (Galvani, 1791) and the pioneering work of Davy and Faraday in the early 19th century. However, it was not until the turn of the nineteenth century that major developments took place, opening up a new field of analytical science. In particular, the definition and determination by Nernst and Riesenfeld of the potential of the membrane–solution boundary which is fundamental to the understanding of electrode interactions, was a turning point in the usefulness of this branch of science. This work led to the rapid development of techniques, such as voltammetry and potentiometry, which have become standard tools in analytical laboratories worldwide (Wang, 1994).

Electroanalysis has been enjoying something of a renaissance recently. The techniques have always been competitive with other methods, such as atomic absorption spectroscopy, but were often cumbersome to use. The advent of the microprocessor and rapid advances in software development have led to a new generation of instrumentation, which is often modular and software driven. These analysers are user friendly and have a large array of electroanalytical functions built in. Data storage, retrieval and editing are straightforward, using 'Windows' type software. It is, however, prudent to mention a word of warning at this point: modern electrochemistry equipment is simple to use, but it is vital to have a good understanding of the subject and to be careful during analysis to obtain meaningful results. This chapter aims to provide an understanding of the fundamental concepts of electrochemistry and practical details of some relevant applications.

10.2 Electrochemistry – fundamentals

Electrochemistry concerns the interaction between electricity and chemistry. Measured parameters include voltage, current and charge and these are related to the activity or concentration of chemical moieties in solution. Some of these terms are briefly described below.

Voltage is the driving force behind an electrochemical reaction and is also a separation of charge. It is sometimes referred to as electromotive force (emf) or potential difference (pd). It has units of volts (V) and the symbol E. Current is a flow of electrons. It has units of amperes (A) and the symbol I (or i). Unlike conventional electronic applications, in electrochemistry, a positive flow of electrons is a positive current. Resistance is the opposition to the voltage, preventing a flow of current. It has units of ohms (Ω) and the symbol R.

Occasionally, the reciprocal of resistance may be used. This is conductance and has units of siemens (S) and the symbol G. Resistance and conductance are not inherent properties of a system; resistivity (ρ in Ω m) and conductivity (κ in $S\,m^{-1}$) are. These properties are related by the following equations:

$$R = \rho l/a \quad \text{and} \quad G = \kappa a/l$$

where l is the distance between any two planes under consideration and a is the area of the planes.

The coulomb (C) is the unit of charge. An electron carries a charge of $1.602\,10 \times 10^{-19}\,C$. This can be related to the current by

$$Q = It$$

Faraday's constant (F) is the charge carried by a mole of electrons. This has a value of

$$F = 6.022\,52 \times 10^{23} \text{ electrons mol}^{-1} \times 1.602\,10 \times 10^{-19} \text{ C/electron}$$

$$= 96\,487\,C\,mol^{-1}$$

Electrochemistry occurs at junctions where electricity interacts with a chemical environment. In other words, this is a region where conduction changes from electronic to ionic. The interaction takes place in an interphase region close to the electrode surface, which often behaves in a similar way to an electronic capacitor. Such a component has the ability to store electrical charge. Capacitance (C) is measured in farads (F) and is defined as charge divided by voltage:

$$C = Q/V$$

Also, total charge

$$Q = \int i(t)\,dt$$

Therefore if a capacitor is charged by a variable current $i(t)$,

$$V = (1/C)\,Q = (1/C)\int i(t)\,dt$$

Alternatively,

$$C = I/(dV/dt)$$

As we shall see later, this has some important implications for the current flow in electrochemical systems.

It is important to realise that at least two electrodes are required for current to flow or for a voltage to be created in an electrochemical system. In its simplest form, a cell can be formed from two electrodes in solution (half-cells). Electrodes are heterogeneous and are often very different from each other in size, structure and chemical nature.

10.2.1   Reference electrodes and standard electrode potentials

Any electrochemical half-cell (an electrode in solution) has a potential associated with it. In isolation, it is not possible to measure this; another half-cell is always required. To simplify the calculation of potential differences, an arbitrary point has been chosen, corresponding to the potential of the standard hydrogen electrode (SHE), which by definition has a potential of 0 V. This electrode consists of a platinum wire in a $1.0\,M$ solution of H^+ ions, over which is bubbling a stream of hydrogen gas at atmospheric pressure and a temperature

Table 10.1 *Some standard electrode potentials at 298 K*

Electrode	Electrode reaction	$E°$ (V)
$Pt\|F_2\|F^-$	$F_2 + 2e \rightleftharpoons 2F^-$	+2.87
$Ag^+\|Ag$	$Ag^+ + e \rightleftharpoons Ag$	+0.799
$Pt\|O_2\|H_2O_2$	$O_2 + 2H^+ + 2e \rightleftharpoons H_2O_2$	+0.682
$Pt\|H^+\|H_2$	$2H^+ + 2e \rightleftharpoons H_2$	0
$Pb^{2+}\|Pb$	$Pb^{2+} + 2e \rightleftharpoons Pb$	−0.126
$Cd^{2+}\|Cd$	$Cd^{2+} + 2e \rightleftharpoons Cd$	−0.403
$Cr^{3+}\|Cr$	$Cr^{3+} + 3e \rightleftharpoons Cr$	−0.744
$Zn^{2+}\|Zn$	$Zn^{2+} + 2e \rightleftharpoons Zn$	−0.763
$K^+\|K$	$K^+ + e \rightleftharpoons K$	−2.925

of 298 K. Using this reference point, tables of standard electrode potentials can be drawn up, as illustrated in Table 10.1.

More comprehensive standard electrode potential tables are available in numerous specialist electrochemical texts and reference sources (Weast and Astle, 1981), but for many practical applications they are not particularly important since they only apply to standard conditions, which rarely occur in real systems. Also, despite its use as the absolute standard, the hydrogen electrode is not a very convenient reference electrode and is rarely used, owing to its complexity. In most real situations, either a silver/silver chloride or saturated calomel electrode is used:

$$Ag, AgCl \mid Cl^- (3.8\,M, aq) \parallel \quad E_{1/2} = +0.2046\,V \text{ versus SHE}$$

$$Hg, HgCl_2 \mid Cl^- (3.8\,M, aq) \parallel \quad E_{1/2} = +0.242\,V \text{ versus SHE}$$

Both types of electrodes are usually constructed within a glass outer body, containing a porous frit at the bottom and an electrolyte refill hole near the top. The frit allows electrical contact between the electrode and the outside environment, while retaining the electrolyte. Some of the practical features of these reference electrodes are that the half-cell potential ($E_{1/2}$) is reproducible and does not drift; they are robust, simple to use and cheap to manufacture. In addition, they exhibit a low temperature response (especially the Ag/AgCl electrode) and are fairly insensitive to interference (particularly the calomel electrode). As can be seen from the above half-cells, chloride ions are required and are usually supplied as a saturated solution of KCl. For a more detailed discussion of reference electrodes, the reader is directed towards the text by Hamman *et al.* (1998).

10.3 Potentiometry

The measurement of the variation of the potential of a cell with the activity of a particular ion is, in principle, the simplest electrochemical technique. A pH electrode is a rather special example of a device which operates by this principle with almost perfect selectivity. This analytical approach evolved from the turn of the century but has advanced dramatically over the past 30 years owing to development of sensitive and selective electrodes and improved electronic instrumentation.

An ion-selective electrode (ISE) measurement system consists of a high-impedance voltmeter across which are connected two electrodes. One of the electrodes is a reference device such as a silver/silver chloride or a saturated calomel electrode, while the other is selective towards a particular ion. The two electrodes can often be found in a single housing and are referred to as a combination electrode. It is possible to use a multi-purpose voltmeter to make measurements, but it is more common to use specialist instrumentation, which displays ion activity directly. A more detailed description of this subject is given by Evans and James (1988).

In the simplest method of measurement, the pair of electrodes is immersed in the sample solution. There follows a process during which an electrothermodynamic equilibrium is set up. Ions move across the electrode–solution interface, driven by their activity in solution. Since ions carry a charge, this creates a difference in the electrical potential, which opposes further motion. The equilibrium voltage is therefore directly related to the activity of the ions in solution. In most instances, a steady state is reached within a few seconds, although it may take considerably longer if the ion activity is low.

An ISE is a potentiometric probe whose output potential is proportional to the activity of a particular ion in solution. Ideally, the behaviour of such a system is governed by the following version of the Nernst equation:

$$E = E_0 + (RT/zF) \ln a_i = E_0 + 2.303(RT/zF) \log a_i$$

where E is the potential (V), R the universal gas constant ($8.134 \, \text{J} \, \text{mol}^{-1} \, \text{K}^{-1}$), T the absolute temperature (K), z the ionic charge of the ion of interest and a_i its activity (M). The constant E_0 depends on a number of factors influenced by the design of the electrodes.

As can be seen, a plot of E against $\log a_i$ will yield a straight line with an intercept of E_0 and a slope of $2.303RT/zF$. In practical terms, this means that, for a monovalent ion at 298 K, a slope of 59.1 mV per decade change in concentration occurs. This is often referred to as a Nernstian response. In practical applications, this theory is usually implemented via a calibration step. On many pH meters, there is often only one calibration point, which assumes that ideally Nernstian behaviour is exhibited. Compensation for temperature effects is also built into most meters these days and is carried out either automatically via a temperature probe or by manually entering the temperature. On more sophisticated instrumentation, calibration is carried out at two, or sometimes more, points, and the slope of the response and the intercept of the calibration plot are automatically calculated.

It is worth stressing at this point that potentiometric devices respond to activity rather than concentration. There are a great many ionic interactions which occur and account for the difference between activity and concentration. The two values can be related via an activity coefficient γ_i:

$$a_i = \gamma_i \, c_i$$

The activity coefficient depends on the ionic strength (I) of the solution, which can be calculated as follows:

$$I = 0.5 \sum_i c_i z_i$$

where z_i is the charge number of the ion i.

Using this, the activity coefficient (at 298 K) can be calculated from the Debye–Hückel equation:

$$\log \gamma_i = -0.5z_i^2\sqrt{I}/(1 + \sqrt{I})$$

As can be seen, the activity coefficient approaches unity in dilute solutions, and activity and concentration are, hence, virtually the same. Furthermore, the effect of the activity coefficient is greater as the charge on the ion increases.

10.3.1   Ion-selective electrode characteristics

There are a number of characteristics of ISEs which are important for their operation. Some of these concern the electrode itself, while others involve the interaction of the electrode and the measuring solution.

10.3.1.1   *Response curve*

As described earlier, potentiometric devices exhibit a log-linear relationship between ion activity and voltage, typically characterised by the Nernst equation. This behaviour usually occurs over a fairly wide range of activities, but eventually the sensitivity reduces asymptotically to zero. The useful measuring range of these devices is thus limited to a region between the linear portion of the response curve and the ultimate limit of the response. It is usual to define the limit of detection as either the intercept of the two asymptotes of the response curve or the point at which the response decays to 30% of its initial value. It is clear, however, that measurements in this range will be less reliable.

In practice, calibration of electrode responses is carried out frequently and, in sophisticated instrumentation, compensation for the loss of sensitivity at low activities can be automatically carried out by using a multi-point calibration. In most applications, however, ion-selective devices are operated in their linear range and a simple one- or two-point calibration, using standard solutions, is sufficient.

10.3.1.2   *Interference*

So far, we have assumed that the electrodes have been absolutely specific for the ion of interest. In practice, this is never the case, although certain ISEs such as the ubiquitous pH probe and the fluoride ISE exhibit remarkable selectivities. The actual response can be calculated from a variation of the previously mentioned Nernst equation, the Nikolsky–Eiscnman equation:

$$E = E_0 + (2.303RT/zF)\log(a_i + K_{i,j}\, a_j^{zi/zj})$$

where $K_{i,j}$ is the selectivity coefficient, which is a measure of the relative effect of an interfering ion. A low value indicates that an electrode is highly selective for ion i compared with the interfering ion j. Values of selectivity coefficients are usually enclosed with commercially available ISEs.

10.3.1.3   *Temperature effects*

As can be seen from the Nernst equation, temperature has an influence on the slope factor $(2.303RT/zF)$. In most cases, electrode responses are quoted at 25 °C, but measurement and compensation are necessary to account for the increased value of the slope with temperature. Many modern instruments have a temperature probe connected to them and perform this automatically. A further effect of raising the temperature is that it reduces the detection limit, so it may be advantageous to operate at low temperatures when measuring dilute samples.

10.3.2   The glass pH electrode

Because of its widespread use and remarkable properties, the glass pH electrode will be treated separately from other ISEs here. Most laboratories contain at least one pH electrode and in addition various designs of device can be found in industrial plant. Because of the compact nature of portable pH meters, they are also widely used in field analysis of many sites. They are cheap, simple to use, reliable (if used and looked after properly) and extremely selective for H^+ ions.

The functional part of a glass pH electrode consists of a thin glass membrane, which establishes an electrochemical equilibrium with hydrogen ions in solution. The mechanism of this response is complex but is based on the ion-exchange properties of the glass surface. The theory of this process has been thoroughly discussed in a book by Eisenman (1976).

Although the pH electrode is a remarkable device, with by far the widest range and best selectivity of all ion-selective devices, it is important to realise that it is not perfect. The two greatest faults of these electrodes are the alkali and acid errors encountered at extremes of pH. The alkali error is usually the most serious and is caused by the response of the electrode to alkali metal ions, in particular sodium. This results in the recorded pH being lower than the true value. Modern glass formulations include lithium oxide instead of sodium oxide and have reduced interference effects, but the user is recommended to be wary of the accuracy of measurements made at a pH of above 11 or so. The acid error occurs at very low values of pH (<0.5) owing to activity effects. Subsequently, values higher than the true pH are recorded.

10.3.3   Methods of analysis

There are a number of methods by which potentiometric devices can be used. Electrodes can be simply calibrated, as discussed previously, and placed in a test solution to measure the activity of a particular ion, or they can be used in a variety of more complicated procedures, such as flow-injection analysis (FIA) or titrations.

It is common to add a buffer to a sample prior to potentiometric analysis. These solutions are usually of high ionic strength and are known as ionic strength adjusters (ISAs). The primary purpose of these buffers is to maintain, as near as possible, the total ionic strength of the solution. This minimises changes in the activity coefficient of the ion being sensed. Other properties of ISAs can include pH adjustment, decomplexing ability and the means to remove interferences. An example of an ISA is total ionic strength adjustment buffer, which is used for the determination of fluoride. Many other examples are commercially available for other ions.

In situations where large numbers of different samples are being analysed, it is often more convenient not to perform a complete calibration prior to analysis. In these situations, standard (or incremental) addition may be preferable. The electrode potential of the unknown solution is first measured. This is followed by the addition of a known volume of a standard solution, followed by re-measurement of the electrode potential. The activity of the unknown solution can then be calculated from the following:

$$a_1 = a_2 \left[V_1/(V_2 - V_1) \right] \left\{ 10^{\Delta E/s} - \left[V_1/(V_1 + V_2) \right] \right\}^{-1}$$

where a_1 and V_1 are the activity and volume of the unknown solution, a_2 and V_2 are the activity and volume of the standard solution, ΔE is the change in the measured potential and s is the slope value of the electrode. As can be seen from this equation, the slope value of the electrode must be known in advance. Furthermore, it is helpful if an estimate of the

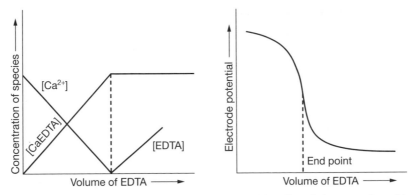

Figure 10.1   Concentration and electrode potential profiles during potentiometric titrimetric analysis of Ca^{2+} using EDTA

activity of the ion in the unknown solution can be made, since the technique is most accurate if the unknown and the standard are of the same order of magnitude. The previous equation can be simplified if the same volumes of standard and sample are always used. Thus a plot can be made of a factor X (such that $a_1 = Xa_2$) against ΔE. This information is often tabulated as a standard addition table.

Standard subtraction works on a similar principle to standard addition, except that the added solution decreases the activity of the measured ion by precipitation or complexation. The above equation for standard addition can still be used provided that the stoichiometry of the reaction is known. This technique is particularly useful for measuring toxic ions such as cyanide, since it minimises preparation and handling of dangerous standard solutions.

10.3.4   Titrimetric operation

Improvements in measurement accuracy can be readily achieved by the use of a titrimetric procedure in which a stoichiometric reaction between two species is followed using an ISE. The end point of such a reaction corresponds to the total disappearance of one species or the emergence of a product. The net result of such a procedure is that there is a sudden change in the electrode potential.

To illustrate this procedure, let us consider the titrimetric analysis of a solution of calcium ions. For this we would use a calcium ISE and a titrant containing a known concentration of EDTA. As the titrant is added, a calcium–EDTA complex is formed, which cannot be detected by the ISE. Figure 10.1 illustrates the concentration profiles and the electrode potential exhibited by the electrode. Once the end point has been determined, it is a simple calculation to determine the initial concentration of calcium from the stoichiometry of the reaction:

$$Ca^{2+} + EDTA^{2-} \rightarrow CaEDTA$$

A further advantage of this procedure is that it allows the determination of species for which no ISEs are currently available, by applying the above principle in reverse. For example, aluminium could be analysed by complexometric titration using a fluoride ISE and adding a standard solution of sodium fluoride.

10.3.5 Potentiometric flow-injection analysis

FIA has become a well-established and widely used technique. It is based on the injection of a liquid sample into a flowing carrier stream of a suitable solution. The sample is carried towards a detector (in this case an ISE) which detects the passage of the sample through the flow cell. The technique is based on three principles: injection of the sample, controlled dispersion of the sample zone and reproducible timing of the movement of the sample to the detector. The resultant signal is recorded as a series of peaks as samples are sequentially injected.

The simplest form of FIA involves the measurement of peak height, although peak area and width may also be used. This subject is relatively new but has evolved rapidly into a very wide area. A more detailed description of FIA is outside the scope of this article, but interested readers may find the text of Kolev (1999) useful.

10.3.6 Other potentiometric devices

As we have discussed, ISEs are available for a very wide range of ions and come in many forms. Glass electrodes are widely used for measuring pH and other cations and solid state devices exist for fluoride and a number of other ions. Liquid membrane electrodes based on ion exchange and neutral carriers are also widespread. Devices of these types are all ISEs. Table 10.2 illustrates this with some examples of commercially available devices.

A more recent introduction to this area is the ion-selective field effect transistor (ISFET), which was first described by Bergveld (1970). This is a solid-state device, based on silicon manufacturing technology, and hence offers the potential for mass production and miniaturisation. The biggest advantage of these devices, however, is that they avoid the use of delicate glass membranes and are consequently much more robust and more suitable for use in remote sites and for field use. Although it is nearly 30 years since the invention of the ISFET, it has been slow to reach the market, owing to technical difficulties during production. However, the pH ISFET has now been introduced and, doubtless, others will soon follow.

Table 10.2 *Some commercially available ion-selective electrodes*

Target ion	Electrode type	Operating range (M)
Barium	Liquid ion exchange	$10^0 – 5 \times 10^{-5}$
Bromide	Solid state	$10^0 – 5 \times 10^{-6}$
Calcium	Liquid ion exchange	$10^0 – 5 \times 10^{-5}$
Chloride	Solid state	$10^0 – 10^{-5}$
Copper	Solid state	$10^0 – 10^{-6}$
Cyanide	Solid state	$10^{-2} – 10^{-6}$
Fluoride	Solid state	$10^0 – 10^{-6}$
Iodide	Solid state	$10^0 – 10^{-5}$
Nitrate	Liquid ion exchange	$10^0 – 5 \times 10^{-5}$
Potassium	Neutral carrier	$10^0 – 5 \times 10^{-5}$
Silver	Solid state	$10^0 – 10^{-6}$
Sodium	Glass	$10^0 – 10^{-6}$
Sulphide	Solid state	$10^0 – 10^{-5}$

10.4 Voltammetry

So far, we have only looked at the voltage dependence of a circuit containing two electrodes, in which (effectively) zero current is flowing. In other words we have been monitoring an equilibrium situation. The following section of this chapter will look at what happens when a voltage is imposed on a system and a current allowed to pass.

10.4.1 Faradaic processes

In order that a reaction occurs at an electrode surface, the applied potential must exceed a particular value. It is sometimes useful to view the voltage as an electron pressure, which forces a chemical moiety to gain or lose an electron. Thus, electrochemical reduction or oxidation of any electroactive species can be achieved. The required voltage is a function of the electrode material, the solution (ionic strength, pH, temperature) and the chemical species reacting. Standard electrode potentials, as discussed previously, may be of some help in determining the required applied potential, but it is rare for standard conditions to apply, so their use is somewhat limited.

The current, which flows as a result of the transfer of electrons during electrochemical oxidation or reduction reactions, is a measure of the rate of the reaction and can be related to the concentration of a particular analyte. Thus, the following general reaction can be followed:

$$O + ne^- \rightleftharpoons R$$

where O and R are the oxidised and reduced forms of the redox couple.

The current resulting from a change in the oxidation state of this couple is termed the Faradaic current, since it obeys Faraday's law:

$$N = Q/nF$$

where N is the number of moles of reactant, Q is the total charge passed (coulombs), n is the number of electrons involved in the reaction and F is Faraday's constant ($96\,487\,C\,mol^{-1}$). The electrode reaction rate is dependent on a number of factors. These include mass transfer to the electrode surface, electron transfer across the interface and transport of products away from this surface. The net rate of the reaction and hence the resultant observed current may therefore be dependent on the mass transport or the electron transfer. When the rate is controlled solely by the mass transfer process, the reaction is referred to as Nernstian or reversible. The processes can be (and usually are) more complicated than this simplified description. For further details, the reader is directed towards the text by Hamman et al. (1998).

10.4.2 Non-Faradaic processes

In addition to the Faradaic current, there are a number of other currents which can flow during an electrochemical reaction. An electrolyte has an electrical resistance and hence there will be a background current due to a voltage being applied across a resistor. The current resulting from the application of a voltage in such a system is referred to as the Ohmic potential drop. This can be minimised by the use of high concentrations of an electrochemically inert buffer. Particular problems can occur in non-aqueous solutions, where the conductivity can be very low.

The excess charge on the electrode in voltammetric reactions causes the formation of an electrical double layer close to the electrode surface. Since the interface must be neutral,

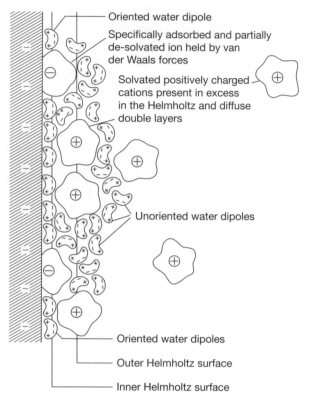

Oriented water dipole

Specifically adsorbed and partially de-solvated ion held by van der Waals forces

Solvated positively charged cations present in excess in the Helmholtz and diffuse double layers

Unoriented water dipoles

Oriented water dipoles

Outer Helmholtz surface

Inner Helmholtz surface

Figure 10.2 Molecular structure of electrical double layer at electrode surface showing ions and water dipoles

a counterlayer forms of ions of opposite charge to that of the electrode. The double layer consists of several regions, as illustrated in Figure 10.2. The inner Helmholtz plane is closest to the electrode and consists of solvent molecules and adsorbed ions. The outer Helmholtz plane lies just outside this layer and is at a distance of the centre of a plane passing through the layer of solvated ions closest to the electrode surface. Together, these planes are known as the compact layer. Beyond this, towards the bulk solution, lies the diffuse layer, containing scattered ions. This structure resembles a parallel-plate capacitor, as described at the beginning of this chapter, and is responsible for the charging current observed during controlled potential analysis. It manifests itself as a charging current, which is a non-Faradaic process, since electrons are not transferred across the solution–electrode interface. In potential step experiments, such as chronoamperometry, an exponentially decreasing background current is observed with time.

10.4.3 Voltammetric cells

In most voltammetric applications, one is only interested in the reactions that occur at one electrode, known as the working electrode, which is coupled to an ideal, non-polarisable electrode of known potential. However, a three-electrode cell is usually used:

reference electrode   Used to give a fixed reference point
counter electrode   Acts as a source or sink of electrons
working electrode   Where electrochemical reaction of interest takes place

A third electrode is used owing to the desirability of maintaining a fixed reference potential. This potential relies on an equilibrium reaction, as illustrated by that of a silver/silver chloride electrode:

$$AgCl \rightleftharpoons Ag^+ + Cl^-$$

If a current is passed through this electrode, the equilibrium will be upset and the potential will alter. To overcome this problem, a third (counter, or auxiliary) electrode is used to donate or accept electrons in the system. It is usual for the surface area of this electrode to be larger than that of the working electrode, so that the reaction rate, and hence the current, is controlled by the working electrode. It is also prudent to ensure that the counter electrode does not produce electrolysis products that will reach the working electrode surface, leading to interference. For highly accurate work, the two electrodes are often placed in separate compartments, separated by a sintered plug. The reference electrode is placed in the working electrode compartment, with its tip located close to the working electrode surface, so that the voltage drop due to the resistance of the solution (iR_s) is minimised.

It is important that the voltmeter used to measure the potential between the working and reference electrodes has a high input impedance, so that negligible current is drawn through the reference electrode (for the same reasons as discussed for potentiometry). This arrangement further minimises the effect of the iR_s contribution of the solution. For further accuracy, compensation for any remaining potential drop can be carried out. Modern instrumentation frequently has this facility built in.

10.4.4   Chronoamperometry

Chronoamperometry involves poising the working electrode at a fixed value, such that a (chosen) oxidation or reduction reaction occurs at the surface. This is carried out with a static electrode in unstirred solution. The resultant current is determined by diffusion of the reacting species to the electrode and is given by the Cottrell equation:

$$I(t) = nFACD^{1/2}/\pi^{1/2}t^{1/2}$$

As can be seen, the current is proportional to $t^{-1/2}$, which is termed Cottrell behaviour.

Analytical applications of chronoamperometry are limited, although it is a very common technique in the specialised field of biosensors. Its more common uses are for calculating the surface area of an electrode or for measuring the diffusion coefficient of an electroactive compound.

10.4.5   Linear sweep and cyclic voltammetry

Because of its widespread use, voltammetry at a dropping mercury electrode is usually referred to as polarography. However, the principles will be treated in this section as voltammetry, followed by a short description of the dropping mercury electrode itself.

Linear sweep voltammetry involves sweeping the potential at a scan rate (υ) between an initial and a final potential and measuring the resultant current. In order to monitor an oxidation process, the initial potential is chosen to be sufficiently negative for the oxidation reaction to not occur and is then scanned anodically. At a sufficiently positive potential, oxidation of the analyte begins and then rises to its diffusion-limited value. It is worth mentioning here that it is a common mistake to regard oxidation reactions as being those which occur at a positive potential and reductions as those which occur at a negative potential. This is simply not true, since the zero point is arbitrary (based on the reference electrode used and ultimately on the SHE). A common variation of this technique then involves reversing the process; a technique known as cyclic voltammetry. In this approach, the product produced during the forward scan can be interrogated during the reverse scan.

Analytical applications of these techniques are limited, owing to the relatively poor resolution of the redox potentials being examined. Nevertheless, useful information about electron transfer kinetics, reaction mechanisms and reversibility can be obtained.

10.4.6   Pulse voltammetric techniques

Pulse techniques involve a sequence of potential steps followed by some form of current sampling and subsequent signal processing. Following each voltage step, the charging current (as discussed earlier) rapidly decays, allowing more accurate and sensitive measurements of the Faradaic currents to be made. Hence, these techniques allow significantly lower detection limits to be achieved and have largely superseded other techniques in analytical electrochemistry laboratories.

10.4.6.1   *Normal pulse voltammetry*

A series of potential pulses with a linearly increasing amplitude are applied to the electrode at a set time interval (towards the end of each drop lifetime for the mercury electrode). As well as the advantage of high Faradaic:charging current ratio, this ensures that the diffusion layer thickness is small, allowing a high analyte flux to the electrode surface.

10.4.6.2   *Differential pulse voltammetry*

This has become one of the most commonly used electrochemical techniques for the routine analysis of a very wide range of species, ranging from trace levels of organic compounds to heavy metals. The process involves the superposition of a series of fixed magnitude potential pulses on a linearly ramped voltage. The current is measured just before the application of the pulse and again towards the end of the pulse (allowing the charging current to decay). The current difference, which is proportional to the concentration of the electroactive species, is then plotted against the voltage, appearing as a series of peaks.

Variables include the scan rate (usually in the range $1-10\,\mathrm{mV\,s^{-1}}$) and the pulse amplitude (often $10-100\,\mathrm{mV}$), the selection of which depends on the sensitivity and speed restrictions. The main benefit is that peak resolution is much improved over conventional voltammetric techniques.

10.4.6.3   *Square-wave voltammetry*

This is a similar technique to differential pulse voltammetry, except that a symmetrical square waveform is imposed on a ramped (staircase) potential. The amplitude of the square wave is high, so that the reverse pulse reverses the reaction occurring on the forward pulse. The current is again measured twice for each wave: at the end of the forward pulse and at the end of the reverse pulse. The difference between the two currents, which is larger than either component, can be plotted against the base potential. The technique is extremely rapid and several times more sensitive that differential pulse voltammetry. As a result, it is becoming more widely available on standard electrochemical instrumentation.

10.4.7 Mercury electrodes

Despite the fact that mercury is toxic and a liquid over the usual analytical temperature range (i.e. $0–100\,°C$), it is still a very widely used electrode material. There are a number of good reasons for this, among which are the following:

1. It provides a smooth, homogeneous surface.
2. It has a high hydrogen overvoltage, greatly extending the cathodic potential window (compared with other commonly used electrodes).
3. Mercury forms amalgams with many metals (discussed in the next section).
4. It has an easily renewable surface.

A mercury drop electrode consists of a mercury reservoir connected to a glass capillary tube. Modern electrodes have microprocessor control of the droplets, such that the drop diameter and its detachment can be controlled. Many of the early problems associated with gravity feed reservoirs are therefore eliminated. It is, however, essential to ensure the cleanliness of the capillary for accurate and reproducible work.

10.4.8 Stripping analysis

So far, we have only looked at direct methods of analysis. In many applications, these techniques are perfectly adequate, but, for maximal sensitivity, it is useful to have a pre-concentration step. This can be achieved by electrochemically oxidising or reducing the analyte at the electrode, where it then undergoes some means of attachment to the electrode surface, followed by stripping the analyte out by ramping the potential in such a way that the analyte is re-reduced or re-oxidised. All of the voltammetric techniques discussed earlier are applicable following the pre-concentration step. Pre-concentrations of the order of 100–1000-fold are common and, hence, detection limits of around $10^{-10}\,M$ are the norm. Much of the work involving these techniques is performed using dropping mercury electrodes because of the ease with which pre-concentration can be achieved.

10.4.8.1 *Anodic stripping and cathodic stripping voltammetry*

Anodic stripping voltammetry (ASV) is the most commonly used electrochemical technique in environmental analysis. It is widely used for the determination of trace levels of metals, which will be used as an example here to illustrate the process. The first stage involves poising the electrode (usually mercury) at a potential which is sufficiently negative to reduce the metal ions from solution at the electrode surface, such that they are electrode-posited. Here, they form an amalgam with the mercury. The deposition time determines the level of preconcentration, but, for practical analyses, $30–180\,s$ is common. This is followed by the determination step, which consists of stripping the deposited species back into solution by scanning the potential anodically. This results in an oxidation current, which is proportional to the amount of the species which was reduced onto the electrode originally. The determination scan can be carried out using any of the techniques discussed earlier, although the differential pulse technique is the most commonly used.

Cathodic stripping voltammetry is the reverse of the above process. An analyte is anodically deposited onto the electrode, followed by cathodic stripping. As with ASV, all of the previously discussed voltammetric techniques can be utilised following deposition. This technique is useful for the determination of a wide range of compounds which are capable of forming insoluble salts with the (mercury) electrode.

10.4.8.2 *Adsorptive stripping voltammetry*

This is a technique whereby a metal complex is formed and adsorbed at the electrode surface and then reduced. The current is proportional to the adsorbed species and, hence, the original metal. As with the other stripping techniques, the sensitivity is dependent on the deposition time. While the method is most commonly used for the determination of metals, it is also possible to measure numerous organic compounds, such as pesticides, using this approach. A further advantage is that it is possible to measure certain non-electroactive compounds by selectively binding them and then measuring the resultant electroactive complex.

10.5   References

Bergveld P 1970 Development of an ion-sensitive solid-state device for neurophysiological measurements, *IEEE Trans. Biomed. Eng.*, **17**; 70–1

Eisenman G 1976 *Glass electrodes for hydrogen and other cations*, Marcel Dekker, New York

Evans A and James A M 1988 *Potentiometry and ion-selective electrodes (analytical chemistry by open learning)*, Wiley, New York

Galvani L 1791 *De viribus electricitatis in motu musculari commentarius*, Bologna

Hamman C H, Hamnett A and Vielstich W 1998 *Electrochemistry*, Wiley-VCH, Weinheim

Kolev S D 1999 *Flow injection analysis*, World Scientific, London

Wang J 1994 *Analytical electrochemistry*, Wiley, New York

Weast R C and Astle M J (eds) 1981 *CRC handbook of chemistry and physics*, CRC Press, Boca Raton, FL, pp D155–60

11

Radiochemical methods and radiation monitoring

11.1 Radioactivity and radioactive decay

Traces of radioactive material are to be found widely distributed in the Earth's crust, in oceans and rivers and in the atmosphere and biosphere. Some is naturally occurring while an increasing amount is man made (Hewitt, 1996). Some of this radioactivity becomes incorporated into living systems as well as being present in inanimate materials which surround us. The human population is therefore subjected to radiation from these sources. Radiation also comes as cosmic radiation from space. Figure 11.1 illustrates the approximate contributions to the total radiation background experienced on Earth. It is important to note that the man-made contribution to the total environmental radioactivity is less than 1% of that which occurs naturally. However, because the biological effects of low-level radiation on the general population are difficult to study and are not well understood, environmental pollution with radioactive substances is considered as potentially detrimental to health. It is important first to examine what is meant by radioactivity before going on to examine the nature of environmental radioactivity. Some general aspects of how radiation interacts with matter will then be discussed leading into an introduction to the methods used for detection and measurement of radiation. This is followed by examples to illustrate their application.

11.1.1 Radioactivity

Radioactivity can be defined as the spontaneous emission of radiation from substances. Only certain intrinsically unstable atomic nuclei show this behaviour. These are termed radionuclides and the majority of such radionuclides, on disintegration, emit a particle which is generally accompanied by electromagnetic radiation. The particle can be an α-particle (a helium nucleus), a β⁻-particle (electron) or a β⁺-particle (positron); in these cases the emission leaves the resulting product nucleus in an excited state. The product nucleus rapidly achieves stability by emission of electromagnetic radiation usually as high-energy γ-radiation. A radionuclide can be characterised by the nature and energy of the emitted radiation and by its half-life which will be discussed in the next section.

11.1.2 Radionuclides

Radionuclides are, with the exception of those of very light elements – particularly hydrogen – chemically identical to their inactive isotopes, e.g. the radionuclide ^{90}Sr is

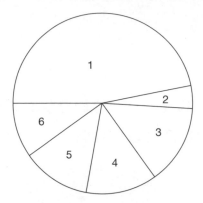
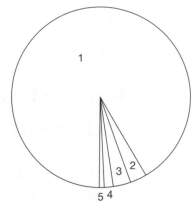

1. 47% radon-222
2. 4% radon-220
3. 14% γ from Earth and buildings
4. 13% from man-made sources
5. 12% internal – food and drink
6. 10% cosmic rays

(a)

1. 91.6% medical
2. 3.0% weapon fallout
3. 3.1% miscellaneous
4. 1.5% occupational exposure
5. 0.8% nuclear discharge

(b)

Figure 11.1   Radiation exposure of a population (UK): (a) contribution to total exposure; (b) contributions to exposure from man-made radioactivity

chemically identical to its stable isotope ^{88}Sr and undergoes all the same reactions and becomes incorporated into the same materials. In addition the similarity between different but neighbouring elements in the periodic table is often a good indicator of behaviour, e.g. Ca and Sr show similar chemical behaviour and are group 2 neighbours in the periodic table. Ca is an important constituent of milk, and any ^{90}Sr in the environment tends to follow the reaction pathways of Ca and can end up in milk.

11.1.3   Radioactive decay

Radioactive decay is the spontaneous disintegration of a radionuclide into a new product nucleus – often referred to as the daughter product. Such decay is detected and measured via the emitted radiation. The rate of decay of all radionuclides is a first-order process and hence for any given radionuclide we have for the rate of disintegration

$$\mathrm{d}N_t/\mathrm{d}t = -\lambda N_t \tag{11.1}$$

where N_t is the number of nuclei of the radionuclide at time t and λ is the decay constant which has a unique value for each radionuclide. Integration of equation (11.1) gives an expression which relates the number of nuclei remaining after time t to the value of t as follows:

$$\ln N_t = \ln N_0 - \lambda t \tag{11.2}$$

or

$$N_t = N_0 \exp(-\lambda t) \tag{11.3}$$

Equation (11.1) expresses the decay in terms of the rate of change of number of nuclei with time, i.e. as dN_t/dt, termed the 'activity' (A_t) of the radioactive source. We note that

$$A_t = \lambda N_t$$

so that multiplication of both sides of equation (11.3) by λ leads to the equation

$$A_t = A_0 \exp(-\lambda t) \tag{11.4}$$

It is practically more convenient to work with the 'activity' of the radionuclide since this can easily be measured.

11.1.4   Activity

Activity is defined as the rate of disintegration of nuclei, i.e. this is equal to A_t discussed above. The unit of activity is the becquerel. This unit is defined so that one becquerel (Bq) is equal to one disintegration per second. A given quantity of radioactivity is therefore most meaningfully expressed in terms of the number of becquerels present.

11.1.5   Radioactive half-life

We note that an important characteristic of exponential decay is that for any arbitrary time zero at which the number of nuclei is N_0, the time taken for decay to a fixed fraction of N_0 is always the same and does not depend on the value of N_0. For convenience the concept of half-life is used. This is defined as the time taken for the number of nuclei, present at a given instant, to decay to one-half that number. Alternatively, this is the same as the time for the activity to decay from A_0 to one-half of A_0. It is denoted by $t_{1/2}$ and is readily shown to be inversely related to the decay constant so that

$$t_{1/2} = 0.693/\lambda$$

Equation (11.4) can then be converted to the form

$$A_n = A_0 (\tfrac{1}{2})^n \tag{11.5}$$

where A_0, as already seen, is the activity at time zero and A_n is the activity remaining after an elapsed time equal to n half-lives.

11.1.6   Decay data

Table 11.1 lists a number of common radionuclides together with their half-lives. Table 11.2 lists values of $(\tfrac{1}{2})^n$ for a range of n values. This is easily employed to calculate the activity remaining as the following example illustrates. If the initial activity is A_0, for a given radionuclide of known half-life, then if the time elapsed is say 3.5 half-lives the required value of $(\tfrac{1}{2})^n$ is at the intersection of row 3.0 and column 0.5, i.e. 0.088, and $A_n = A_0 \times 0.088$.

11.1.7   Radiation energy

It is important to note that, in the case of an α-emitting radionuclide, disintegrations produce α-particles having a small number of discrete energies. These energies are characteristic of the nuclide. In the case of β-emitters the particle energy is not the same for each decaying nucleus. In this case a range of energies can be observed up to a maximum

Table 11.1 *Some radionuclides and their half-lives*

Radionuclide	Half-life	Radionuclide	Half-life
Uranium-238	4.8×10^9 years	Chlorine-36	3.1×10^5 years
Uranium-235	7.04×10^8 years	Cobalt-60	5.27 years
Plutonium-239	2.4×10^4 years	Caesium-137	30.0 years
Radium-226	1.62×10^3 years	Iodine-131	8.04 days
Radon-222	3.8 days	Strontium-90	28.0 years

Table 11.2 *Values of $(1/2)^n$ for integral and non-integral values of n*

n	0	0.1	0.2	0.3	0.4	0.5	0.6	0.7	0.8	0.9
0	1.000	0.933	0.871	0.812	0.758	0.707	0.660	0.615	0.578	0.536
1	0.500	0.467	0.435	0.406	0.379	0.354	0.330	0.308	0.287	0.268
2	0.250	0.233	0.217	0.203	0.190	0.177	0.165	0.154	0.144	0.134
3	0.125	0.117	0.109	0.102	0.095	0.088	0.083	0.077	0.072	0.067
4	0.063	0.058	0.054	0.051	0.047	0.044	0.041	0.039	0.036	0.034
5	0.031	0.029	0.027	0.025	0.024	0.022	0.021	0.020	0.018	0.017
6	0.016	0.015	0.014	0.013	0.012	0.011	0.010	0.010	0.009	0.008

value and the energy spectrum and the maximum energy are characteristic of the decaying radionuclide. Positron (β^+) decay is accompanied by annihilation radiation which results from the destruction of the positron via its interaction with an electron. Electron capture and internal conversion processes create unoccupied inner electron energy levels. Electron transitions then occur with the emission of X-rays. For γ-emitters the γ energies correspond to transitions between nuclear energy levels and are observed as characteristic of the emitting nucleus. We recall that γ emission normally occurs following α or β emission.

11.2   Radionuclides in the environment

11.2.1   Naturally occurring radionuclides

The environment contains radionuclides which occur naturally mainly in uranium- and thorium-bearing rocks. There are 45 such radionuclides and they separate into three radioactive series in which members of a given series are all related through the growth and decay of other members of the series. Each series terminates in a stable and different nuclide of lead. To illustrate, Table 11.3 shows the interrelationships between members of the uranium series which terminates with ^{206}Pb. For such series to exist in nature the half-life of the first member must be comparable with the age of the Earth which is estimated to be about 5×10^9 years. It is noteworthy that the gas ^{222}Rn, which is present as a member of the U series, is of current interest owing to its accumulation in buildings erected in uranium-rich areas. This radionuclide accounts for a large proportion of the radiation background.

Table 11.3 *Uranium decay series*

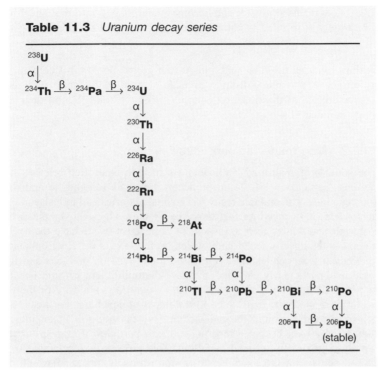

Table 11.4 *Some naturally occurring radionuclides – not members of radioactive decay series*

Radionuclide	Emission	Half-life
Potassium-40	β^-, electron capture	1.28×10^9 years
Rubidium-87	β^-	4.8×10^{10} years
Neodymium-144	α	2.4×10^{15} years
Platinum-190	α	6.9×10^{11} years
Carbon-14	β^-	5.7×10^3 years
Hydrogen-3 (tritium)	β^-	12.3 years

In addition to the radioactive members of these series there are small amounts of a small number of very long half-life radionuclides presumably formed when the Earth was formed. Some of these are given together with their half-lives in Table 11.4.

Further, there are radionuclides which owe their natural occurrence to their continuous production in the upper atmosphere. In particular ^{14}C is produced continuously by the reaction of cosmic-ray neutrons with nitrogen:

$$^{14}_{7}N + ^{1}_{0}n \rightarrow ^{14}_{6}C + ^{1}_{1}p$$

This ^{14}C is oxidised to CO_2 and becomes incorporated into biological material. It is therefore widely distributed in nature. Likewise tritium (^3_1T) is produced via

$$^{14}_{7}N + ^1_0n \rightarrow ^{12}_6C + ^3_1H$$

Tritium oxidises to water and adds to the global pool. Considerable additions to global tritium have also come from man-made sources to be dealt with in the next section. Characteristics of the radiation emissions from ^{14}C and ^3H and their decay half-lives are given in Table 11.4.

11.2.2   Man-made radionuclides

The sources of man-made radionuclides include controlled releases from nuclear power stations (particularly of fission products), hospitals (diagnostic and therapeutic unsealed sources), industrial and university research laboratories and industry in general. Often waste disposal is via licenced waste disposal companies. All establishments which use radioactive materials and which need to dispose of radioactive waste have their releases regulated by a responsible governmental body which, in the UK, is the Environmental Agency.

Nuclear weapons testing has released mainly fission products and radioactive weapons materials, particularly enriched uranium, plutonium and tritium. For atmospheric testing in which the fireball has not touched the ground, the majority of the radioactive material becomes diluted and distributed in the upper atmosphere. For tests in which the fireball has touched the ground the radioactivity mixes with soil taken into the fireball. Much radioactivity then becomes incorporated into particles, a large proportion of which may rapidly return to the ground as fallout. This process is accelerated by precipitation. The fallout contains a large number of fission products, many of which decay very rapidly, i.e. in minutes to hours. Important radionuclides of moderate-to-long half-life are ^{137}Cs, ^{90}Sr and ^{131}I. Some decay characteristics are given in Table 11.1.

11.3   Radiation characteristics

11.3.1   Importance in detection and measurement

Any device used to detect, measure and characterise radiation must be capable of allowing the radiation:

(a)   to enter the detector, and
(b)   to lose most of its energy within the detector.

For such a system it is then possible to measure both the quantity of radiation entering and its energy characteristics. The type and characteristics of the radiation to be measured and the way in which the radiation interacts with matter have an important bearing on the choice of detector and some salient points will now be considered. All detection methods rely on the fact that radiation emitted by radionuclides interacts with matter and transfers energy to it. The precise way in which energy is transferred will now be outlined.

11.3.2   Charged particles

α- and β-particles lose energy by collisions with atoms which cause:

(a)   atomic or molecular excitation – subsequent de-excitation usually produces visible radiation;

(b) ionisation, which involves ejection of an orbital electron from an atom, so creating an ion pair.

Kinetic energy is transferred at each collision until after hundreds of thousands of collisions the charged particle is 'stopped', i.e. it has insufficient energy to cause further ionisation or excitation.

The stopping power of materials, which is described as the energy loss, by the charged particle, per unit distance travelled, is proportional to

$$(z/v)^2 NZ$$

where z is the number of electronic charges on the incident charged particle having velocity v, N is the number of atoms per unit volume and Z the atomic number of the atoms of the irradiated matter. Consideration of this expression shows that:

α-particles are more easily stopped than β-particles of the same energy;
low-energy particles are more easily stopped than high-energy particles.

Since stopping power is directly proportional to electron density and since the number of electrons per unit mass is roughly the same for all elements except hydrogen, then equal weights of different materials cause the same attenuation of a beam of charged particles.

11.3.3 γ- and X-radiation

γ-Rays lose energy by three different interaction processes with electrons:

(a) **Photoelectric absorption.** The γ-photon transfers all its energy to a bound inner orbital electron of the matter. This electron is ejected and causes further ionisation. This process predominates at low energies up to ~10 keV.
(b) **Compton scattering.** The γ-photon interacts with a 'loosely' bound outer electron. Sufficient energy transfer occurs to eject the electron at high speed and the photon moves away with lower energy, but will undergo further similar interactions until it is incapable of ejecting more electrons. This predominates for energies ~0.5–10 MeV.
(c) **Pair production.** A γ-photon having energy greater than 1.02 MeV can, under the close influence of the nucleus of an atom, create a positron and an electron. These move away in opposite directions at high speeds. This predominates for energies greater than ~10 MeV.

All these processes result in the production of large numbers of ion pairs. Unlike for charged particles the absorption of γ-photons follows an exponential law:

$$I_x = I_0 \exp(-\mu x)$$

where I_x and I_0 are the intensity at a distance x into the material and the incident intensity respectively and μ is the linear absorption coefficient of the material. Table 11.5 provides a summary of the properties of commonly occurring radiation including approximate ranges.

11.4 Radiation detection

The above discussion on the effects of radiation on matter suggests that the presence of radiation can be detected and measured either by measuring the current which is carried by the ions produced or, where applicable, by measurement of emitted light. Devices employed to detect radiation can be divided into three different types, namely gas ionisation,

Table 11.5 *Properties of commonly occurring radiation*

Radiation	Type	Charge	Mass (AMU)	Energy	Approximate range (cm)	
					Aluminium	Air
α	Particle	+2	4	4–8 MeV	0.0025–0.0075	3–8
β⁻	Particle	−1	1/1830	0–4 MeV	0–1	0–1000
β⁺	Particle	+1	1/1830	0–4 MeV	0–1	0–1000
X	Electromagnetic radiation	–	–	10–100 keV	0.04–4[a]	500–10 000[a]
γ	Electromagnetic radiation	–	–	1–2 MeV	14–19[a]	24 000–38 000[a]

[a] Exponential attenuation; values are for attenuation to one-tenth of the initial intensity.

scintillation and semiconductor ionisation. Before outlining the principles of detector operation we first consider what is meant by detector efficiency.

11.4.1   Efficiency of detector systems

For any detection system the relationship between the true activity and the observed count rate, of a radioactive source, is called the counting efficiency (E). This efficiency is expressed as

E = observed counting rate from the detector/disintegration rate of the source

There are numerous factors which affect E and these have been extensively dealt with by previous workers (Evans and Muramatsu, 1977). It is sufficient to note that they include:

(a)   The geometrical position of the source with respect to the detector.
(b)   Absorption of radiation by air or by any material which separates the source and detector.
(c)   Self absorption of radiation by the source such that not all the radiation emitted within the source escapes from the source, e.g. solid samples containing α- or low-energy β-emitters.
(d)   The range of radiation within the detector.
(e)   Backscatter radiation reflected from the source support.
(f)   Dead time losses – immediately following detection of a radiation pulse the detector system is momentarily out of action owing to the finite time taken for the system to recover. Radiation entering the detector during this 'dead time' will not be counted although it is possible to make a correction for lost counts if the dead time (paralysis time) of the instrument is known.

An obvious requirement for radiation detection is that the radiation of interest should enter the detector. Many detectors incorporate thin windows whose function might be to contain a gas or in some cases to act as a shield for a light-sensitive component. Such windows are capable of absorbing low-energy β-radiation and α-radiation, thus preventing their detection. It is clearly essential to choose a detector system suitable for the radiation to be measured.

For some measuring systems the radioactivity measurement is made with the source external to the detector but placed close to it. For such an arrangement, since a source emits radiation in all directions, only a proportion of the emitted radiation can enter the detector and the actual proportion depends on the position of the source relative to the detector window, i.e. on the counting geometry.

For other measuring systems the source to be assayed is actually incorporated into the detector. This is the case with liquid scintillation counting in which instance the detector is a liquid containing radiation-responsive components. The source to be assayed is either dissolved or dispersed in this liquid. This arrangement can give highly efficient detection of the emitted radiation.

We shall begin by discussing detection methods dependent on current flow and which depend on gas ionisation.

11.5 Radiation detection using gases

11.5.1 Gas ionisation detectors

The detector is a gas-filled chamber, often a cylinder, in which an axial wire is fixed in and insulated from the cylinder. A potential difference is applied between the wire and the cylinder such that the wire is made the positively charged anode and the cylinder is the negatively charged cathode. Each time a charged particle passes through the space between the electrodes molecules of gas are ionised. Each gas molecule ionised gives rise to an ion pair, i.e. a positively charged molecular ion plus an electron. This, on average, requires an energy of approximately 30 eV per ion pair. Thus the number of ion pairs produced per charged particle is dependent on the particle energy. Electrons flow towards the anode and the molecular ions towards the cathode under the influence of the potential gradient. However, because of their much smaller mass the electrons travel very much faster and essentially carry all the observed current. Clearly the observed current will be dependent on the potential applied across the electrodes but the relationship is not generally ohmic. For a constant radiation source Figure 11.2 shows the general way in which current varies with applied potential difference.

At zero voltage recombination of electrons and positive ions occurs and no current flows. At small voltages the electrons and positive ions will be attracted respectively to the anode and cathode. Some recombination will take place so that not all the initially produced electrons and ions reach the electrodes. As voltage is further increased the fraction of these electrons and ions which reach the electrodes increases up to a limiting value of unity. This corresponds to the saturation current. Further moderate increases in voltage cause no increase in the current. Typical values of voltage are in the range from 50 V to 150 V and the current in the range 10^{-15}–10^{-12} A. These conditions correspond to those for operation of ion chambers and will not be considered further here. Further increase in the voltage causes the current to increase approximately in proportion to the increase in voltage; this is therefore called the proportional region. Here the high potential gradient in the vicinity of the axial anode imparts such high acceleration to the ion pair electrons nearby that these cause further ionisation of the gas. As a result the number of ion pairs per initial ionisation is increased and as a consequence so is the observed current. The ratio of total ion pairs per initial ion pair is known as gas multiplication. Gas multiplication can vary from unity to 10^6. It increases proportionately with the voltage but this behaviour does reach an upper limit. This limit is controlled by the total number of ion pairs which can

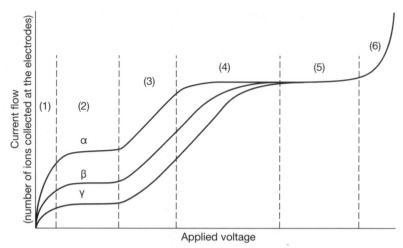

Figure 11.2 Dependence of current on applied voltage for radioactive sources of equal source strength (schematic)

be produced. This is limited by the gas concentration and by the counter dimensions. At higher voltages a region is reached in which all initial ionisations, no matter how small or large, will produce the same final ion concentration and hence current. This corresponds to the plateau known as the *Geiger–Müller* region, i.e. region 5. Further increase in applied voltage beyond this region causes continuous discharge in the absence of radiation and this can damage the detector. We now consider proportional and Geiger–Müller counters in a little more detail.

11.5.2 Proportional counters

Such counters operate in region 3. Here there is proportionality between the number of primary ion pairs, created by the incoming radiation, and the total number of ion pairs, resulting from, the chosen operating conditions. Figure 11.2 illustrates the effect of radiation type on the response of the counter. We note the consequences of the stopping power expression discussed in Sections 11.3.2 and 11.3.3. The ionisation produced by an α-particle is much more intense than for a β-particle which is more intense than for a γ-ray all having the same initial energy. The differences observed are a consequence of the fact that a decreasing proportion of the energy loss is inside the detector gas and this is least for γ-radiation. Such counters can therefore be used to distinguish between these types of radiation using a signal discriminator. They can also be used to give an energy spectrum for a given source. For this to be possible the radiation must lose all its energy in the detection medium. In this case, because the medium is a gas, only α-radiation, low-energy β-radiation, e.g. from hydrogen-3, carbon-14 and sulphur-35, and low-energy γ- and X-radiation will be effectively absorbed. The counter gas is typically argon plus a quencher gas, e.g. ethanol or chlorine (see Section 11.5.3). The counter can be operated as a sealed unit having a thin mica window to allow entry of low-energy β-particles, e.g. for measurements on carbon-14. Alternatively it can be operated as a windowless counter for measurement of α and very low-energy β-radiation such as that from tritium. Dead times for such a counter are much shorter, ~250 ns, than for Geiger-Müller counters so that proportional counters can be used for high count rate measurements.

11.5.3   Geiger–Müller counters

In region 5 gas multiplication produces a pulse of uniform size irrespective of the number of ions formed by the passage of the primary radiation. If a pure gas is used in this type of detector a sequence of secondary ionising events occurs before the full process is complete so that the dead time can be ~200 μs, which is considerably longer than for the proportional counter. The most effective way to reduce or quench these secondary events is to add a quenching gas to the main gas filling. Quench gases can be either organic, e.g. ethyl alcohol, or a halogen, e.g. chlorine or bromine. Organic quenchers are decomposed during the quenching process so that Geiger–Müller tubes employing these have a life limited by the decomposition rate. For halogen-quenched tubes the halogen molecule is reformed during the quenching process so that tube life is limited by gas leakage from the detector seal. The main gas filling is typically argon.

11.6   Radiation detection using solids and liquids

11.6.1   Solid and liquid state counters

The much greater densities of solids and liquids, compared with gases, provide opportunities for more efficient absorption, and hence detection, of high-energy β- and γ-radiation. Solid and liquid state detectors are of two main types, namely those which depend on the measurement of light flashes or scintillations and those which depend on the measurement of electric current. The former make use of the properties of insulators whereas the latter employ the properties of semiconductors.

11.6.1.1   *Electronic structure and behaviour of insulators and semiconductors*

The ability of solid materials to conduct electricity can be understood in terms of their electronic energy levels. In the case of an insulator the inner electron levels are fully occupied as are the valence levels and there is a large energy gap between the highest occupied valence level and the next unoccupied level or conduction band. In this case no electrons are available to give electron conductivity, and, except in the presence of potential gradients high enough to cause dielectric breakdown, such materials do not conduct electricity. However, incident radiation can lead to the formation of charge carriers. These are electrons promoted from the valence level to the conduction band together with corresponding positive holes remaining in the valence level. With imposition of an applied potential it is possible to measure a current. However, for certain materials it is more advantageous to allow electron–hole recombinations to occur since these give rise to photon emissions having energies similar to the band gap. Such materials can be used to provide the detector stage in scintillation counting systems. For inorganic scintillators the electron band structure is dependent on the three-dimensional structure of the solid. For organic scintillators this is no longer the case and it is the molecular unit which determines the energy conversion process. For this reason organic scintillators, unlike inorganic scintillators, are able to function either in the solid state or in solution.

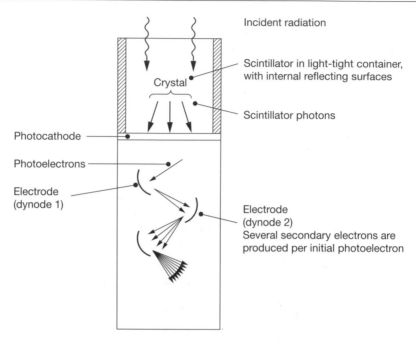

Incident radiation

Scintillator in light-tight container, with internal reflecting surfaces

Crystal

Scintillator photons

Photocathode

Photoelectrons

Electrode (dynode 1)

Electrode (dynode 2)
Several secondary electrons are produced per initial photoelectron

Figure 11.3   Crystal scintillator connected to a photomultipler tube

11.6.2   Scintillation methods

These methods, as indicated, depend on the fact that certain substances are capable of absorbing radiation whose energy is then transformed into a burst of photons, i.e. a scintillation. The intensity of the scintillation depends on the energy of the radiation absorbed. The scintillations are detected by a photomultiplier tube which is a photoelectric device capable of converting the scintillation into an electrical signal which is then given considerable amplification. The primary detection medium in such counters is the scintillator or phosphor and this can be solid or liquid. The sequence of processes involved in scintillation counting is illustrated schematically in Figure 11.3. The signal amplification occurs in the photomultiplier tube. This is achieved by applying a potential, typically 100 V, across the electrodes so that electron multiplication occurs at each electrode. The overall multiplication factor, which can be up to 10^6, depends on the number of dynode stages and the applied potential. At the end of the dynode chain the electrons are collected by an anode and the current is fed into an electronic circuit. The pulse counting rate is the same as the initial rate of scintillations and this is proportional to the activity of the radiation source. The magnitude of the signal output, or pulse amplitude, is proportional to the initial number of scintillation photons and this is proportional to the energy of the incident radiation. The scintillation counter can therefore be used as a spectrometer.

11.6.2.1   *Scintillators*

The most common solid material used is sodium iodide activated with thallium, designated as NaI(Tl). This is well suited for γ-ray detection and spectrometry. However, since it is hygroscopic it has to be sealed into a metal can. The encapsulated crystal can be made as a simple cylinder on which the source is placed. A special beaker known as a Marinelli beaker designed to surround the crystal, and to give good counting efficiency, can be used to contain radioactive solutions. Alternatively the crystal can be drilled to give a cylindrical recess, or well, into which a container, holding a prepared radioactive source, is inserted for counting. The latter arrangement also improves the counting efficiency. Such a counter has a dead time of ~250 ns and very high count rates can be measured. Organic scintillators are typically aromatic compounds such as naphthalene, anthracene and terphenyl. Terphenyl is also used combined in a polymeric matrix to give a solid plastic scintillator.

11.6.3 Liquid scintillation counting

The liquid scintillator is a cocktail of solvent, which is often an alkyl benzene, e.g. toluene or xylene, together with a scintillator solute, e.g. p-terphenyl at ~0.5 g l^{-1}. A range of commercial scintillation cocktails is available for counting aqueous and non-aqueous samples. An important practical and sometimes expensive problem is the need to dispose of large volumes of used cocktail. Recent developments in the design of new efficient cocktails have taken account of the need for environmentally friendly products which can accept both aqueous and non-aqueous samples. As indicated, the sample for measurement is incorporated either by dissolution or by suspension. For the latter it is important that the suspension should be stable during the measurement period or counting efficiency will not be constant. The solvent absorbs energy from the radiation and for efficient detection this energy must be rapidly transferred to the scintillant which de-excites via scintillation emission. A secondary solute is usually incorporated in order to give a wavelength shift. Its function is to capture the emission from the primary scintillant and emit at a lower wavelength which gives more efficient transfer of light to the photomultiplier tube.

The major application is in the counting of low-energy β-emitters such as tritium and carbon-14. Because the sample to be assayed is incorporated into the detector medium very high counting efficiencies can be obtained, e.g. for tritium >50% and for carbon-14 >90%. α-Emitters and high-energy β-emitters can also be measured as well as low-energy X-ray and γ-emitters. Several types of sample vial are commercially available and can be obtained with low ^{40}K background and to give very low leakage of organic components.

Several factors influence counting efficiency. These include colour and chemical quenching in the sample plus cocktail. Dissolved substances present in the scintillant solution interfere with the energy transfer processes and so reduce light output and hence counting efficiency. This is called chemical quenching. Substances which cause strong quenching include, for example, aliphatic iodides, bromides, thiols and ketones. Apart from chemical quenching the presence of coloured or opaque substances can absorb scintillation light before it reaches the photomultiplier. This has the effect of reducing the output pulse amplitude. Secondary solutes can improve performance by shifting the emission to a region in which the sample is more transparent. There is an extensive literature on the technique (Birks, 1964; Dyer, 1974).

11.6.4 Semiconductor methods

In the earlier discussion on the electronic behaviour of insulators and semiconductors we briefly referred to the large energy gap between the highest fully occupied valence level and the conduction band. In the case of diamond, a very poor conductor, this band gap is 6 eV. Other group 4 elements, particularly silicon and germanium, have crystal structures

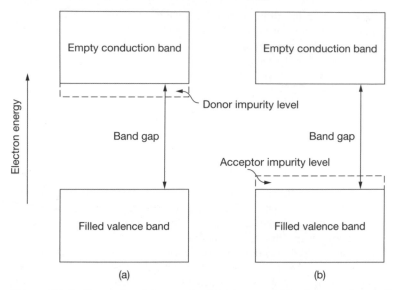

Figure 11.4   Energy band diagram for (a) n-type and (b) p-type semiconductors

similar to diamond but the band gaps are only 1.1 eV and 0.7 eV and at room temperature a small fraction of electrons have sufficient energy to enter the conduction band. As a consequence a small current can be made to flow in an applied field. Silicon has an electrical conductivity intermediate between those of a good conductor and an insulator so is referred to as a semiconductor. The conductivity of a pure semiconductor can be considerably increased by doping with small concentrations of impurities. By incorporation, i.e. doping, of the As atom $(4s^2 4p^3)$ into Si $(3s^2 3p^2)$ or Ge $(4s^2 4p^2)$ four of the electrons from As become involved in four covalent bonds to Si whereas the fifth electron requires only ~0.05 eV in Si and ~0.01 eV in Ge for excitation into the conduction band. Thus the As 'impurity' provides new energy levels known as donor levels which an electron must occupy for conduction to take place. The Si–As mixture is called an n-type semiconductor since the current is carried by negative charges. Similarly, doping of Si with Ga $(4s^2 4p^1)$ leaves vacancies called holes in the electronic structure. When an applied field is applied an electron can move into a hole, thus creating a hole elsewhere, and flow of current is generally described in terms of the movement of holes which behave like positive charges. Such semiconduction is known as p type. In this case the presence of Ga provides new energy levels – acceptor levels. The two types are shown schematically in Figure 11.4. The operation of most semiconductors devices depends on the properties of junctions made between n- and p-type material. By applying reverse bias, holes and electrons are drawn away from the junction, leaving a depleted layer (Figure 11.5). In the early days of semiconductors it was noticed that such materials having very low concentrations of charge carriers had good sensitivity to radiation. The depleted layer discussed above is particularly sensitive to creation of new charge carriers induced by radiation and this is the basis of operation of semiconductor detectors. The requirement for detection is that the radiation should lose its energy by creation of electron–hole pairs in the depleted region. At room temperature this requires 3.62 eV/ion pair in silicon and 2.96 eV/ion pair in germanium. Both electrons and holes have to be collected to produce an output pulse proportional to

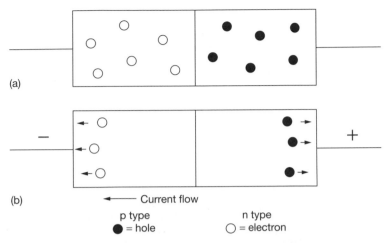

Figure 11.5 Formation of a depleted layer near an n–p junction: (a) no bias; (b) reverse bias

the total energy of the incident radiation. This can be achieved since electrons and holes have similar mobilities.

The width of the depletion region should ideally be optimised to the range of the radiation because the sensitive width should be sufficient to completely absorb all the energy of the radiation. The width is determined by the reverse bias and the conductivity of the base crystal. Modern semiconductor detectors employing lithium-doped silicon, Si(Li), and germanium, Ge(Li), allow wide depletion widths to be produced. The former can be operated at room temperature whereas the latter requires storage and use at liquid nitrogen temperatures (77 K). The more recent hyperpure Ge detectors must also be operated at 77 K but can be stored without degradation at room temperature. The advent of modern semiconductor detectors has revolutionised the measurement of charged particles and γ-radiation as is seen for example by the great improvements in energy resolution, speed of response and close linearity of response (Delaney and Finch, 1992).

11.7   Measurement of environmental radioactivity

Measurement of the radioactivity in water, air, many foods and animals is done routinely. Natural waters contain small amounts of radioactivity. This comprises members of the natural decay series plus man-made radionuclides which include those resulting from accidents such as that at Chernobyl in 1986. γ-Spectrometry is extensively used to measure a very wide range of radionuclides. Guidance on good practice procedures which include preparation and storage of samples, optimisation of equipment and counting procedure, and treatment of data is available (HMSO, 1989).

Detailed protocols for the determination of specific radionuclides in various media have been agreed by the Association of Official Analytical Chemists International and have been published. These include the following:

(a)   tritium in water using liquid scintillation counting;
(b) ^{90}Sr in water using a low-background thin-window Geiger–Müller counter to count β-particles;

(c) ^{89}Sr and ^{90}Sr in milk using a low-background thin-window Geiger–Müller counter to count β-particles;

(d) ^{131}I, ^{140}Ba and ^{137}Cs in milk and other foods counted, as a liquid, in a Marinelli beaker using a solid state scintillation detector for γ-ray spectroscopy (Association of Official Analytical Chemists International, 1998).

11.8   Radioanalytical methods for detection of non-radioactive analytes as pollutants

11.8.1   Introduction

Until now the objective has been to examine radioactive substances considered as pollutants and to consider methods for their detection based on their emitted radiation. Our emphasis now changes to methods which use radioactivity to detect and measure the presence and concentration of non-radioactive substances and which capitalise on the sensitivity of available detection techniques. These will be discussed under the headings of radiotracer methods, activation analysis and other radioanalytical methods.

11.8.2   Radiotracer methods

In this method a radioactively labelled atom in a substance is used to follow the behaviour of the substance through a series of physical and/or chemical and/or biological processes. A vital requirement in this method is that the tagged atom must not exchange with similar atoms in other substances. The method can be highly specific and very sensitive allowing the measurement of very small amounts of material, e.g. $<10^{-18}$ g for optimum choice of radionuclide and counting conditions. This means that for analyses where high dilution of material is involved a radiotracer method might be suitable, e.g. as it has been employed in some large-scale environmental studies of river flow and to assess the dispersion of wastewater in coastal waters (HMSO, 1984).

On a smaller scale many potential pollutants, e.g. pesticides, are released into the environment. Their fate can be tracked using a radiolabelled version of the compound. Such substances are subject to atmospheric washout and solubility then becomes an important property. Solubility studies with radiotracers have been undertaken using ^{14}C-labelled γ-benzene hexachloride, pp′ DDT and dieldrin (Atkins and Eggleton, 1971). The detailed procedure may involve separation and concentration of components for which conventional analytical techniques are often quite suitable.

11.8.3   Activation analysis

11.8.3.1   *Basic principles*

In activation analysis the analyte sample is bombarded with particles which interact with and induce radioactivity in atoms of the sample. The induced radiation is measured and can then be used to identify and quantify the amount of radionuclide present and also the original atoms in the sample. The state of combination of the atoms in the sample is unimportant. Irradiation of a sample generally produces a number of radionuclides which may require physical and chemical separation, prior to radioassay, in order to simplify the radiochemical analysis. For simple mixtures it may be possible to avoid destructive treatment of the sample and to conduct the radiation measurements directly on the irradiated material. Neutron irradiation is the most commonly used activation process.

Activation analysis relationships   The activation process involves the conversion of an inactive atom, I, to a radioactive atom, R:

$$I \rightarrow R$$

The radionuclide R then decays into a daughter product D:

$$R \rightarrow D$$

Straightforward mathematical treatment of these simultaneous processes leads to an expression for the number of radioactive nuclei N_R after irradiation time t:

$$N_R = N_I \, \phi \, \sigma[1 - \exp(-\lambda t)]/\lambda \qquad (11.6)$$

where N_I is the number of target atoms, λ is the decay constant for R, ϕ is the particle flux (number of particles $cm^{-2} s^{-1}$) and σ is the activation cross-section (cm^2) for the target atom, I, for particles of a given energy. Comprehensive tables of cross-sections are available (Hughes and Harvey, 1955; Mughabghab, 1984). We note that an element may have a range of isotopes all of which will normally be present in their natural abundances. Each has a different capture cross-section and will give a different product.

The activity of R at time t, A_t, is given by

$$A_t = \lambda N_R \qquad (11.7)$$

so that

$$A_t = N_I \, \phi \, \sigma[1 - \exp(-0.693t/t_{0.5})] \qquad (11.8)$$

where the decay constant has been replaced by $0.693/t_{0.5}$ for R.

We note that for 'long' irradiation, i.e. for times such that $\lambda t \gg 1.0$,

$$A_\infty = N_I \, \phi \, \sigma \qquad (11.9)$$

where A_∞ is the saturation activity and for a given concentration of the element I it has a value determined by the available particle flux. Equation (11.8) can be written

$$A_t = A_{sat}[1 - \exp(-0.693t/t_{0.5})] \qquad (11.10)$$

The term in square brackets is known as the fraction of saturation. An important consequence of this relationship is that, for any given particle flux, irradiation for a time equal to one half-life of R will achieve 50% of the saturation activity. For two, four, eight and ten half-lives the fractions of saturation will be 75%, 93.7%, 99.6% and 99.9% respectively and there is little point in continuing with irradiation beyond about eight half-lives. If it is known which elements are present then an appropriate irradiation time can be chosen. However, this will not always be known and preliminary exploratory measurements may be necessary. Also the sample for analysis will generally contain several elements all of which may undergo activation to give different fractions of saturation for each for a fixed irradiation time.

The above equation can give a useful indication of the feasibility of the method as applied to a problem. In practice it is not usual to apply the method in an absolute way but to make comparative measurements using a sample and standard which are irradiated under near identical conditions.

Neutrons are the most commonly used particles and these can be thermal or slow neutrons having energy of ~0.025 eV. In this case the most important nuclear reaction is one in which the neutron is captured and this is rapidly followed by a 'prompt' γ-ray. These reactions are therefore called nγ reactions. Many nuclides have high capture cross-sections for this type of reaction.

Sources of thermal neutrons are:

(a) **Fission reactors.** These provide high fluxes in the range 10^{12}–$10^{15}\,\mathrm{n\,cm^{-2}\,s^{-1}}$.

(b) **'Natural' neutron sources.** These make use of a nuclear reaction such as an α,n reaction, e.g. in a mixture comprising ^{241}Am and ^{9}Be. The α-particles from americium induce neutron emission in the beryllium. Neutron intensity is normally quoted as $\mathrm{n\,s^{-1}\,TBq^{-1}}$ and values of $\sim 10^{7}$–10^{8} are typical. Emitted neutrons have energies up to ~ 10 MeV and require moderation to give thermal energies. ^{226}Ra–^{9}Be is a similar source. The advantage of these types of source is that they do not require sophisticated operating equipment. However, because of their γ emission they require to be shielded and a severe disadvantage is their low neutron flux.

(c) **Neutron generators.** Neutrons, with energy of 14 MeV, are produced by bombardment of a tritium target with deuterons. Instruments are available which can generate fluxes in the range 10^{9}–$10^{12}\,\mathrm{n\,cm^{-2}\,s^{-1}}$.

11.8.3.2 *Analysis procedures and applications*

Following irradiation the induced activity decreases at a rate determined by the half-life of the product. The radionuclides are measured and ideally are identified via the nature and type of radiation emitted and half-life. Where products of widely differing half-lives are produced simplification of the analysis can sometimes be achieved by activity measurements over a suitable period of time, so allowing short-lived products to decay almost completely. For complex mixtures these may require chemical separation before activity measurements are made. For solid samples any β-radiation may experience considerable self-absorption and because of this γ-ray spectrometry, using crystal scintillation or semiconductor detectors, is used to identify the radioactive constituents. This necessitates measurement of the number, energies and intensities of the γ-peaks and peak decay. The smallest amount of element which can be determined by the method, i.e. sensitivity, clearly depends on the lowest activity which can be measured with sufficient precision. As indicated in equation (11.8), high flux, high capture cross-section and a half-life long enough for convenient handling of the sample will give high sensitivity. For realistic count rates sensitivities are routinely $<10^{-1}\,\mu$g and can extend to $10^{-6}\,\mu$g (Dean, 1995). We now consider two applications to illustrate the use of the method.

River and rainwater analysis By employing evaporation techniques to covert the liquid sample to a small quantity of solid residue it is possible to measure a wide range of elements which includes non-metals (Br, Cl, Se, As, I) and metals (Ag, Co, Cr, Cs, Fe, K, Mn, Na, Rb, Sb, Sc, Zn, Al, Au, Ca, Cd, Ce, Cu, and others). Combined with local knowledge this information can be used to identify the source (HMSO, 1984). Methods have been described for the analysis of trace metals in coastal waters (Salmon, 1975; Schnier and Karbe, 1976) and also to illustrate the use of pre-concentration methods prior to irradiation (Van der Sloot and Luten, 1976).

Airborne particulate analysis A great deal of work has used thermal neutron activation analysis to analyse airborne particulates. The method generally involves trapping the particulates by drawing a known volume of air through a filter. A large variety of filters have been investigated. Filter choice may be made to facilitate subsequent processing which can involve chemical treatment. For convenience particulates and filter can be irradiated and counted together; it may then be important to ensure there is little activity induced in the filter material in order to maintain a low background. The method has been used to identify and investigate abnormal levels of elements for source tracing. Comparison of city and pollution 'free' environments has been studied. Examination of air particulates present in heavy industrial areas has shown the significant contribution from aerosols and flyash from coal combustion. In some cases particle size information has been combined with elemental analysis to assist in identification of particle sources. Most of this type of work uses neutron activation followed by γ-spectrometric analysis (IAEA, 1976; Nuclear Analytical Methods, 1993).

11.8.4   Other radioanalytical methods

Other techniques which have found use in general chemistry, biochemistry and related fields have been less used in environmental analysis. These include isotopic dilution analysis (IDA) and related methods. The direct application of IDA is, in principle, very simple and it is possible to analyse for a given pollutant in a complex mixture which can be solid, liquid or gas. Let the substance of interest be A, and suppose that a mass of A equal to m_A is present in a total mass of mixture m. The method requires a sample of pure radiolabelled A whose specific activity, i.e. activity per unit mass, is known or can be measured. Let this be S_1. A weighed amount of this radiolabelled A, w_A, is added to the mixture and thoroughly mixed in to give a uniform concentration. A sample of A is recovered, purified and weighed and its activity is measured to give the new specific activity S_2. Quantitative separation of A is not necessary but the sample recovered for measurement must be pure. The observed change in specific activity is due to dilution of the added radiolabelled A with inactive A in the sample for analysis. From the initial and final specific activity the amount of A in the mixture can be calculated as follows. If necessary corrections are made for radioactive decay then total activity added to the mixture is conserved and we can write

$$w_A S_1 = (w_A + m_A)S_2$$

so that

$$m_A = w_A(S_1/S_2 - 1)$$

Clearly, for each substance to be analysed, a sample of the pure radiolabelled version is required. Unless these substances are readily available the cost of custom syntheses may be prohibitive unless a large programme of work is being undertaken. Very comprehensive coverage of this and related radioanalytical methods has been given (Tolgyessy and Kyrs, 1989; Tolgyessy and Bujdoso, 1991).

11.9   References

Association of Official Analytical Chemists International 1998 *Official methods of analysis*, 16th edn

Atkins D H F and Eggleton A E J 1971 *Nuclear techniques in environmental pollution*, IAEA Conference, Vienna

Birks J B 1964 *The theory and practice of liquid scintillation counting*, Pergamon, Oxford

Dean J A 1995 *Analytical chemistry handbook*, McGraw-Hill, New York, Section 11

Delaney C F G and Finch E C 1992 *Radiation detectors – principles and applications*, Oxford Science Publications, Oxford

Dyer A 1974 *Introduction to liquid scintillation counting*, Heydon, New York

Evans E A and Muramatsu M (eds) 1977 *Radiotracer techniques and applications*, Volume 1, Marcel Dekker, New York, Chapter 1

Hewitt C N 1996 Radiation in the environment. In R M Harrison (ed) *Pollution – causes, effects, control*, 2nd edn, Royal Society of Chemistry, London, Chapter 17

HMSO 1984 *Four essays on the application of radiation measurements in the water industry*, London

HMSO 1989 *Determination of radioactivity in water by γ-ray spectrometry*, London

Hughes D J and Harvey J A 1955 *Neutron cross-sections*, USAEC, McGraw-Hill, New York

IAEA 1976 *Measurement, detection and control of environmental pollutants*, Conference, Vienna, 15–19 March, several papers

Mughabghab S F 1984 *Neutron cross-sections*, Volume 1, *Neutron resonance parameters and thermal cross-sections*, Part B, Academic Press, New York

Nuclear Analytical Methods 1993 *Abstracts from international conference on nuclear analytical methods in the life sciences*, Prague, 13–17 September

Salmon L 1975 *Instrumental neutron activation analysis in environmental studies of trace elements*, AERE Report R7859, HMSO, London

Schnier C and Karbe L 1976 *Measurement, detection and control of environmental pollutants*, Conference, Vienna, 15–19 March, paper 206/31, p 415

Tolgyessy J and Bujdoso E 1991 *CRC handbook of radioanalytical chemistry*, CRC Press, Boca Raton, FL

Tolgyessy J and Kyrs M 1989 *Radioanalytical chemistry*, Volumes 1 and 2, Ellis Horwood, Chichester

Van der Sloot H A and Luten J B 1976 *Measurement, detection and control of environmental pollutants*, Conference, Vienna, 15–19 March, paper 206/40, p 435

Appendix 1

Priority pollutants

	Substance	Merck Index number	CA Registry number	Molecular weight	Melting point (°C) (boiling point (°C))	LD_{50} (mg kg^{-1})	LD_{50} comments
1	Aldrin	219	309–00–2	364.93	104–105.5	39,60	m,f rat, oral
2	Atrazine	886	1912–24–9	215.7	173–175	1750.0	Mice, oral
3	Azinphos-methyl	926	86–50–0	317.3	≈73	11.0	f rat, oral
4	DDT	2832	50–29–3	354.5	108	113,118	m,f rat, oral
5	1,2-Dichloroethane	3754	107–06–2	99	–35.4 (83.5)	770	Rat, oral
6	Dichlorvos	3069	62–73–7	220.98	(120)	80,56	m,f rat, oral
7	Dieldrin	3093	60–57–1	381	176–177	46	Rat, oral
8	Endosulfan	3529	115–29–7	406.95	106	43,18	m,f rat, oral
9	Endrin	3533	72–20–8	380.90	≈200	18,7.5	m,f rat, oral
10	Fenitrothion	3922	112–14–5	277.2	(140–145)	250	Rat, oral
11	Hexachlorobenzene	4600	118–74–1	284.80	227–229 (322–326)	–	–
12	Hexachlorobutadiene	–	87–68–3	261	–19 to –22 (210–220)	–	–
13	γ-Hexachlorocyclohexane (lindane)	5379	58–89–9	290.85	112	88,91	m,f rat, oral
14	Malathion	5582	121–75–5	330.36	2.85 (156–157)	1375,1000	m,f rat, oral
15	Pentachlorophenol (and its compounds)	7059	87–86–5	266.35	188–191 (310 decomposes)	146,175	m,f rat, oral
16	Simazine	8485	112–04–9	201.7	225–227	5000	Rat, oral
17	Trifluralin	9598	1582–09–8	335	48.5–49 (139–140)	500	Rat, oral

m,f indicate male,female respectively. Where gender is not specified the data for each sex are not available and the value should be assumed to represent an equal proportion of both sexes.

Appendix 1   Priority pollutants

The following substances are also on the Red List but are generic and chemical and toxicity data cannot be given:

	Substance
18	Cadmium and its compounds
19	Mercury and its compounds
20	Polychlorinated biphenyls (PCBs)
21	Tributyltin compounds
22	Trichlorobenzenes
23	Triphenyltin compounds

The data below show some of the original EU limits for some of the Red List compounds.

The Dangerous Substances Directive (76/464/EEC) list 1 (Red or Black List) substances include the following.

Compound	Level ($\mu g\,l^{-1}$)
Carbon tetrachloride	12.0
Chloroform	12.0
1,2-Dichloroethane (DCE)	10.0
Trichloroethylene (TRIC)	10.0
Perchloroethylene (PER)	10.0
DDT, p-DDT	0.01
Total 'drins'	0.03
Endrin	0.005
Aldrin	0.01
Dieldrin	0.01
Isodrin	0.005
Hexachlorobenzene (HCB)	0.03
Hexachlorobutadiene (HCBD)	0.01
Hexachlorocyclohexane HCH or lindane	0.11
Pentachlorophenol	2.0
Trichlorobenzene (TCB)	0.1

A number of these limits are under almost constant review and subject to updates on a regular basis. The preferred analytical method for many of these compounds is chromatography.

The following are Orange List hazardous compounds whose environmental effects are under review for addition to the Red List:

2-Amino-4-chlorophenol
Anthracene
Biphenyl
Chloracetic acid
4-Chloro-2-nitrotoluene
Cyanuric chloride
Dementon O
1,4-Dichlorobenzene
2-Dichloroethanol

1,1-Dichloroethylene
2,4-Dichlorophenoxyacetic acid
Dimethoate
1,3-Dichloropropan-2-ol
Ethylbenzene
Hexachloroethane
Linuron
Mevinphos
Pyrazon
1,1,1-Trichloroethane

Typical EC directives specifying limits on organic compounds to be measured in several water supplies are listed below. Note that the preferred analytical method is HPLC or GC.

Number	Title	Organic components	Level ($\mu g\,l^{-1}$)
75/440/EEC	Surface water directive	Cyfluthrin	0.001
		Hydrocarbons	50–1000
		Permethrin	0.01
		Pesticides	1–5.0
		Phenols	1–100
		Polycyclic aromatic hydrocarbons	0.2–1.0
76/659/EEC	Quality standards for freshwater required to support fish	Cyfluthrin	0.001
		Flucofuron	1.0
		PCSDs	0.05
		PAD	0.05
		Permethrin	0.01
		Sulcofuron	25.0
80/778/EEC	Quality of water intended for human consumption (Drinking Water Directive)	Pesticides individually	0.1
		Pesticides total	0.5
		Benzo-3,4-pyrene	0.01
		Tetrachloromethane	3.0
		Trichloroethane	30.0
		Tetrachloroethane	10.0
		Trihalomethanes	100.0
		Phenols	0.5
		Polyaromatic hydrocarbons	0.2

Appendix 2

Occupational exposure levels

The following texts are extensive sources of occupational exposure levels data:

Materials Safety Data Sheets
Published by Sigma Aldrich Fluka
1001 West St. Pauls Avenue
Milwaukee
WI 53233
USA

Sax's Dangerous Properties of Industrial Materials
Published by Wiley VCH
PO Box 101161
69451 Weinheim
Germany

Both of the above databases are available in printed hardback versions and compact discs for automatic searching.

Also available is a publication, EH 40, *Occupational Exposure Levels*, from the UK Health and Safety Executive, which is amended every year for many of the key compounds widely used in commerce and identified in the various Health and Safety at Work acts and amendments. Also available are a series of publications which detail hazards and safe handling procedures for a number particularly toxic materials.

Publications are available from:

Health and Safety Executive Books
PO Box 1999
Sudbury
Suffolk CO10 6FS
UK

The Royal Society of Chemistry produces *The Dictionary of Substances and their Effects* (DOSE) which provides data for hazard and risk assessment for 4000 chemicals which are known to be pesticides, food carcinogens and endocrine disruptors. It is available from:

Royal Society of Chemistry
Thomas Graham House
Science Park
Cambridge CB4 0WF
UK

Similar documents are available from the EU commission under the Environmental Health
Commission directorate, and in the USA from a number of agencies, particularly the
National Institute of Health and the Environmental Protection Agency.

Appendix 3

SI units, definitions, conversion tables and physical constants

Units

The quantities of parameters are expressed in the internationally accepted 'Système International d'Unités', abbreviated as SI units. Names and symbols of the base units and those of some parameters derived from these units are given (alphabetic order) as follows:

Parameter	Name of unit	Symbol
Base and supplementary units		
Amount of substance	Mole	mol
Angle (plane)	Radian	rad
Angle (solid)	Steradian	sr
Electric current	Ampere	A
Length	Metre	m
Luminous intensity	Candela	cd
Mass	Kilogram	kg
Temperature	Kelvin	K
Time	Second	s
Derived units		
Acceleration (linear)	Metre per second squared	$m\,s^{-2}$
Acceleration (angular)	Radian per second squared	$rad\,s^{-2}$
Area	Square metre	m^2
Capacitance	Farad	$F = A\,s\,V^{-1}$
Electric field strength	Volts per metre	$V\,m^{-1} = N\,C^{-1}$
Electric resistance	Ohm	$\Omega = V\,A^{-1}$
Electrical potential	Volt	V
Energy (work, quantity of heat)	Joule	J
Entropy	Joule per kelvin	$J\,K^{-1}$
Force	Newton	$N = kg\,m\,s^{-2}$
Frequency	Hertz	$Hz = s^{-1}$
Illuminance	Lux	lx
Inductance	Henry	$H = V\,s\,A^{-1}$
Luminance	Candela per square metre	$cd\,m^{-2}$
Luminous flux	Lumen	lm
Magnetomotive force	Ampere	A
Magnetic field strength	Ampere per metre	$A\,m^{-1}$

Parameter	Name of unit	Symbol
Magnetic flux	Weber	$Wb = V\,s$
Magnetic flux density	Tesla	$T = Wb\,m^{-2}$
Power	Watt	$W = J\,s^{-1}$
Pressure (mechanical)	Pascal	$Pa = N\,m^{-2}$
Quantity of electricity	Coulomb	$C = A\,s$
Radiant intensity	Watt per steradian	$W\,sr^{-1}$
Radiation activity (source)	Becquerel	$Bq = s^{-1}$
Radiation dose	Gray	$Gy = J\,kg^{-1}$
Specific heat capacity	Joule per kilogram per kelvin	$J\,kg^{-1}\,K^{-1}$
Thermal conductivity	Watt per metre per kelvin	$W\,m^{-1}\,K^{-1}$
Velocity (linear)	Metre per second	$m\,s^{-1}$
Velocity (angular)	Radian per second	$rad\,s^{-1}$
Viscosity (kinematic)	Square metre per second	$m^2\,s^{-1}$
Viscosity (dynamic)	Newton second per square metre	$N\,s\,m^{-2}$
Wavenumber	Per metre	m^{-1}

Useful physical constants (CODATA recommended values)

Electron charge (e)	$1.6022 \times 10^{-19}\,C$
Electron mass (m_e)	$9.1094 \times 10^{-31}\,kg$
Neutron mass (m_n)	$1.6749 \times 10^{-27}\,kg$
Proton mass (m_p)	$1.6726 \times 10^{-27}\,kg$
Mass of a molecule (m_m)	$M \times 1.67 \times 10^{-27}\,kg$ (M is the relative molecular mass)
Volume of 1 mol of a gas at NTP	$2.24 \times 10^{-2}\,m^3$ (mole = gram molecule = M in g)
Number of molecules per cm^3 at NTP	2.6871×10^{19}
Avogadro's number (N_a)	$6.025 \times 10^{23}\,mol^{-1}$
Velocity of light (c)	$2.9979 \times 10^8\,m\,s^{-1}$
Planck's constant (h)	$6.62607 \times 10^{-34}\,J\,s$
Boltzmann's constant (k)	$1.38066 \times 10^{-23}\,J\,K^{-1}$
Gravitational constant (G)	$6.6726 \times 10^{-11}\,N\,m^2\,kg^{-2}$
Gas constant (R)	$8.3145\,J\,mol^{-1}\,K^{-1}$
Permittivity of free space (ε_0)	$8.854 \times 10^{-12}\,C^2\,N^{-1}\,m^{-2}$
Permeability of free space (μ_0)	$4\pi \times 10^{-7}\,Wb\,A^{-1}\,m^{-1}$
Acceleration due to gravity (g)	$9.7805\,m\,s^{-2}$ (sea level at equator)
Absolute zero temperature (T_{abs})	$0\,K$ ($-273.15\,°C$)

Source: Lisle D R (ed) *Handbook of Chemistry and Physics*, 77th edn, CRC Press, Boca Raton, FL

Appendix 3 SI units, definitions, conversions and constants

Conversion of units

Name of unit (Length)	Metre	Name of unit (Time)	Second	Name of unit (Power[a])	Watt	Name of unit (Mass[b])	Kilogram
Kilometre (km)	10^3	Second (s)	1	Exawatt (EW)	10^{18}	Kilogram (kg)	1
Metre (m)	1	Millisecond (ms)	10^{-3}	Petawatt (PW)	10^{15}	Gram (g)	10^{-3}
Decimetre (dm)	10^{-1}	Microsecond (μs)	10^{-6}	Terawatt (TW)	10^{12}	Milligram (mg)	10^{-6}
Centimetre (cm)	10^{-2}	Nanosecond (ns)	10^{-9}	Gigawatt (GW)	10^9	Microgram (μg)	10^{-9}
Millimetre (mm)	10^{-3}	Picosecond (ps)	10^{-12}	Megawatt (MW)	10^6	Nanogram (ng)	10^{-12}
Micrometre (μm)	10^{-6}	Femtosecond (fs)	10^{-15}	Kilowatt (kW)	10^3	Picogram (pg)	10^{-15}
Nanometre (nm)	10^{-9}			Dekawatt (dW)	10^1		
Angstrom (Å)	10^{-10}			Watt (W)	1		
Picometre (pm)	10^{-12}			Milliwatt (mW)	10^{-3}		
				Microwatt (μW)	10^{-6}		
				Nanowatt (nW)	10^{-9}		
				Picowatt (pW)	10^{-12}		

[a] 1 hp (horse power) = 550 ft lb s^{-1} = 745.7 W, 1 Btu h^{-1} (British thermal unit per hour) = 0.2931 W.
[b] 1 u (atomic mass unit) = 1.6605 × 10^{-27} kg; 1 tonne (metric ton) = 10^3 kg.

Energy

$$1 \, eV = 1.6 \times 10^{-12} \, erg = 10^{-7} \, J$$

$$1 \, cal = 4.2 \, J$$

$$1 \, eV \rightarrow 1.24 \times 10^3 \lambda^{-1} \, (\lambda, \text{ wavelength in nanometres})$$

Pressure

$$1 \, mbar = 0.75 \, Torr = 100 \, Pa \, (N \, m^{-2})$$

$$1 \, atm = 760 \, mmHg = 1.013 \times 10^5 \, Pa \text{ (standard atmospheric pressure)}$$

Gas law:

$$PV = nRT$$

where P is the pressure $(N\,m^{-2})$, V the volume (m^3), n the number of moles of the gas, R the gas constant $(8.31\,J\,mol^{-1}\,K^{-1})$ and T the temperature (K). The number of molecules in 1 mol of the gas is N_A (Avogadro's number).

Temperature

0 K (kelvin) = $-273.15\,°C$ (degree Celsius)

°F (degree Fahrenheit) = $°C \times 9/5 + 32$

$°C = (°F - 32) \times 5/9$

Optical quantities

Photon energy

$$E = h\nu = hc/\lambda$$

where h is Planck's constant, ν = frequency of the light (electromagnetic wave) and λ is the wavelength.

1 einstein $\equiv$ 1 mol of photons = 6.023×10^{23} photons

Absorption of light

A (absorbance) = $\log_{10}(1/T)$

T (transmittance) = transmitted intensity/incident intensity

$$= \exp(-kcx)$$

where k is defined as the absorption cross-section (m^2), c the concentration (m^{-3}) and x the path length (m). For a uniformly absorbing medium, $kc = \alpha\,(cm^{-1})$ defined as the absorption coefficient.

$T = 10^{-A}$ and $A = 0.434\,kcx$

OD (optical density) = $\log_{10}(1/T)$

Throughput of a monochromator

TP (throughput) = $f^2 w/(1 + 4F^2)$

where f is the focal length of the monochromator and w, the angular width of the slit, is equal to 1/RP (where RP is the resolving power).

F (f number) = f/A

where f is the focal length and A is the aperture (effective diameter) of the monochromator mirror.

Appendix 3   SI units, definitions, conversions and constants

Spectral bands

100–280 nm	UV-C (far ultraviolet)
280–315 nm	UV-B (ultraviolet)
315–400 nm	UV-A (near ultraviolet)
380–780 nm	Visible (subjective)
	Violet 360–450 nm
	Blue 450–500 nm
	Green 500–570 nm
	Yellow 570–590 nm
	Orange 590–610 nm
	Red 610–830 nm
0.78–2.5 μm	IR-A (near infrared) (μm = micrometre = 10^3 nm)
2.5–50 μm	IR-B (mid-infrared)
50–1000 μm	IR-C (far infrared)

Useful mathematical formulae

Circumference of a circle	$2\pi r$ (*r* is the radius)
Area of a circle	πr^2
Volume of a sphere	$(4/3)\pi r^3$
Volume of a cone	$(1/3)\pi r^2 h$ (*h* is the height from the base)
Volume of a cylinder	$\pi r^2 h$ (*h* is the length)
Area of a triangle	$\frac{1}{2}$ base × height
Volume of a pyramid	$(1/3)$base area × height

$\log_{10} x = \ln_e x / \ln_e 10 \sim 0.4343 \ln_e x$

$\log A \times \log B = \log(A + B); \quad \log A / \log B = \log(A - B); \quad \log A^n = n \log A$

$\pi \sim 3.14$, e ~ 2.72

Equation of a straight line: $y = mx + c$, *y* is the independent axis, *x* is the variable, *m* is the slope and *c* is the value of *y* at $x = 0$ (*y*-intercept)

Equation of a circle of radius *r*: $y^2 = r^2 - x^2$

Appendix 4

Standards for environmental diagnostics

Very few of the analytical methods used in environmental diagnostics are absolute. In most cases the method may need to be calibrated with a standard. Two types of **standard** are used: **primary** and **secondary**. Primary standards are those which can be sampled and prepared by mass while secondary standards are calibrated with primary standards using a stoichiometric chemical reaction by volumetric or other methods. Their accuracy is limited by the nature of the reaction used in the calibration. Some standards use a physical property of the material whereas for other standards the concentration of the chemical is the standard. Our discussion of standards will only give a brief description of typical standards used in the various analytical methods with an indication of some of the sources of discrete chemical compounds and possible sources of cocktails of a number of chemicals of one type, i.e. standard herbicides or pesticides. Some of the requirements for primary standards will be briefly discussed.

In order to be considered as a primary standard a chemical should meet the following criteria:

(a) It must be readily obtainable in a pure (>99.95%) form which is easy to dry and maintain in a pure state without degradation or alteration on storage.
(b) The material must not react with the oxygen, water or carbon dioxide in the atmosphere during handling.
(c) Established tests for known or likely impurities must be readily available.
(d) The reaction between the standard and any other material used in the calibration should be rapid and stoichiometric.

Additional useful attributes are ready solubility in common solvents and a relatively high molecular weight so that weighing errors are reduced to a minimum.

A secondary standard is a chemical which can be used in calibrations and standardisations after its active component has been determined by the use of a primary standard in a stoichiometric chemical reaction usually a titration process. Our discussions of analytical methods will concentrate on primary or certified standards.

Analytical determination standards

Gravimetric methods of analysis are unique in that they are absolute and require no calibration, but for other methods calibrations with standards are required and these will be considered below.

Volumetric analysis

Acids and bases

The classical standard acid is obtained by boiling a solution of analytical grade hydrochloric acid until a constant boiling temperature is obtained when the compositions of the solution and the vapour is the same. After measuring prevailing atmospheric pressure, as well as the boiling temperature, the composition of the acid can be read from a calibration table. This is the procedure adopted to produce the 'Volucon', certified standard acid solution concentrate, which only requires a suitable dilution with distilled water to provide working solutions. More convenient standard acids are potassium hydrogen phthalate $KH(C_8H_4O_4)$, potassium hydrogen iodate $KH(IO_3)_2$ (notice the high molecular weight of this compound) and benzoic acid $C_6H_5CO_2H$. The last of these is also used as a calibration compound in calorimetric studies.

It should also be noted that the British standard, 1647, for a buffer solution is a 0.05 M solution of potassium hydrogen phthalate in distilled and deionised water which has a pH of 4.001 at 20 °C. This is particularly important for the calibration of the glass electrode in pH determinations. This standard has been adopted by other bodies such as IUPAC and ASTM.

Standard bases are more difficult to obtain and only disodium tetraborate, $Na_2B_4O_7$, and sodium carbonate, Na_2CO_3, are widely used. The latter requires careful manipulation since it is mildly hygroscopic and cakes over a period of time, but can be readily air dried in an oven at 120 °C.

Oxidation–reduction systems

Standard oxidising agents are potassium dichromate, $K_2Cr_2O_7$ (also used as a spectrophotometric standard), and the potassium halates: bromate, $KBrO_3$, and iodate, KIO_3. Reducing agents are sodium oxalate $Na_2C_2O_4$, arsenic(III) oxide, As_2O_3, and ultra-pure iron metal.

Precipitation and complex formation systems

Usually silver metal, silver nitrate, sodium or potassium chlorides, potassium bromide are used for precipitation standards. For complex formation standards, spectroscopic grade metals zinc, magnesium and copper are added to the list.

Conductimetric methods

Usually absolute conductivities of solutions are not required since determinations based on conductance look for changes in conductance but when absolute values are required a standard 0.1 M solution of 'Analar' or 'Spectroscopic' grade potassium chloride (7.4191 g in 1000 g of solution) has a conductance of $0.012\,856\,\Omega^{-1}\,cm^{-1}$.

Potentiometric methods

The potential of the pure hydrogen electrode containing hydrogen gas at unit pressure in contact with hydrogen ions at unit activity on a platinum black electrode is the absolute standard, assigned a potential of 0 V at all temperatures. However, for ease of maintainance and use, the calomel and silver electrodes are used routinely. Mercury in contact with a saturated solution of mercurous and potassium chlorides has a standard potential of 0.2444 V compared with the standard hydrogen electrode and silver in contact with silver chloride and unit activity hydrochloric acid has a potential of 0.222 V. Usually the glass indicator electrode, in which the calomel or silver electrode is included as the standard, is used in pH measurements by potentiometric methods.

Spectrophotometric and spectroscopic systems

Two standards are used for spectrophotometry, one to calibrate the wavelength and the second to verify the absorbance level. The wavelength calibration is provided by a rare earth neodymium or gadolinium glass filter. The absorption by these materials is characterised by a number of narrow absorption bands. The absorbance standard, as defined by the Beer–Lambert law, is usually a 0.01 M solution of analytical grade potassium dichromate $K_2Cr_2O_7$ in 0.01 M HCl or H_2SO_4. For the dispersive emission spectrometer the wavelength calibration is based on the light emission from a sodium vapour discharge source. The two lines seen as a yellow doublet by a dispersive spectrometer have wavelengths of 5889.950 and 5895.924 Å.

Sources of standard materials

A number of organisations and institutes as well as chemical suppliers will provide primary standard materials or analysed systems as secondary standards for calibration purposes. Major sources of certified standards are:

UK and Europe
> Bureau of Analysed Samples, Cleveland, UK
> National Physical Laboratory, UK
> National Chemical Laboratory, UK
> Bundesanstalt für Materialforschung und Prüfung (BAM), FRG
> Centre Technique des Industries de la Fonderie, France
> Institut de Recherches de la Siderurgie Française, France
> Community Bureau of Reference, Belgium
> Swedish Institute for Metal Research, Sweden

USA and Canada
> National Bureau of Standards (Department of Commerce), Washington
> Canadian Centre for Mineral and Energy Technology

Asia
> National Institute for Environmental Studies, Japan

Other certified standards containing mixtures of chemical such as pesticides can be obtained from the Environmental Protection Agencies (USA and UK) and many of the chromatographic suppliers. Usually the samples contain measured quantities of one particular type of pesticide, e.g. carbamates or organophosphates. In many environmental samples several types of pesticide will be present and thus individual standards may be best.

Example

The simplest sources of certified standard solutions available for wet chemical processes are the major chemical suppliers. A typical example is the 'Volucon' series supplied by

BDH (Merck) Limited of Poole, Dorset, UK. An accurately prepared solution is provided, in a sealed container, which is pierced by downward action of a sharpened non-porous rod, thus delivering the contents into a volumetric flask, via a filter funnel. The container is easily rinsed with water to ensure complete transfer of the contents vial into the flask. Simple dilution of the supplied solution with pure water in the calibrated flask provides a recognised standard solution suitable for system calibration as part of a legal investigation.

Appendix 5

Further reading for Chapter 9

Cellular methods

Babich H and Borenfreund E 1991 Cytotoxicity and genotoxicity assays with cultured fish cells: a review. *Toxicol. in Vitro*, **5**; 91–100

Bitton G 1983 Bacterial and biochemical tests for assessing chemical toxicity in aquatic environments. A review. *CRC Crit. Rev. Environ. Control*, **13**; 51–65

Blum D J and Speece R E 1990 Determining chemical toxicity to aquatic species, *Environ. Sci. Technol.*, **24**(3); 284–93

Buikema A L Jr, Geiger J G and Lee D R 1980 *Daphnia* toxicity tests. In A L Buikema Jr and J Cairns Jr (ed) *Aquatic invertebrate bioassays*, ASTM STP 715, American Society for Testing and Materials, Philadelphia, PA, pp 46–69

Bunch R L and Chambers C W A 1967 Biodegradability test for organic compounds. *J. Water Pollut. Control Fed.*, **39**; 181–7

Centeno M D, Brendonck L and Persoone G 1993 Cyst-based toxicity tests. III. Development and standardization of an acute toxicity test with the freshwater anostracan crustacean *Streptocephalus proboscideus*. In A Soares and P Calow (ed) *Progress in standardization of aquatic toxicity tests*, CRC Press, Boca Raton, FL, pp 37–55

Department of the Environment 1978 Standing Committee of Analysis. In *Amenability of sewage sludge to anaerobic digestion 1977*, HMSO, London

Dutka B J 1986 Method for determining acute toxicity in water, effluents and leachates using *Spirillum volutans. Toxic. Assess.*, **1**; 139–49

Elnabarawy M T, Robideau R R and Beach S A 1988 Comparison of three rapid toxicity test procedures: Microtox, Polytox, and activated sludge respiration inhibition. *Toxic. Assess.*, **3**; 361–70

Erasin B R, Turner A P F and Wheatley A D 1994 A fixed film bioassay for the detection of micropollutants toxic to anaerobic sludges. *Anal. Chim. Acta*, **298**; 1–10

Gotvajn A Z and Zagorc-Koncan J 1996 Comparison of biodegradability assessments for chemical substances in water. *Water Sci. Technol.*, **33**; 207–12

Janssen C R, Persoone G and Snell T W 1994 Cyst-based toxicity tests. VIII. Short-chronic toxicity tests with the freshwater rotifer *Brachionus calyciflorus. Aquat. Toxicol.*, **28**; 243–58

Kameya T, Murayama T, Kitano M and Urano K 1995 Testing and classification methods for the biodegradabilities of organic compounds under anaerobic conditions. *Sci. Total Environ.*, **170**; 31–41

307

Keddy C J, Greene J C and Bonnell M A 1995 Review of whole-organism bioassays: soil, freshwater sediment, and freshwater assessment in Canada. *Ecotoxicol. Environ. Safety*, **30**; 221–51

Kilroy A and Gray N F 1995 'Treatability, toxicity and biodegradability test methods. *Biol. Rev.*, **70**; 243–75

Livingstone D R 1993 Biotechnology and pollution monitoring: use of molecular bio-markers in the aquatic environment. *J. Chem. Technol. Biotechnol.*, **57**; 195–211

Ma T H, Xu Z, Xu C, McConnell H, Rabago E V, Arreola G A and Zhang H 1995 The improved Allium/Vicia root tip micronucleus assay for clastogenicity of environmental pollutants. *Mutat. Res.*, **334**; 185–95

McFeters G A, Bond P J, Olson S B and Tchan Y T 1983 A comparison of microbial bioassays for the detection of aquatic toxicants. *Water Res.*, **17**; 1757–62

Novotny V and Olem H 1994 *Water quality, prevention, identification and management of diffuse pollution*, Van Nostrand Reinhold, New York, pp 817–60

Nyholm N 1996 Biodegradability characterisation of mixtures of chemical contaminants in wastewater – the utility of biotests, *Water Sci. Technol.*, **33**; 195–206

Persoone G and Janssen C R 1993 Freshwater invertebrate toxicity tests. In P Callow (ed) *Handbook of ecotoxicology*, Blackwell, Oxford, pp 51–65

Santojanni A, Gorbi G and Sartore F 1995 Prediction of mortality in chronic toxicity tests on *Daphnia magna*. *Water Res.*, **29**; 1453–9

Snell T W, Moffat B, Janssen C R and Persoone G 1991a Acute toxicity bioassays using rotifer. III. Effects of temperature, strain and exposure time on the sensitivity of *Brachionus plicatilis*. *Environ. Toxicol. Water Qual.*, **6**; 63–75

Snell T W, Moffat B, Janssen C R and Persoone G 1991b Acute toxicity bioassays using rotifer. IV. Effect of cyst age, temperature and salinity on the sensitivity of *Brachionus calyciflorus*. *Ecotoxicol. Environ. Safety*, **21**; 308–17

Stroud K C G and Jones D B 1975 Development of a cross pollution detector: field trials. *Water Treat. Exam.*, **24**; 100–19

Van Steertegem M and Persoone G 1993 Cyst-based toxicity tests. V. Development and critical evaluation of standardized toxicity tests with the brine shrimp *Artemia* (Anostraca, Crustacea). In A Soares and P Calow (ed) *Progress in standardization of aquatic toxicity tests*, CRC Press, Boca Raton, FL, pp 81–95

Williamson K J and Nelson P O 1983 *Bacterial bioassay for level one toxicity assessment*, EPA-600/3-83-017, US Environmental Protection Agency Environmental Research Laboratory, Corvallis, OR.

Molecular methods

Immunoassays

Despande S S 1996 *Enzyme immunoassay from concept to product development*, 1st edn, Chapman and Hall, London pp 1–450

Gould R 1996 *ELISA – a new name in environmental analysis*, Croner's Environmental Special Report, August 1996, pp 1–15

Hammock B D and Gee S J 1995 *Impact of emerging technologies on immunochemical methods for environmental analysis*, ACS Symposium Series 586, Immunoanalysis of Agrochemicals, ACS Press, pp 1–22

Harlow E and Lane D 1988 *Antibodies: a laboratory manual*, Cold Spring Harbor Laboratory, pp 1–726

Hermanson G T 1996 *Bioconjugate techniques*, 1st edn, Academic Press, New York, pp 120–75

Marco M-P, Gee S and Hammock B C 1995 Immunochemical techniques for environmental analysis, 1. Immunosensors. *Trends Anal. Chem.*, **14**(7); 341–50

Selim M I, Achutan C, Starr J M, Jiang T and Young B S 1997 *Comparison of immunoassay to high-pressure liquid chromatography and gas chromatography–mass spectrometry analysis of pesticide in surface water*, ASC Symposium Series 657, American Chemical Society, pp 234–44

Sherry J P 1992 Environmental chemistry: the immunoassay option. *Crit. Rev. Anal. Chem.*, **23**(4); 217–300

Genetics

Lewin B 1997 *Genes VI* Oxford University Press, Oxford

Molecular biology techniques

There are many laboratory manuals covering general molecular biology and also manuals on specific topics, e.g. the polymerase chain reaction (PCR). Examples include:

Griffin H G and Griffin A M (ed) 1994 *PCR technology current innovations*, CRC Press
White B A (ed) 1993 *PCR protocols, current methods and applications*, Humana Press

Also useful are the books in the Practical Approaches Series, edited by D Rickwood and B D Hames, published by Oxford University Press. The series includes books on all aspects of molecular biology including use of gene probes and PCR.

DNA-based diagnostics in the environment

Many of the companies producing DNA-based diagnostic kits (e.g. Affymetrix, Pharmacia, Murex, Qualicon, Perkin Elmer) have web sites where information on their kits is available. The journal *Applied and Environmental Microbiology* is a useful source of articles describing the application of molecular methods in the environment.

Amann R I, Ludwig W and Schleifer K H 1995 Non-culture methods of cell identification. *Microbiol. Rev.*, **59**; 143–69

MacNeil L, Kauri T and Robertson W 1995 Molecular techniques and their potential application in monitoring the microbiological quality of indoor air. *Can. J. Microbiol.*, **41**; 657–65

Sayler G and Layton A 1990 Environmental application of nucleic acid hybridisation. *Annu. Rev. Microbiol.*, **44**; 625–48

Steffan R J and Atlas R M 1991 Polymerase chain reaction: applications in environmental microbiology. *Annu. Rev. Microbiol.*, **45**; 137–61

Extraction of nucleic acids from environmental samples

Moran M A, Torsvik V L, Torsvik T and Hodson R E 1993 Direct extraction and purification of rRNA for ecological studies. *Appl. Environ. Microbiol.*, **59**; 915–18

Ogram A, Sayler G S and Barkay T 1988 DNA extraction and purification from sediments. *J. Microbiol. Methods*, 7; 57–66

Steffan R J, Goksayr J, Bej A K and Atlas R M 1988 Recovery of DNA from soils and sediments. *Appl. Environ. Microbiol.*, **54**; 2908–15

Tsai Y-L, Park M J and Olson B H 1991 Rapid method for direct extraction of mRNA from seeded soils. *Appl. Environ. Microbiol.*, **57**; 765–8

Appendix 6

Standard methods for biodegradability and toxicity testing

EC 1981 *Modified semi-continuous activated sludge test*, Directive 79/831/EEC, Method DGX 1/718/81, Commission of the EC, Brussels

EC 1982 Council directive amending directive 73/405/EEC on the approximation of the laws of member states relating to the methods of testing the biodegradability of anionic surfactants, Commission of the EC, Brussels

ISO 1982 *Water quality – determination of the inhibition of the mobility of Daphnia magna Straus (Cladocera, Crustacea)*, ISO 6341, International Standardisation Organisation, Geneva

ISO 1984 *Water quality – evaluation in an aqueous medium of the 'ultimate' aerobic biodegradability of organic compounds – method by analysis of DOC*, ISO 7827

ISO 1986a *Water quality – test for the inhibition of oxygen consumption by activated sludge*, ISO 8192

ISO 1986b *Water quality; test for inhibition of oxygen consumption by activated sludge*, Standard method 8192, Part A, pp 1–10

ISO 1989 *Water quality – determination of the inhibition of the mobility of Daphnia magna Straus (Cladocera, Crustacea)*, 2nd edn, ISO 6341, International Standardisation Organisation, Geneva

ISO 1991a *Soil quality – effects of pollutants on earthworms (Eisenia fetida). Method for the determination of acute toxicity using artificial soil substrate*, ISO/TC190/SC4/WG2 N20, ISO, Paris

ISO 1991b *Soil determination of the effect of chemical substances on the reproduction of earthworms*, ISO/TC190/SC4/WG2 N27, ISO, Paris

OECD 1971 *Pollution by detergents: determination of the biodegradability of anionic synthetic surface active agents*, OECD, Paris

OECD 1976 *Proposed method for determination of the biodegradability of surfactants used in detergents*, OECD, Paris

OECD 1981 *Guidelines for the testing of chemicals*, OECD, Paris

OECD 1984 *Guidelines for the testing of chemicals: activated sludge respiration inhibition test*, Method 209, OECD, Paris

OECD 1992 *Guidelines for the testing of chemicals*, fifth addendum, OECD, Paris

Painter H A and King E F 1978 *WRC porous-pot method for assessing biodegradability*, Technical report TR70, Water Research Centre

UK SCA 1979 *The instrumental determination of TOC, TOD and related determinants. Methods for the examination of waters and associated materials*, HMSO, London

UK SCA 1981a *Assessment of biodegradability. Methods for the examination of waters and associated materials*, HMSO, London

UK SCA 1981b *Biochemical oxygen demand. Methods for the examination of waters and associated materials*, HMSO, London

UK SCA 1982 *Methods for assessing the treatability of chemicals and industrial wastewaters and their toxicity to sewage treatment processes. Methods for the examination of waters and associated materials*, HMSO, London

UK SCA 1988 *Examining biological filters, toxicity to bacteria, effect of SRT and temperature. Methods for the examination of waters and associated materials*, HMSO, London

UK Standing Technical Committee on Synthetic Detergents 1966 Ministry of Housing and Local Government, supplement to the 8th progress report, Appendix II

US Environmental Protection Agency 1985 *Methods for measuring the acute toxicity of effluent to freshwater and marine organisms*, EPA 600/4-85/013, US EPA, Washington, DC

US Environmental Protection Agency 1991 *Technical support documents for water quality-based toxics control*, EPA 505/2-90-001, Office of Water, US EPA, Washington, DC

Index

Index

chemical oxygen demand, COD, 55
chemical screening, 82
chloride ion determination, 52
chromatography, 41
   applications in environmental
      analysis, 150
   efficiency of columns for, 149
   optimisation of column
      performance, 149
   theory of, 148
chromophores table of, 70
Clark oxygen electrode, 229, 234
colloid formation, 59
competitive immunoassay, 245, 246
composite sampling, 25, 27
Compton scattering, 279
conductance, 260
conductimetric titrations, 50
confidence limit,
   in error analysis, 18
contaminated land, 5
continuous and semi-continuous
      methods, 238
continuous sampling, 25, 26, 33, 37
Cottrell equation, 269
Coulomb, 260
crustaceans, 226
current, 259
   Faradaic, 267
cyanide, 32

data set,
   population of, 13, 15, 18
   sample of, 12, 13, 18
dead time, 280, 282, 285
Debye-Huckel Equation, 262
decay of radioactivity, 274
   activity, 275
   daughter product, 274
   decay constant, 274
   disintegration rate, 274
   half life, 273, 275–7
detection limit,
   in atomic spectroscopy, 132
detection of radiation, 279 *et seq.*
detectors for gas/liquid
      chromatography,
   katharometers, 155
   flame ionisation, 155
detectors for liquid chromatography,
      169
   fluorescence, 171
   electrochemical, 171
   spectroscopic, 169

diffraction,
   Bragg's law of, 138
   Fraunhofer, 94
dipstick test, 231
direct biodegradability testing, 237
direct methods, 237
dispersion,
   Cauchy's equation of, 91
displacement immunoassay, 246,
      247
dissolved oxygen 56, 57
dithizone, 76
Doppler broadening, 126
double layer, 267
DNA, *see* nucleic acid-based tests,
      247–58
Drager tubes, 85

ecosystems, 1, 4
efficiency of detection, 280, 285
electrochemical cell, 260
   half-cell, 260
electrode,
   combination, 262
   counter, 269
   ion-selective, 262
   mercury, 271
   pH, 261, 264
   reference, 260, 269
   standard hydrogen, 260
   working, 268–9
electromagnetic wave,
   field of, 88
   Maxwell's theory of, 88, 133
electromotive force (emf), 259
electron capture detectors, 155
electron orbital, 100–3
   bonding, 104, 108
   non-bonding, 108
electron spin,
   Pauli's principle of, 127
electro-thermal furnace for
      atomization, 130
elution chromatography, 145
emission spectroscopy, 79
energy, 1, 3
environment,
   factors affecting, 194
   parameters of, 193
environmental management, 7
environmental standards, 8
enzyme assays, 240
enzyme bioassays, 230
equilibrium constants, 45, 148

ethylenediamminetetraacetic acid,
      53
   table of cations determined by, 54
external conversion,
   in energy deactivation, 111

Faraday's constant, 260
Fermat's principle, 90
field test kits, 41, 240
finger print matching
   in fluorescence spectroscopy, 124
fish activity monitor, 227
fish tests, 227, 228
flame for atomization, 131
flame photometer, 79
   table of elements determined by,
      80
   sensitivity of, 80
flame photometric detectors
flow-injection analysis, 264, 266
fluorescence, 126, 197, 219
   quantum yield of, 111, 112
   quenching of, 113
fluorescence spectroscopy, 67
Fourier analysis, 135, 119–22, 214
Fourier transform infra red/
      chromatography
   connections, 185
freshwater rotifer, 225

gamma radiation, 273, 276, 279,
      280, 282, 283
gamma ray spectrometry, 287, 288,
      290, 291
gas ionisation, 281
gas/liquid chromatography, 151
   detectors for, 154
   applications to environmental
      analysis, 157
gas multiplication, 281, 283
gas phase chromatography, 151
gated charge integrator,
   for signal-to-noise improvement,
      24, 119
Gaussian,
   optics, 92
   curve, 24, 17
Geiger-Muller counter, 282, 283,
      287, 289
Geiger-Muller region, 282
geometrical signal compression,
      212
geosphere, 1, 2
germanium-lithium detectors, 287

314

Index